Recovery of Lithium from Secondary Resources

Recovery of Lithium from Secondary Resources: Recycling Technologies of Spent Lithium-Ion Batteries presents a state-of-the-art review of recent advances in the lithium recovery from spent lithium-ion batteries (LIBs). It examines the recovery of lithium from secondary sources and provides an introduction to the classification and structure of LIBs. It explains the development of LIBs for electric vehicles and addresses the potential impact of spent LIBs in the environment. Further, it also addresses the multiple treatment protocols for the recycling of LIBs and discusses the high value-added products from these processes. This book provides an essential resource for professionals, researchers, and policymakers in academia, industry, and governments around the globe.

Recovery of Lithium from Secondary Resources
Recycling Technologies of Spent Lithium-Ion Batteries

Muammer Kaya

CRC Press
Taylor & Francis Group
Boca Raton London New York

CRC Press is an imprint of the
Taylor & Francis Group, an **informa** business

Designed cover image: Shutterstock

First edition published 2025
by CRC Press
2385 NW Executive Center Drive, Suite 320, Boca Raton FL 33431

and by CRC Press
4 Park Square, Milton Park, Abingdon, Oxon, OX14 4RN

CRC Press is an imprint of Taylor & Francis Group, LLC

© 2025 Taylor & Francis Group, LLC

ISBN: 978-1-032-47097-9 (hbk)
ISBN: 978-1-032-47100-6 (pbk)
ISBN: 978-1-003-38455-7 (ebk)

DOI: 10.1201/9781003384557

Typeset in Times
by codeMantra

Contents

Introduction

This book provides an up-to-date review of the recycling technology of S-LIBs, which are rich in critical and valuable metals and non-metals. From both the economic and environmental perspectives, the efficient recycling of S-LIBs is of great importance for secondary sources of critical and strategic metals. Nowadays, population increases, rapid economic growth, technological innovations, living standard improvements, shortened life spans of electronic and electrical equipment, and consumer attitude changes have resulted in a significant increase in the amount of waste LIBs that needs to be safely managed. In the last decade, conventional and advanced recycling technologies for S-LIB recycling have been introduced in the world.

Li is a rare metal with many uses, and as the popularity of hybrid electric vehicles (HEVs), portable electronics, and electrical and electronic equipment (EEE) grows, so will the demand for Li. Li can be extracted from minerals and clays, but doing so involves expensive mining and energy use, whereas doing so from brine and bitterns/ seawater requires a lengthy evaporation process. As a result, these processes need to be appropriately modified in order to produce efficiency and better financial results.

Electric vehicles (EVs) and portable electronics frequently use lithium-ion batteries (LIBs) with high power densities. Since their applications have grown so rapidly in recent years, there are more LIBs being used. Heavy metals and toxic hazardous materials found in end-of-life (EoL) LIBs pose a serious threat to the environment and public health and must be properly managed. For used LIBs (S-LIBs), recycling is a crucial option because it not only reduces the pollution of toxic components but also protects primary natural resources. This technology book provides an overview of battery structures as well as information on the recycling status of waste LIBs today. Additionally, recent developments in hydrometallurgy, pyrometallurgy, and direct recycling are thoroughly examined on both a research and an industrial level. For scientists and engineers who might benefit from applying the theory to the examples outlined in the thorough book, this scientific document will serve as a helpful handbook and reference source.

Management hierarchy possibilities of S-LIBs include prevention, reuse, recycling, recovery, and disposal in descending order of environmental desire. In this book, emphasis will be given only to recycling options.

Recycling should be avoided in favor of remanufacturing and repurposing. The best option for S-LIBs is remanufacturing, which maximizes battery value while minimizing emissions and energy use. However, the strict battery quality requirements present a significant obstacle in this case. Direct recycling of LIBs from first-life usage (in vehicles) is less preferable due to the lack of benefits and resulting energy and material losses. On the other hand, recycling has benefits because, rather than being rejected, EoL LIBs join the circular economy, which reduces the need for the extraction of new resources in part. Remanufacturing or repurposing only delays recycling, which is all LIBs' ultimate fate.

The chemistry of a battery's constituent materials affects the extraction procedures used with secondary resources like batteries. The majority of procedures involve disassembling LIBs, separating cathode and anode components, and leaching expensive metals such as cobalt, lithium, nickel, and manganese. Although expensive, $LiCoO_2$ (LCO), which is used in portable electronic devices, performs better. It is replaced by safer alternatives like NMC, which is used in EVs and portable electronics, LMO, which is used in EVs and electric bikes and where Mn provides structural stability, and C-coated LFP ($LiFePO_4$). Metals are separated and recovered from the solutions by solvent extraction, ion exchange, and precipitation from the cathode material in various mineral and organic acids. Currently, there are no commercial methods for extracting Li from LIBs. Therefore, there is enough room to both streamline the currently used processes and increase the effectiveness of metal extraction and separation, including Li and Co recoveries. Process intensification to save energy and improve the kinetics of leaching while addressing the selectivity issue as well as the use of untested/synergistic solvents with an ultra-high metal loading capacity to reduce the number of process steps, are some possible considerations. The extraction of all valuable metals from their primary and secondary resources requires the development of technology that can overcome the limitations of the current processes.

Hydrometallurgy, pyrometallurgy, and direct recycling are the three main pathways used to recycle S-LIBs. On an industrial scale, recycling methods have some pros, cons, and challenges. Hydrometallurgy has a high recovery rate, high purity product, low energy consumption, less waste gas, and high selectivity advantage. More wastewater generation and long processes are disadvantages. Wastewater treatment and optimization of the process are the main challenges of the hydrometallurgical route. The pyrometallurgical process has the advantages of simple operation and short flow; feeding of mixed and all-size material LIBs; and high efficiency. Disadvantages of pyrometallurgy are unrecovered Li and Mn; high energy consumption; low recovery efficiency; more waste gas; and the cost of waste gas treatment. Challenges of the pyrometallurgical process are reduction of energy consumption and pollution emissions; reduction of environmental hazards, and combination with hydrometallurgy route for Li and Mn recovery. Direct recycling route has the following advantages: short recovery route; low energy consumption; environmentally friendly; and high recovery rate. However, direct recycling has the disadvantages of high operational and equipment cost requirements and incomplete recovery rates. The challenges for direct recycling will be lowering the costs, lowering the category of requirements, and further optimizing product performance. For LIB recycling, pyrometallurgical and hydrometallurgical treatments are industrially used, but direct recycling is in lab-scale demonstration.

Fe can be recycled by mechanical processes and thermal treatment. Al can be recycled by mechanical processes, NaOH leach, and chemical precipitation. Cu can be recycled by mechanical processes. Graphite can be recycled by mechanical processes and thermal treatment. Binder (PVDF) can be recycled by thermal treatment. Electrolytes ($LiPF_6$, $LiBF_6$, $LiClO_4$, etc.) can be recycled by thermal treatment and solvent extraction. Ni can be recycled by mechanochemical process; chemical precipitation; and electrochemical process, and Mn can be recycled by chemical precipitation. Co from cathode active material can be recovered by mechanochemical process; dissolution process; thermal treatment; conventional acid leaching; bioleaching; solvent extraction; chemical precipitation; and electrochemical process.

Li can be recycled by mechanochemical process; thermal treatment; dissolution process; traditional acid leaching; bioleaching; and solvent extraction.

Preliminary treatment processes for LIBs include discharging and pretreatment. Discharging can be achieved with saturated salt solutions, which have an ionic contamination disadvantage. Pretreatment can be carried out in four ways: Thermal pretreatment (calcination, oxygen-free roasting, vacuum environment, and vacuum pyrolysis); mechanical and physical pretreatment (grinding, sieving, ultrasonic washing, and flotation); chemical pretreatment (electrolyte dissolution and binder dissolution); and mechanochemical pretreatment (EDTA and PVC addition during grinding).

H_2SO_4 and HCl are chemically cheap inorganic leaching acids. H_2SO_4 has the disadvantage of solubility limitation of sulfate ions, while equipment corrosion and Cl contamination are disadvantages of HCl. HNO_3 has a high extraction rate on Co and Li. However, dissolution of Al, oxidation of Mn and Co, and discharge of NO_x are the disadvantages. H_3PO_4 has low acid consumption, but phosphate has very limited solubility. Organic acids and bioleaching are environmentally friendlier, but extraction rates are lower than mineral acids. Bioleaching has low kinetics.

Energy storage, LIB materials, LIB types, battery evolution, how LIBs work?, lifecycle, life expectancy, handling, transportation, recycling capacity and capabilities are focused in Chapter1. In Chapter 2, lithium-ion batteries and their market are described. The advantages and disadvantages of LIBs are presented. The chemical composition of LIBs, LIB demands in the world, Li compound prices, and the EV battery market are covered in detail. LIB anatomy, anode materials, and their properties, anode manufacture, cathode materials and their properties, Li versus S as cathode, Li-metal technology, solid-state electrolytes, and batteries are extensively given. Testing LIBs, and three LIB production case studies from Australia, America, and China are summarized.

In Chapter 3, LIB structure, assembly, and functioning are covered. Firstly, the history of LIB, how LIBs work, LIB cathode, anode, electrolyte, separator, and additive types are described in detail. The need for recycling S-LIBs is explained. Comprehensive current and novel S-LIB indirect and direct recycling technologies are described in Chapter 4. Pretreatment methods, discharging, dismantling, mechanical and mechanochemical separation processes, Al and binder removal processes, and evaluation of various LIB recycling technologies are covered. Finally, the challenges and future outlook of LIB recycling are explained.

In Chapter 5, hydrometallurgical recycling of LIBs using inorganic acids, ammonia compounds, and alkaline solutions is reviewed extensively. Optimum leaching conditions are summarized with the best metallurgical results. In Chapter 6, organic acid leaching and bioleaching results of S-LIBs are summarized with optimum results.

Industrial S-LIB recycling technologies are summarized in Chapter 7 with flowsheets and product types. Challenges and future outlook of EVs as well as the limits and opportunities of the urban mines are also covered in this chapter. Chapter 8 covers the solution purification technologies after leaching. Fe, Cu, Mn, and Al removal methods are described. Chemical Co and/or Li precipitation from different PLS are described. Solvent extraction of Ni, Li, and Co metals, sol-gel method, electrochemical process, and some of the previous SX studies are summarized in this chapter.

About the Author

Muammer Kaya was born in Eskisehir, Turkey, in 1960. He obtained his B.Sc. from the Mining Engineering Department of Eskisehir Osmangazi University (ESOGU) in 1981. Dr. Kaya received his M.Sc., and Ph.D. from the Metallurgical Engineering Department of McGill University in Montreal, Canada in 1985 and 1989, respectively. He has over 79 national and international (SCI) publications, 85 conference papers/presentations, published 17 books, and contributed 33 book/encyclopedia chapters to edited international books.

Dr. Kaya is interested in mineral processing, flotation, leaching, solvent extraction, precipitation, paper/e-waste/LIB/REE recycling, and environmental protection. He has been working as a full Professor at ESOGU since 1999. He has worked as the Director of ESOGU Technological Research Center (TEKAM) for 15 years. Dr. Kaya also established a Vocational School (ESOGU-EMYO) in Eskisehir. Prof. Kaya has been the Dean of ESOGU Engineering and Architectural Faculty since 2019. Prof. Kaya has been ranked among the top 2% of scientists in the field of Mining & Metallurgy according to the H index, total publications, and citations for the last three years. Prof. Kaya was a member of the Minerals Metals & Materials Society (TMS), the Canadian Institute of Mining, Metallurgy, and Petroleum (CIM), and the American Institute of Mining, Metallurgy, and Petroleum Engineers (AIME).

This book is the third book of Prof. Kaya. The first book was *Electronic Waste and Printed Circuit Board Recycling Technologies* (2019; ISBN 978-3-030-26592-2) and the second book was *Recycling Technologies for Secondary Zn-Pb Resources* (2022; ISBN: 978-3031146848).

Eskisehir, Turkey
March 17, 2024

Acknowledgments

I would like to thank my kind wife Yasemin Kaya for her sacrifice, patience, motivation, and support during writing my third book. Also, I would like to thank my previous postgraduate students (Assoc. Prof. Dr. Sait Kursunoglu (Batman University), Dr. Shokrullah Hussaini (Nevada University), Ph.D. students Angela Manka Tita & Umut Kar (Nevada University) and Hossein Delavandani (ESOGU) for their continuous collaboration, help, and encouragement. It would be impossible to complete this book without their support.

I would like to dedicate this book to my role model decedent Prof. Dr. Andrea Laplante (McGill University, Canada), Prof. Dr. Rıfat Bozkurt (ESOGU, Turkey), and Prof. Dr. Ersan Cogulu (Ottawa University, Canada). I am glad and lucky to be a student of them.

1 Lithium-Ion Batteries

ABBREVIATIONS

ABTC	American Battery Technology Company
CE	Circular economy
CEs	Consumer electronics
DEC	Diethyl carbonate
DMC	Dimethyl carbonate
EC	Ethylene carbonate
EMC	Ethyl methyl carbonate
EoL	End-of-life
HEV	Hybrid electric vehicle
LAB	Lead acid battery
LBM	Lithium metal batteries
LE	Linear economy
LIB	Lithium-ion battery
MFR	Materials recovery facility
PE	Polyethylene
PM	Precious metal
PP	Polypropylene
PVC	Polyvinyl chloride
R&D	Research and Development
SS	Stainless steel

1.1 PERFECT ENERGY STORAGE WITH LIBs

LIBs have

- twice the battery life,
- 50% less space consumption,
- no maintenance, and
- 60% less recharge time.

Electric vehicles (EVs) stand for the advancement of green transportation. In this regard, energy storage is a field of technology that is quickly developing. Since a few years ago, EVs and consumer electronics products that require rechargeable batteries have mostly used lithium-ion batteries (LIBs). In addition to research on improving battery performance (i.e., capacity, safety, cost, etc.), there is a global race to develop a market for mass-produced, economically viable, and environmentally benign LIBs.

Owing to its many benefits over traditional battery systems, LIB is growing in popularity. It is extensively employed in EVs and it is also used as a kind of energy storage in solar-powered products including home inverters, street lights, lighting systems, etc. LIBs can be charged/discharged 2,000 times in their lifetime. The charging time of LIB is 4 hours. The battery comes with a warranty of 6 months to 5 years. The battery has a life of 10 years.

Materials, such as Cobalt (Co), lithium (Li), and graphite (C), are regarded as key critical minerals and are used to make LIBs. Raw resources known as critical minerals have a high risk of supply disruption, which are strategically and economically significant, but lack simple alternatives. We completely lose these valuable resources when we throw away these batteries. Instead of being thrown away or placed in municipal recycling bins, LIBs, or those found in electronic devices, should be recycled at authorized battery electronics recyclers that take batteries. Figure 1.1 shows 60 years of R&D summary on LIBs.

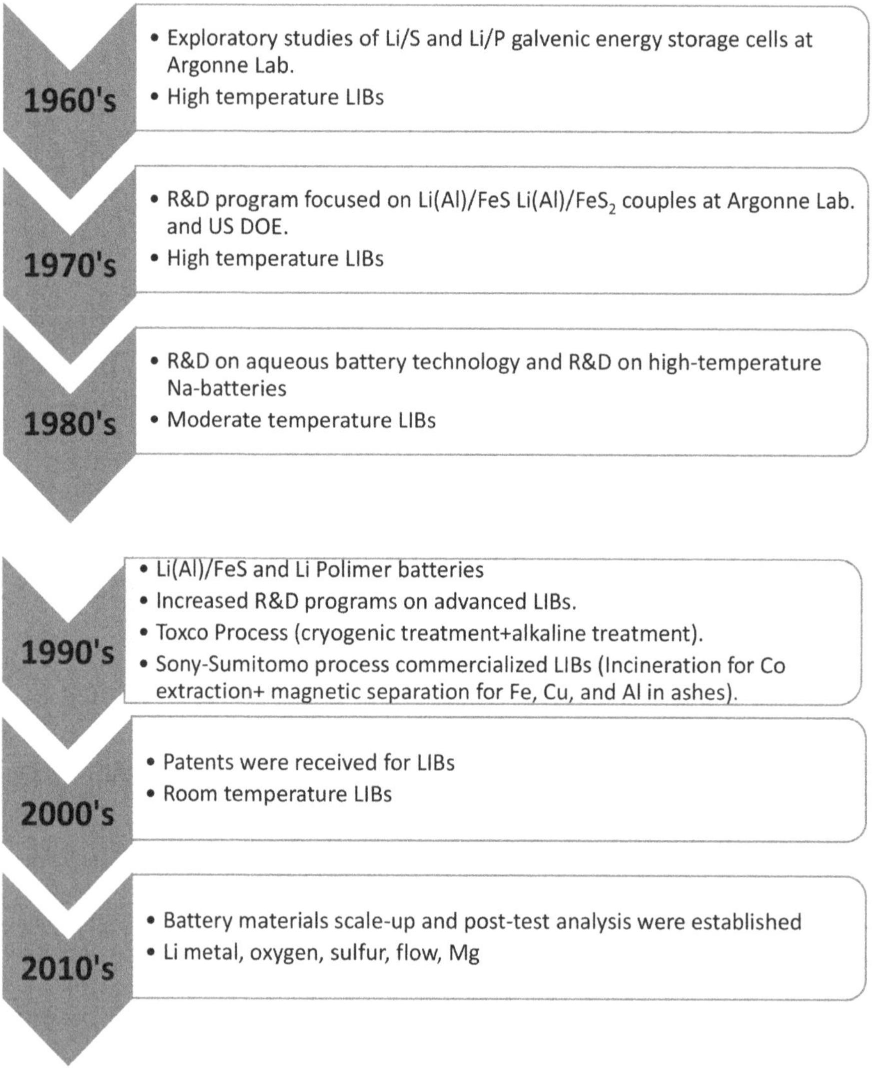

1960's
- Exploratory studies of Li/S and Li/P galvenic energy storage cells at Argonne Lab.
- High temperature LIBs

1970's
- R&D program focused on Li(Al)/FeS Li(Al)/FeS$_2$ couples at Argonne Lab. and US DOE.
- High temperature LIBs

1980's
- R&D on aqueous battery technology and R&D on high-temperature Na-batteries
- Moderate temperature LIBs

1990's
- Li(Al)/FeS and Li Polimer batteries
- Increased R&D programs on advanced LIBs.
- Toxco Process (cryogenic treatment+alkaline treatment).
- Sony-Sumitomo process commercialized LIBs (Incineration for Co extraction+ magnetic separation for Fe, Cu, and Al in ashes).

2000's
- Patents were received for LIBs
- Room temperature LIBs

2010's
- Battery materials scale-up and post-test analysis were established
- Li metal, oxygen, sulfur, flow, Mg

FIGURE 1.1   Sixty years of R&D on LIBs.

Material Discovery
Models, Synthesis

Material Characterization
In Situ, Operando

System-level Analysis
Vehicle, Grid, Techno-Economic

Electrode and Cells
Modeling, Characterization

**BATTERY
RESEARCH
AT ARGONNE**
Li-ion, Li-metal,
flow batteries,
multivalent systems

Recycling
Life Cycle, Processing

**Material Process
R&D and Scale Up**
Organic, Inorganic

Cell Diagnostics and Modeling
Performance, Degradation

Large Format Devices
Pouch, 18650

Standardized Testing
Vehicle, Grid

FIGURE 1.2 LIB battery program across the value chain.

Since being successfully commercialized by SONY in the 1990s, LIBs have replaced other commercial energy storage devices as the primary power source for CEs due to their high energy density, extended lifespan, and minimal self-discharge. Figure 1.2 shows the battery program of LIB across the value chain (Spangerberger, 2018).

1.2 LIB MATERIALS

Cobalt (Co), lithium (Li), graphite (C), nickel (Ni), and manganese (Mn) are the five essential components of today's LIBs. A few countries dominate the world's production, which is unequally divided (Table 1.1). The Congo provides more than half of the Co used. About 80% of the Li is produced in Australia and Chile, and 70% of the graphite is produced in China.

LIBs are classified as category 9 hazardous materials due to their unstable thermal and electrical properties and because of the risk of thermal runaway if poorly handled during transportation by road, sea, or air. LIBs must undergo and pass a suite of national and/or international tests prior to shipment. Having established local and mature recycling facilities close by has various advantages (economic, access to crucial materials, etc.) over exporting batteries to nations with less stringent regulations and transport laws. China dominates the refining and processing of key LIB materials in the world (Figure 1.3).

TABLE 1.1
Global Mine Production of Main LIB Materials

	Li	Co	Ni	Mn	Graphite
Chile	34%				
China				16%	**67%**
Australia	**44%**	5%	9%	14%	
Argentina	13%				
S. Africa				33%	
Brazil					8%
Russia		5%	9%		
D.R. Congo		**59%**			
Canada			10%		
India					13%
Philippines			11%		
Others	9%	31%	**61%**	**37%**	12%

Note: Bold figures are the biggest producers.

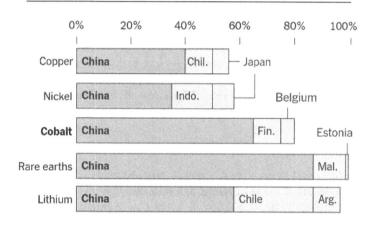

FIGURE 1.3 Key energy element refining and processing.

1.3 TYPES OF LIBs

There are two types of LIBs that consumers use and need to manage at the end of their useful life: single-use, non-rechargeable Li metal batteries (LMBs) and rechargeable LIBs (Table 1.2) (https://www.epa.gov/recycle/used-lithium-ion-batteries#single-use).

1.4 EVOLUTION OF BATTERIES WITH TIME

Energy density has changed for different types of batteries. Figure 1.4 shows the comparison among battery types according to the energy density (Wh/L vs. Wh/kg). Energy density has increased through time with different chemistries in the following order: LAB < NiCd < NiMH < LIB = PLiON < Li metal (which is unsafe).

TABLE 1.2
LIB Types and Use Areas

Single-use (non-rechargeable Li-metal batteries)	Products such as cameras, watches, remote controllers, handheld games, and smoke detectors are frequently made with Li metal.
Li batteries use Li-metal anodes (usually non-rechargeable)	Although the sizes of these batteries may be hard to differentiate from those of standard alkaline batteries, they can also come in unique shapes (such as button cells or coin batteries) for particular devices, such some types of cameras: To identify them, search for the word "lithium" on the battery.
Rechargeable LIBs	Commonly found in tablets, e-readers, e-cigarettes, small and large appliances, laptops, power tools, digital cameras, cell phones, and children's toys.
LIBs use graphite or other material	While some LIBs cannot be easily removed from the items they power, others can.

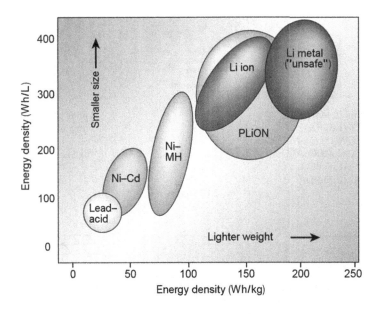

FIGURE 1.4 Battery evolution based on the energy density.

As the energy density increases, batteries become smaller in size and lighter in weight. New batteries are small and light. There is so much interest in Li metal due to increase in Wh/kg for metal anodes.

1.5 HOW LIB WORKS

Figure 1.5 shows the structure of LIBs such as current collects (Cu-anode, Al-cathode), electrolyte, and charging and discharging directions. Cathode and anode chemistries are also given in the figure along with cell module and pack LIB shapes.

The two electrodes are separated by an inert insulating layer. The chemical reactions in the two electrodes can be simply expressed as follows:

The cathodic reaction: $6C + Li^+ + e^- \leftrightarrow C_6Li$ (1.1)

The anodic reaction: $LiCoO_2 \leftrightarrow CoO_2 + Li^+ + e^-$ (1.2)

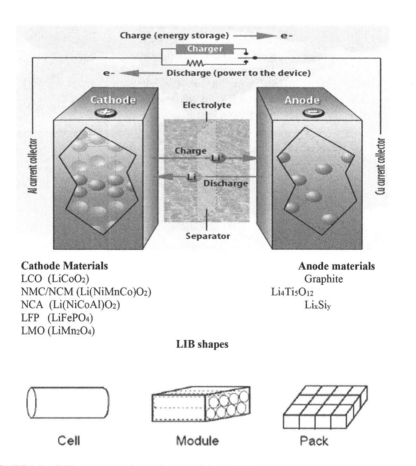

Cathode Materials
LCO ($LiCoO_2$)
NMC/NCM ($Li(NiMnCo)O_2$)
NCA ($Li(NiCoAl)O_2$)
LFP ($LiFePO_4$)
LMO ($LiMn_2O_4$)

Anode materials
Graphite
$Li_4Ti_5O_{12}$
Li_xSi_y

LIB shapes

Cell   Module   Pack

FIGURE 1.5   LIB structure, electrode materials, and shapes.

1.6   BATTERY TYPES

Laptops and EVs use cylindrical-type LIBs. Pouch cell design, which was manufactured by stacking layers of electrodes in a foil envelope, is used in LIBs. Rolled layers of anode, separators, and cathode layers are constructed in a steel casing. LIBs are much more complex in structure and chemistry than lead acid batteries (LABs), and every LIB producer owns different chemistries in their LIBs, which further complicates recycling. Pure graphite flakes (>97%) and $LiCoO_2$ (97%) are used with top sizes of 44 and 30 µm, respectively, in LIBs production.

Because of their high specific energy and energy density, LIBs are used in modern portable CEs such as cellphones, laptops, and tablets. As the most intriguing battery technology for pure and hybrid EVs, LIBs are also widely used in both private and industrial processes. An economic necessity emerges because Ni and Co are expensive transition metal oxides that are used in cathode compositions. Table 1.3 summarizes the advantages and disadvantages of waste LABs and LIBs for EVs and the materials used in the components (compiled from Huang et al., 2018; Zhang et al., 1998).

1.7   LIB LIFECYCLE STAKEHOLDERS

Relevant stakeholders for LIB's lifecycle include LIB recyclers, battery collectors, hazardous and municipal waste managers, household hazardous waste managers, LIB manufacturers, device and EV manufacturers, insurance agents, non-profit organizations, trade associations, academia/research, emergency management, and state, local, and federal government officials.

The E.U. ELIBAMA (European Li-Ion Batteries Advanced Manufacturing) initiative, which will last for 3 years, aims to strengthen and speed up the development of a robust European automotive battery industry built around commercial enterprises already committed to the mass manufacturing of Li-ion cells and batteries. To achieve this target, innovative electrode and cell manufacturing processes were developed up to a high TRL to guarantee drastic cost reductions and significantly enhanced environment-friendliness across the value chain of battery production.

1.8   LIFE EXPECTANCY OF LIBs

LIB technology dominates energy storage solutions for EVs, CEs, and renewable energy grids (such as solar systems and wind). The life expectancy of these batteries is between 2 and 10 years. After that, they become hazardous waste. In 2040, about 340,000 t/y LIBs from EVs will be available for recycling. In LIB production, scarce Co and Li are critical metals. LIBs are classified as hazardous materials under environmental regulations in many countries. LIBs can catch up to the fire and cause an explosion if they are short-circuiting. LIBs disposal in landfills will have a detrimental effect on the environment and human health. In recent years, much scientific research and manuscripts have been devoted to the environment and lifecycle impacts of LIBs recycling and have been focused on developing efficient recovery methods for the materials found in S-LIBs. However, compared to LABs,

TABLE 1.3

Comparison of LABs and LIBs from Production and Recycling Point of View

	LABs	LIBs
Advantages	Well-known technology, Cheap (Pb is roughly seven times cheaper than Li), Pb reserve: 2 billion tons, Complete recycling system (99%), 60% of battery mass is Pb, Life: 2 years, 100% of automobile batteries, and 90% of electrical bicycle batteries are recycled.	Exceptional electrochemical performance, High density of energy (120 Wh/kg), High average battery voltage of the LIBs is 3.6 V, which is three times greater than the voltage of NiCd battery or the NiMH battery, Long lifespan (500–2,000 cycles), Low self-discharge, Higher portability, A wide range of temperatures (−20 to +60°C), Safe and environmentally responsible, Contain high-value metals (Co (5%–15%), Li (2%–7%), Mn, Al, Cu) than natural ores, LIBs recycling supplies multiple metals (Li, Co, Ni, Mn, etc.)
Disadvantages	Hazardous heavy metals (Pb), Contain corrosive acids.	Hazardous heavy metals (Co), Flammable organic electrolytes, Fluorine-containing Li salts. Chemistry is complex, Diverse electrode materials are used, Li is more expensive than Pb, Variety of shapes and sizes, Have residual energy, Only 1% of LIBs is recycled, Li reserve: 89 million tons
Recycling purposes	Resource conservation, Environmental protection.	Resource conservation, Environmental protection
Cathode current collector	PbO_2/Pb	$LiMO_2$ (M = Co, Ni, Mn), $LiCoO_2$, $LiFePO_4$, NMC/Al
Anode current collector	Pb/Pb	Graphite/Cu
Electrolyte	H_2SO_4	$LiPF_6$ + organic solvent (EC, DMC, EMC, DEC, etc.)
Separator	PE or PVC w/silica	PE/PP
Case	PP	Al-plastic film, Al, SS

DEC, diethyl carbonate; DMC, dimethyl carbonate; EC, ethylene carbonate; EMC, ethyl methyl carbonate; PE, polyethylene; PP, polypropylene; PVC, polyvinyl chloride; SS, stainless steel.

FIGURE 1.6 In Morris, Illinois, a warehouse with roughly 9,000 kg of LIBs caught fire in July 2021. More than 5,000 locals had to leave.

LIB recycling is far less developed because the recycling network and technology are still in their infancy. The USA has the lowest S-LIB recycling rate.

A virtually exponential pace of growth is being observed in the need for LIBs to power consumer devices and EVs. The risk of fires from inappropriate battery disposal, especially about consumer devices, is growing as usage increases. LIBs may short-circuit and ignite if they are damaged. This risk can result in thermal runaway, explosion-like occurrences, and the ignition of combustible items close to damaged batteries. These fires and associated incidents are occurring more frequently in waste management facilities, materials recovery facilities (MRFs), and during transportation. The fires brought on by thermal runaways can be disastrous at MRFs, where towns manage a range of recyclable materials, many of which are flammable. These circumstances could affect the recycling system infrastructure in the USA (ERG Inc. and JSE Associates Reports, 2022). Figure 1.6 shows a LIB fire in Illinois in 2021.

1.9 CREATING A CIRCULAR ECONOMY FOR SUSTAINABILITY

Getting manufacturers to consider recyclable materials at the product's conception is one strategy to make recycling commonplace. Recently, the concept has gained popularity: producers and recyclers collaborate to make money while producing as little waste as possible. When a battery runs out of power in a linear economy (LE), it is disposed away in a landfill. Batteries in a circular economy (CE) begin a new life as raw materials and are immediately reintroduced into the production process. We should effectively maintain these metals in that loop eternally after they have been extracted once. This implies that all the linked businesses may, in theory, continue to make money while wasting little to no resources.

Recycling facilities must produce as much as manufacturing facilities do for a circular battery economy to function. Although the manufacturing sector is expanding very quickly, there are no recycling facilities that operate on a commercial scale. In comparison to mining, recycling would provide a steady supply, reduce costs, and maybe have a smaller environmental impact. The development of relationships at all points in the supply chain, from refineries to vehicle manufacturers to battery recyclers, is essential to achieving the recycling goal. The United States Advanced Battery Consortium, made up of General Motors, Ford, Stellantis, and the Department of Energy, awarded American Battery Technology Company (ABTC) a $2 million contract last year to assist in this endeavour. The grant gives funds for more than 2 years to show that making batteries from recyclable materials are better for the environment and the economy. It also means ABTC will be working with a cathode producer and battery recycler as well as a cell technology developer. If there is anything to be learned from LABs, it is that you can always return one. Making the best decision possible has worked well for LABs; the same strategy must be used for LIBs. Compared to a half-century ago for LABs, LIB innovation is still in its infancy, with significant advancements occurring in little more than the last decade. Although ABTC has a lofty timeline in mind, it might take another 10 years for solutions to reach the required size (https://americanbatterytechnology.com/projects/usabc-project/). There are three issues with the availability of battery metal in the USA:

- **Security of Supply:** Currently, less than 1% of the world's manufacturing capacity for each of the major battery metals—Li, Ni, Co, and Mn—is located in the USA. All four minerals are regarded by the federal government as being "essential to the economic or national security of the US."
- **Cost of Supply:** In recent years, manufacturing and importing battery metals have become more expensive due to an increase in demand that has outpaced the ability of new supplies to reach the market.
- **Environmental Impact of Supply:** The extraction of battery metals using traditional methods may lead to significant emissions of greenhouse gases, air pollutants, contaminated water, and soil.

1.10   WHY RECYCLE?

Waste management is one of the main factors influencing LIB recycling, yet it is incorrect to say that all batteries produced today are destined for landfills. When a LIB reaches the end of its useful life, it still has around 80% of its charge left; while this is insufficient for an EV, it is sufficient for many other uses, like energy storage. For at least 10 years, these discarded batteries could be put to use. Because they are so valuable and recycling is expensive, this form of repurposing is always preferred to recycling. A Tesla battery, for example, has a material value of about $1,500, but the market value is between $10,000 and $15,000. It must be disassembled, crushed, and put through some sort of recycling process in order to recover that material's worth. The remaining funds are merely a few hundred dollars.

The world's largest tower operator, China Tower, plans to switch practically all of the 2 million tower base stations' LAB backup batteries for second-hand LIBs, demonstrating the high demand for these battery packs. That is equivalent to 2 million batteries or 54 GWh of battery storage. Reuse can help create value in this way by giving the recycling sector time to put together the infrastructure required to recycle batteries at scale. However, by 2040, the overall amount of LIBs is anticipated to reach 7.8 mt/y. The demand for EoL batteries in second-life applications is expected to be greater than their global supply by a significant margin. Not all batteries will also be reusable; those with questionable origins or those involved in a crush will need to be recycled straight away. Batteries do, of course, eventually die permanently, so in the long term, recycling is crucial for waste management and building up key material reserves. Recycling should be all LIBs' ultimate destination after their initial and maybe second life.

Since LIBs often contain valuable materials, such as Co, Ni, and Mn, to eventually discard them in landfills as waste volumes rise would be a waste of precious resources. If the purpose of having EVs is to reduce CO_2 emissions, then extracting these raw materials from the ground to produce the EVs goes against the purpose of having them. A strong recycling infrastructure would also prevent the depletion of the critical Co reserves necessary to manufacture these batteries—it's expected that by 2030, recycling could provide Europe with 10% of its Co needs.

As a result, the sector gains the additional benefit of being less dependent on unreliable sources: at least 60% of the world's Co is mined in the Democratic Republic of the Congo, where part of the most vulnerable in society bear the brunt of its extraction—Co mining in the Democratic Republic of Congo is linked to human rights abuses, such as child labor and armed conflict.

1.11 BATTERY HANDLING AND TRANSPORTATION

Compliant materials handling and convenient transportation with licensed hazardous waste transportation via approved carriers are compulsory for LIBs. Figure 1.7 shows a fire on a mobile phone battery and swallowed batteries. Table 1.4 shows the dos and don'ts for LIB storage and transportation.

1.11.1 UNDAMAGED BATTERIES FROM ELECTRONIC DEVICES

Regular Packing: If contacts are connected, cut off the connecting wires and isolate them with insulation tape. Batteries that have not been damaged should not be wrapped in foil or plastic. Batteries should be stored in a barrel (200 L barrel or 400 kg

FIGURE 1.7   Damaged LIBs on fire and swallowed.

TABLE 1.4

Dos and Don'ts for LIB Storage and Transportation

Dos	Don'ts
Before continuing with the material recovery, make sure batteries are removed from the devices and stored separately.	Dispose of LIBs with "regular" waste.
Approach an expert to remove batteries if they cannot easily be separated and make sure the expert processing facility has the necessary permits.	Handle the batteries in a way that could cause them to be crushed, punctured, thrown, or have electrodes touch and short circuit.
To prevent short circuits, insulate the battery terminals or cables.	Combine damaged and non-damaged batteries
Pack batteries in UN-approved barrels or crates (depending on the kind), dividing layers with vermiculite (for other battery types) or dry sand (for LIBs).	Avoid grouping numerous batteries together without sufficient segregation because this increases the risk of a fire.
For damaged packages, follow the aforementioned UN-approved barrel process, and additionally wrap leaking or swollen batteries with plastic.	Place batteries, or products containing batteries, in any process that is not specifically designed to accommodate batteries.
	Keep used LIBs inside since they pose a significant risk of short-circuiting and igniting.

pallet box) that has been certified by the UN and has an inner plastic lining. The bottom should be covered with a 10 cm layer of dry sand (for LIB) or vermiculite (for other types of battery). The barrel can then have a layer of batteries added to it, followed by a 10 cm layer of dry sand or vermiculite, and so forth (https://www.simslife-cycle.com/blog/2019/guide-how-to-responsibly-dispose-of-lithium-ion-batteries/).

1.11.2 DAMAGED, SWOLLEN, OR LEAKING BATTERIES FROM ELECTRONIC DEVICES

Regular Packing + Plastic Bag/Foil and Extra Sand: Cut attached cables. Put the battery in a clear plastic bag or cover it in clear plastic foil. Batteries must be kept in an UN-approved 200 L barrel with an inner plastic lining. The bottom should be covered with a 10 cm layer of dry sand (for LIB) or vermiculite (for other battery types). The barrel can then have a layer of batteries added to it, followed by a 10 cm layer of dry sand or vermiculite, and so forth. Batteries with and without damage should never be combined. Each type must be kept in a barrel that has been certified by the UN. Sand and batteries should make up one-third of the weight. Batteries that are harmed, leaking, or swelling ought to be packaged in plastic.

Devices with Swollen Batteries: Place the device in a clear plastic bag or wrap it in clear plastic foil before adding the products to a 200 L barrel that has been certified by the UN and is lined with plastic. The bottom should be covered with a 10 cm layer of dry sand (for LIBs) or vermiculite (for other types of battery). The barrel can then have a layer of products added to it, followed by a layer of 10 cm dry sand or vermiculite, and so forth. Batteries with and without damage should never be combined. Each type must be kept in a barrel that has been certified by the UN. Sand and batteries should make up one-third of the weight. Batteries that are harmed, leaking, or swelling ought to be packaged in plastic.

Batteries Not Allowed to Be Processed Further: Battery scrap cannot be coupled to any other battery scrap or object, including without limitation a plastic, steel, or aluminum foundation, and cannot contain radioactive components, products, or materials. For example, no entire laptop computers or other CEs are permitted. Each of the three categories (1, 2, and 3) described above shall be put in a separate 200 L barrel that has been certified by the UN. It is completely banned to combine several types in one barrel. The following information should be printed on a sticker that is placed on each UN-approved barrel:

- Cat. 1. Undamaged batteries, mentioning also: "LIBs" or "other battery types"
- Cat. 2. Damaged, swollen, or leaking batteries, mentioning also: "LIBs" or "other battery types"
- Cat. 3. Devices with swollen batteries, mentioning also: "LIBs" or "other battery types"

The superior function and performance characteristics of LIBs enhance the performance of our electronic devices. However, more care must be taken when managing LIBs during the EoL phase of electronic goods. Incorrect handling of batteries increases the risk of fire, pollution, and other bad outcomes.

1.12 WHY CAN'T LIBs BE DISPOSED OF WITH NORMAL INDUSTRIAL/CHEMICAL WASTE?

Various compounds can be found in LIBs. Significant consequences of improper disposal include environmental contamination and the depletion of (material) resources. Li is extremely reactive and challenging to manage. A chemical process called an exothermic reaction, which releases energy through light or heat, can occur in the battery as a result of factors including high temperatures, an excessive charging voltage, short circuits, or even an excessive amount of heavy strain. It can quickly catch fire. Because of this, airlines do not permit Li metal batteries (LMBs) and LIBs in checked baggage. A thin polypropylene (PP) sheet that isolates the electrodes and prevents short-circuiting is a component of LIBs. A thermal event, however, may happen if a gadget is damaged or crushed because batteries short circuit when the separator between their positive and negative poles is broken. The thermal reaction increases with battery size. In this manner, a discarded battery may set off nearby combustible materials. A single LIB could catch fire, especially if many cells are gathered together.

1.13 WHY IS PROCESSING LIBs AS E-WASTE CHALLENGING?

Processing LIBs can be problematic for three key reasons. First, because these batteries are frequently linked to the hardware of gadgets, removing batteries might be challenging. The second reason is that it is rather simple to harm the battery while disassembling the device. Because of this, expert separation and recycling typically require specialized equipment, procedures, and controls. Finally, some fire occurrences can be linked to LIBs because of high-temperature exothermic reactions. To lessen the possibility of fire or the production of harmful vapors during processing, high-capacity batteries can be discharged. When batteries are in storage, the risk of a thermal event needs to be categorized as much higher than for normal batteries. In different regions, different rules are in effect, such as Weeelabex and CENELEC; however, waste batteries are covered by European regulations including the WEEE directive and the Batteries directive. The right permits must be in place for treatment facilities to process used batteries.

1.13.1 WHAT IS THE PROPER BATTERY DISPOSAL PROCEDURE?

LIBs can be recycled, but only at approved treatment facilities. When discarded, they must be disposed of at a household e-waste collection point or battery-recycling drop-off location. Undoubtedly, the electrical risk needs to be properly taken into account. High-capacity batteries can present an electric shock hazard as they can provide higher voltages.

1.13.2 HOW SHOULD BUSINESSES AND PROFESSIONAL USERS/ COMPANIES DISPOSE OF BATTERIES?

Make sure LIBs are not left in hot or direct sunlight over an extended period of time. Batteries should not be exposed to moisture or heat. Be careful when handling and packaging these batteries to prevent damage to the safety device (casing). When batteries are piled without a protective covering or put underneath large things, this is a simple possibility. Additionally, make sure that cores cannot come into direct contact with one another and that batteries are expertly depleted prior to further processing. Batteries may explode, catch fire, or rupture if this is not done. The right procedure can recycle the materials found in LIBs. Metal compounds, ferrous metal, aluminum, copper, graphite, and plastic are examples of recyclable materials.

A battery disposal decision tree has been created by Sims Lifecycle Services to assist. This chart provides an overview of the acceptance criteria and processing for three different types of batteries: small industrial batteries, rechargeable (low- and high-grade), and (swollen) LIB. For each of these three groups, pouch cells and batteries without protective covering must be sorted individually out of safety concerns, creating a total of six possible sorting categories.

1.14 RECYCLING CAPACITY AND CAPABILITIES

Plastics, metals, glass, and precious metals (PMs) are used to make electronics. Advanced recycling calls for big, expensive machinery that separates the various materials from one another. Cleanly separated streams of recycled plastic, Fe, steel, Cu, Al, glass, and PMs are the "products" that recycling produces. More ways can be found to use these commodities to create items for the next generation, the better these resources are separated. To create closed-loop systems that lessen pollution and carbon emissions, consume less energy and water, and keep recyclable materials out of landfills, advanced separation is essential. Depending on its complexity, an advanced recycling line may cost between $7M and $14M. Your recycler will need a site of about $13,935 \, m^2$ ($150,000 \, ft^2$) and between 100 and 150 employees if they claim to handle 40 million pounds annually (https://www.simslifecycle.com/resources/tip-sheet-recycling/).

1.15 SUMMARY

LIBs are perfect portable energy storage aids for CEs and EVs with respect to capacity, safety, life span, and cost. Since 1960, there have been many detailed global research studies on LIB materials. Most of the key energy raw materials production and refining come from unstable countries. Rechargeable and non-rechargeable energy batteries evolved with time from LABs to LIBs, which have many advantages. The life expectancy of LIBs is 2–10 years. But LIBs contain hazardous and flammable materials. Building a sustainable CE can be achieved by proper LIB recycling technology from an environmental and economic point of view. LIB handling, transportation, and recycling should be performed by licensed companies. LIB storage and transportation should be carried out according to UN regulations. LIB recycling is a big challenge today for extracting critical raw materials and preventing environmental problems.

2 Lithium-Ion Batteries and Their Market

Great time to be in the Lithium-ion batteries!

LIBs have revolutionized our modern lives since they first entered the market in 1991. M.S. Whittingham, who won the Nobel Prize in Chemistry in 2019, discovered a way to harness the potential energy in the lightest Li metal. He constructed a battery partly made of Li that utilized the element's natural tendency to shed electrons thereby transferring energy.

ABBREVIATIONS

CAM   Cathode active material
CATL Contemporary Amperex Technology
CSPG Coated spherical graphite
EA   Energy absolute
HE   High energy
LAB   Lead-acid battery
LCO $LiCoO_2$
LFP $LiFePO_4$
LIB   Lithium-ion battery
LMO   Lithium manganese oxide
NCA   Nickel cobalt aluminium
NEV   New energy vehicle
NMC   Nickel manganese cobalt
PC   Propylene carbonate
PLA   Protected Li anode
PoM   Put on the market
SEI   Solid electrolyte interphase
SHE   Standard hydrogen electrode
VDLi Vapour deposited lithium

2.1 BATTERIES AND BATTERY MARKET

Table 2.1 shows the comparison of battery parameters of Lead-acid (LAB), NiCd, NiMH, and LIBs. From the four energy types, LIBs have the best properties with respect to energy density, overcharge, toxicity, and voltage point of view.

Table 2.2 shows the typical market share of household waste batteries. Alkaline and/or zinc-carbon batteries are mostly used as household batteries. Eighty-one percent of the used household batteries are of the alkaline or zinc-carbon type.

                                    DOI: 10.1201/9781003384557-2

TABLE 2.1

Comparison of Batteries

Battery types→	Lead Acid				Ni Cd				Ni MH				Li ion			
Battery parameters↓	low	med.	high	v.high	low	med.	high	v.high	low	med.	high	v.high	low	med.	high	v.high
Energy density																☺
Overcharge														☺		
Toxicity																
Cost																
Voltage		2.0 V				1.2 V				1.2 V				3.6 V ☺		

TABLE 2.2

Waste Household Battery Market Share

Battery Type	Market Share
AIkaline/Zinc Carbon	81%
LIB	6%
NiCd	5%
Lead	3%
NiMH	2%
Li Primary	2%
Other Button Cells	1%

LIBs have a share of 6%, and primary Li batteries have a share of 2%. NiCd and NiMH batteries have a market share of 5% and 2%, respectively. Button cells have a market share of 1%.

2.2 LIBs

However, it should be noted that the legal category of industrial batteries, which mostly consists of LIBs for cars, e-bikes, and energy storage, has already surpassed the number of batteries placed on the market (PoM) in the EU. The quantity of batteries in use has already surpassed the sales of residential batteries. As a result, battery collection programs and the recycling industry need to be prepared for a sizable return of trash, mainly LIB waste. Figure 2.1 shows the metal/elements present in both non-rechargeable and chargeable batteries.

2.2.1 ADVANTAGES OF LIBs

- They have a high energy density of approximately 250–670 Wh/L or 100–265 Wh/kg,
- No memory effect,
- A low self-discharge rate of 1.5–2% per month,
- Low maintenance costs,
- The high operating cell voltage of 3.6 V, and
- They are always simple to charge and are largely unaffected by a decreased total amount of charge over time.

FIGURE 2.1   What is inside non-rechargeable and chargeable batteries?

2.2.2 DRAWBACKS OF LIBS

- High cost (approximately 40% more than NiCd),
- Possibility of overheating and damage at high voltages (flammable in extreme circumstances),
- Costs are kept high because of the scarcity of Co and Ni,
- Aging results in a decreased capacity over time and charge-discharge cycles, and
- They perform poorly at very low temperatures (below 0°C).

2.2.3 WHAT IS INSIDE LIBS?

The most difficult battery type to carry, handle, and recycle is Li batteries because of their high energy density, reactive alkali metals, and extremely flammable electrolytes. Additionally, only processes that use selected technologies may meet the standards for optimum material evaluation and recycling efficiency. Figure 2.2 shows the key green energy metals (Li, Co, and Ni) and other elements in the EV batteries. As the demand for LIBs grows, so does the need for higher purity Li. The recovery of valuable key metals from S-LIB cathodes is a sustainability driver for EV markets moving forward. The metal values of Co and Ni represent the bulk value for recycled LIB, with the recovery of Li and Mn contributing additional value. Figure 2.3 presents the battery value chain activities such as mining & refining, battery material/precursor production; battery manufacture & testing; battery deployment, and recycling.

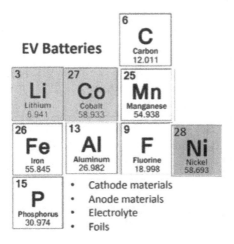

FIGURE 2.2 Element in EV batteries.

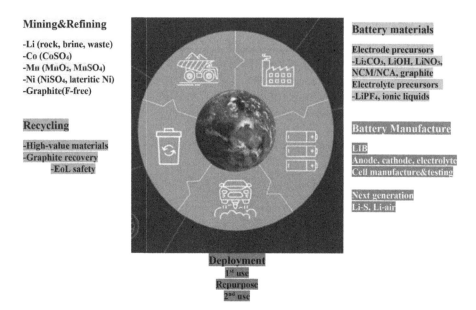

Mining&Refining

-Li (rock, brine, waste)
-Co (CoSO₄)
-Mn (MnO₂, MnSO₄)
-Ni (NiSO₄, lateritic Ni)
-Graphite(F-free)

Battery materials

Electrode precursors
-Li₂CO₃, LiOH, LiNO₃,
NCM/NCA, graphite
Electrolyte precursors
-LiPF₄, ionic liquids

Recycling

-High-value materials
-Graphite recovery
-EoL safety

Battery Manufacture

LIB
Anode, cathode, electrolyte
Cell manufacture&testing

Next generation
Li-S, Li-air

Deployment
1st use
Repurpose
2nd use

FIGURE 2.3   Summary of the battery value chain activities.

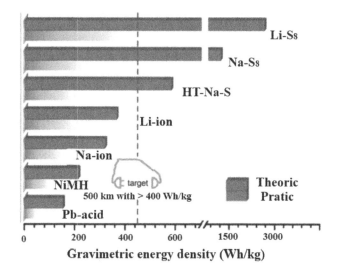

FIGURE 2.4   Practical and theoretical gravimetric energy density comparison of some battery types.

Figure 2.4 shows the practical and theoretical gravimetric energy density comparison of some battery types. For 500 km distance with a car, it requires more than 400 Wh/kg gravimetric energy density. Pb-acid, NiMH, Na-ion, and LIBs are out of this range. Altering the chemistry of LIB electrodes to achieve the theoretical maximum. Revolutionary Li-metal-based chemistries should be explored from now on.

2.3 LIB DEMANDS

According to McKinsey battery demand model 2023, global LIB cell demand (GWh) driven by electrification of transport and energy storage is shown in Figure 2.5.

The demand for LIB will dramatically increase until 2030 and reach 4,700 GWh. Stationary storage and consumer electronics share are very low as compared to mobility in 2025 and 2030. Figure 2.6 shows EV, e-bike, and industrial LIBs consumption in t/a between 2020 and 2035. There will be a significant increase in demand for LIB. EV has the maximum demand for LIBs.

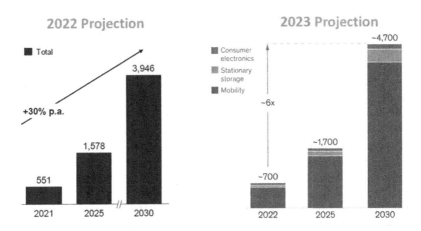

FIGURE 2.5 LIB total and product-based demand in 2022 and 2023 projection.

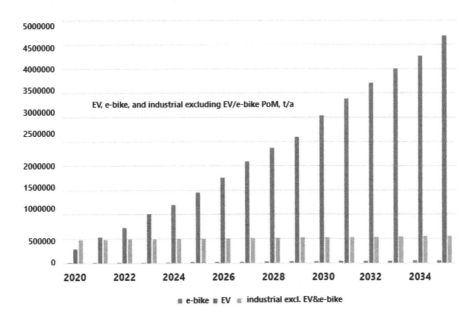

FIGURE 2.6 Electrical vehicle, e-bike, and industrial excl. EV/e-bike PoM.

2.4 Li COMPOUND PRICES

Argus forecasts that global Li consumption will reach 2.78 Mt/y by 2030, mainly driven by demand from the EV battery segment. *Argus* most recently assessed spot prices for battery grade 99.5% lithium carbonate at 290,000–305,000 yuan/t ($40,036–42,107/t) ex-works on July 17, 2023 up by 63% from late April, when prices started to rebound from a downturn. Fluctuations in Li feedstock prices in the past few years have raised concerns about the impact on the uptake of EVs (https://www.argusmedia.com/en//news).

2.5 BATTERY ELECTRIC VEHICLE MARKET

All-electric vehicles usually lack spark plugs, gearboxes, radiators, oil and fuel filters, exhausts, and other components that are exclusive to internal combustion engines. The cost of maintaining an electric vehicle is therefore lower than that of a gasoline-powered vehicle. Regenerative braking, which heavily relies on the electric motor to conduct the brakes, is another feature seen in the majority of electric vehicles. The brake rotors and pads on electric cars last longer since maintenance expenses are reduced. Even if you choose a plug-in hybrid electric vehicle with an internal combustion engine, maintenance costs will be lower. PHEV engines require less oil and coolant and sustain less damage because they operate less frequently than gasoline-powered engines.

In electric vehicles, the LIB is intended to last 6–7 years, maybe only 8 years. After this period, the only alternative left to the owner of an electric vehicle is to buy a new battery, which runs close to 3/4 of the cost of the entire vehicle. This truth is now widely known. The price of batteries will be a major concern for EV owners in the long run because electric cars are still a relatively new technology for the market and consumers.

There are numerous businesses that produce EVs:

- Audi AG
- BAIC Group
- Bayerische Motoren Werke AG (BMW)
- Blue Bird Corporation
- BYD COMPANY LTD.
- Chery Automobile Co., Ltd.
- Ford Motor Company
- General Motors Company (GM)
- Groupe Renault (legally Renault S.A.)
- Hyundai Motor Company
- Kia Motors Corporation
- POLESTAR AB
- Porsche Automobil Holding SE (Porsche SE)
- SAIC MOTOR CORPORATION LIMITED
- Tata Motors Group (Tata Motors)
- Tesla, Inc.
- and many more.

2.6 LIBs ANATOMY

Figure 2.7 shows the electrochemical anatomy of a LIB. Electrons flow from anode to cathode and Li-ions flow from cathode to anode through the separator during discharging and vice versa during charging. Most of the elements in the periodic table can be used as a form of energy storage (Figure 2.8). Elements can be used as conversion anodes, cathodes, or intercalation electrodes. Li and Si have the highest gravimetric capacity (mAh/g). Table 2.3 summarizes recycling processes or treatment methods for the components of LIBs.

Since Co is a rare and precious metal and Ni and Li are expensive, the majority of current research on the recovery and recycling of LIBs is focused on the recovery or recycling of the precious metals such as Co, Ni, and Li of the electrode materials. There are also more studies that need to be done on the recovery or disposal of additional components found in S-LIBs, including electrolytes and graphite.

2.7 ANODE MATERIALS

Because Li metal generates dendrites that can result in short-circuiting, ignite a thermal runaway reaction on the cathode, and ignite the battery, anode materials are essential for LIBs. Additionally, Li metal has a short cycle life. Graphite, LTO, Si, Ge, Sn, and Li_2O (amorphous) were tested as anode materials (Nitta et al., 2015).

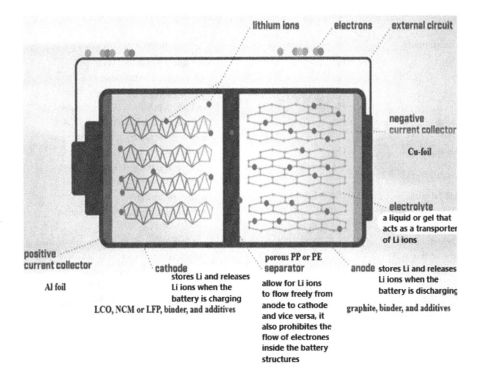

FIGURE 2.7 The most basic anatomy of the electrochemical LIB unit.

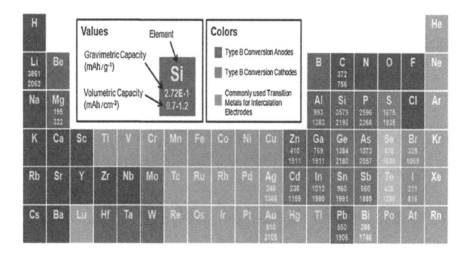

FIGURE 2.8   Gravimetric and volumetric capacities of elements used in energy storage for batteries.

TABLE 2.3
A Summary of the Procedures Used to Recycle or Treat the LIBs' Constituent Parts

Components	Elements	Recycling Processes
Shells	Fe	Mechanical processes (Magnetic separation); thermal treatment (pyrometalurgy)
	Plastics	Mechanical processes (Gravity separation)
Cathode (Aluminum foil)	Al	Mechanical processes (sieving); NaOH leaching; chemical precipitation
Anode	Cu	Mechanical processes (sieving)
	C (graphite)	Mechanical processes (sieving); froth flotation; thermal treatment
Adhesive agent	PVDF binder	Thermal treatment
Electrolyte (organic liquid)	($LiPF_6$, $LiBF_6$, $LiClO_4$)	Thermal treatment; solvent extraction
Cathode active material (CAM)	($LiCoO_2$, $LiNiO_2$, $LiMnO_4$)	Co: Mechanochemical process; dissolution process; thermal treatment; acid leaching; bioleaching; solvent extraction; chemical precipitation; electrochemical process Li: Mechanochemical process; thermal treatment; dissolution process; acid leaching; bioleaching; solvent extraction Ni: Mechanochemical process; chemical precipitation; electrochemical process Mn: Chemical precipitation

Table 2.4 compares possible anode materials for LIBs. Li has a maximum specific capacity and virtually infinite volume change. Li and silicon have a very high energy density, while graphite is stable. The disadvantages of Li are instability and high reactivity. Silicon's capacity fades away due to the large volume capacity and graphite has a low energy density.

Li has the maximum theoretical capacity and minimum electrochemical potential, which makes it the ultimate anode material. The energy density and specific energy of gasoline and various LIB materials are compared in Figure 2.9.

Both specific energy and energy density increases occur in the following order:

Li-ion < Li-LMO < Li-S < Li-Air

Cycled Li is different from fresh Li metal (Figure 2.10). LIB capacity fades up in the following manner (Figure 2.11).

TABLE 2.4

Possible Anode Material Comparison for LIBs

Anode Material	Spec. Cap. (mAh/g)	Volume Change (%)	Advantages	Challenges
Lithium (Li)	3,862	Virtually infinite	Highest energy density	Unsable, highly reactive
Silicon (Si)	3,600	320	Ultra-high energy density	Capacity fade due to large volume change
Graphite (C)	372	10	Stable	Large energy density

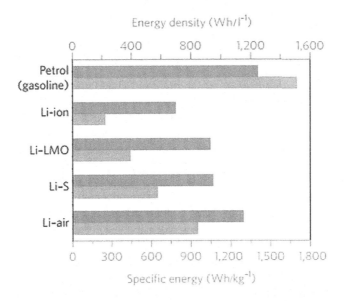

FIGURE 2.9 Comparison of energy density and specific energy for petrol and some LIBs.

FIGURE 2.10   Difference between fresh and cycled Li.

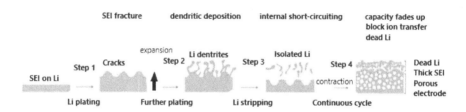

FIGURE 2.11   LIB capacity fades up and dead.

2.7.1 GRAPHITIC AND HARD CARBONS (C)

The C anode, which continues to be the preferred anode material, made the LIB commercially viable more than 20 years ago. Li is intercalated between the graphene planes, which provide strong two-dimensional mechanical stability, electrical conductivity, and Li transport, to produce electrochemical activity in C (Figure 2.12). This allows for the storage of up to one Li atom per six C. Compared to Li, C has a low delithiation potential and a high diffusivity of Li, as well as a strong electrical conductivity and a comparatively small volume change during lithiation and delithiation. C also has inexpensive costs and abundant availability. In comparison to other intercalation-type anode materials, C offers an appealing balance of low cost, abundance, moderate energy density, power density, and cycle life. C has a greater gravimetric capacity than most cathode materials, but the volumetric capacity of commercial graphite electrodes is still small (330–430 mAh/cm^3).

Commercial C anodes fall mostly into two categories. Large graphite granules and almost theoretical charge capacities are characteristics of graphitic Cs. However, due to its low melting point and quick Li transfer, the favored electrolyte is based

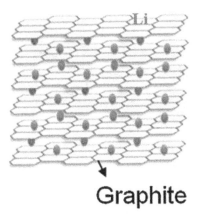

Graphite

FIGURE 2.12   Li in graphite layers.

on propylene carbonate (PC), which graphitic Cs do not interact well with. Between the graphitic planes, PC intercalates with Li^+, causing the graphite to exfoliate and lose its capacity. Li intercalation takes place at the basal planes even in the absence of solvent intercalation, leading to the SEI's preferential formation on these planes as well. Single crystalline graphitic particles experience uniaxial 10% strain along the edge planes during Li intercalation. Such a strong strain may harm the SEI and shorten the cell cycle. Graphitic Cs have recently been coated with a thin layer of amorphous C to protect the vulnerable edge planes from electrolytes and achieve high Columbic efficiency.

Small, disordered-oriented graphitic grains characterize hard carbons, which are far less prone to exfoliation. Nano gaps between these grains also result in reduced and isotropic volume expansion. Additionally, nanovoids and imperfections offer excess gravimetric capacity, enabling capacity greater than the 372 mAh/g theoretical limit. These characteristics make hard Cs a material with a high capacity and long cycle life. The absolute amount of SEI generated is increased by the high fraction of exposed edge planes; however, this lowers the Columbic efficiency in the initial cycles. A full Li-ion cell has a constrained amount of available Li, which severely reduces the amount of capacity that can be achieved. Additionally, the particle density is greatly decreased by the void areas, which lowers volumetric capacity even more. Lastly, contaminants like hydrogen atoms can also provide extra capacity in C-based anodes. However, such electrodes suffer from larger voltage hysteresis, higher irreversible capacity loss, and even lower volumetric capacity, and thus are unlikely to be commercialized (Nitta et al., 2015).

2.7.2   PROPERTIES OF GRAPHITE (C)

Graphite is a non-metallic mineral and one of only two naturally occurring forms of carbon (the other being diamond). Graphite is the strongest and stiffest material by nature, making it a great conductor of heat and electricity. Over a wide variety of temperatures, it is stable. The melting point of graphite, which has a high

FIGURE 2.13   Graphite electrode picture

refractory value, is 3,650°C. One of the lightest of all reinforcing materials, graphite possesses one of the highest inherent lubricities and hydrophobicities. Graphite has great corrosion resistance and is chemically inert. The anode in a Li-ion battery is made of graphite, especially Coated Spherical Graphite (CSPG) (https://westwa-terresources.net/minerals-portfolio/graphite-market/). Figure 2.13 shows a graphite electrode.

China (70%) North Korea (10%), and Brazil (8%) are the three main nations that produce natural graphite. The demand for premium flake graphite is expected to rise over the next few years, but the supply appears to be at best stable, and there are worries about potential export restrictions from China.

2.7.3   GRAPHITE FACTS

- Graphite is a crucial/critical strategic material that has been designated as a supply-critical mineral by both the US and the EU.
- Each electric car contains more than 90 kg of CSPG. The USA currently imports 100% of all graphite consumed.
- Although there are about 200 uses for graphite, LIBs-for which graphite is used to make the anode-have the highest and longest-lasting demand.
- The minimum purity of the graphite needed to create a LIB is 99.95% C, which requires 10–30 times as much as Li.
- Future graphite demand will mostly be fueled by the growing stationary and transportation battery sectors.
- Graphite is the appropriate anode material for LIBs because of its special features; however, clients must use CSPG graphite rather than conventional run-of-mine graphite.
- China has implemented a program of resource nationalization that includes export taxes, permits, and environmental regulations. China controls over 75% of the world's graphite production.
- In comparison to an HF-based process, Alabama Graphite Products uses a patented purification method that offers benefits in terms of chemical safety, graphite purity, sustainability, and environmental process cost.

2.7.4   Drivers of Demand: LIBs Are the Source of Graphite Market Growth. Batteries Represent the Most Significant Demand Driver for Battery-Grade (Coated Spherical or 'CSPG') Graphite

Two types of CSPG graphite are used in LIBs: synthetic (~20,000 USD/t average selling price) and natural (~$8,000–$11,000 USD/t average selling price). Figure 2.14 shows the mineral processing anode value chain of graphite. The anode value chain increases from natural graphite to nano silicon addition graphite. Because of the cost and performance efficiencies, many battery manufacturers are transitioning to natural graphite.

There are three key market segments within the LIB market itself.

- **Transportation Batteries:** To power electric vehicles; electrification of the automobile industry is quickly transitioning from a specialized to a mainstream market (large predicted increase)
- **Stationary Storage Batteries:** They also known as grid-storage batteries, are used to store energy for the electrical grid, as well as for commercial and residential structures (exponentially predicted expansion)
- **Consumer Electronics Batteries:** Cellphones, laptops, tablets, wearable technology electronics, power tools, and other battery-operated devices are expected to expand moderately.

2.7.5   2020–2030 Graphite Consumption and LIB Production Forecasts

According to Roskill's – Natural & Synthetic Graphite: Outlook to 2030 report, the graphite consumption forecast up to 2030 is given in Figure 2.15. There will be a significant increase in both graphite consumption and LIB production. Graphite consumption will reach about 1,800 kt and LIB production to 2,250 GWh in 2030. Natural and synthetic graphite consumption in Li-ion battery production will increase significantly.

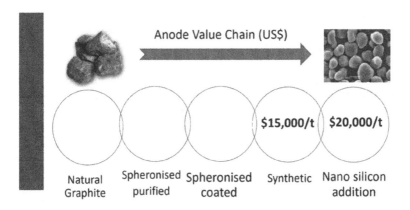

FIGURE 2.14   Anode value chain of graphite.

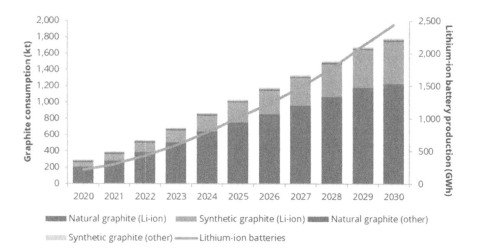

FIGURE 2.15 Graphite consumption and LIB production between 2020 and 2030.

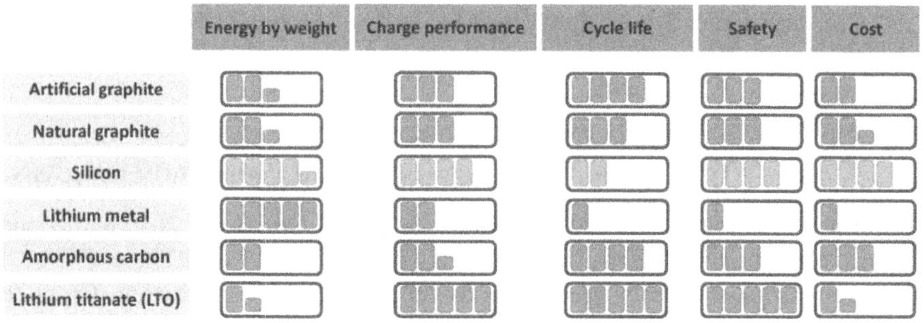

FIGURE 2.16 Comparison of anode materials for LIBs.

2.7.6 ANODE MANUFACTURE

Different anode materials can have very different characteristics across several metrics Silicon and Li metal's cycle life can be enhanced by the use of appropriate enabling technologies. Figure 2.16 compares anode materials concerning energy by weight, charge performance, cell life, safety, and cost point of view.

Li has a theoretical specific capacity of 3,861 mAh/g and 2,061 mAh/cm^3. Silicon stores close to ten times more Li by weight and three times more Li$^+$ by volume. Near-term opportunity to develop cheaper, smaller, and lighter batteries. Silicon anodes can make batteries smaller, lighter, and cheaper. 500$^+$ charge/discharge cycles demonstrated while retaining 90% of the initial capacity in pure silicon anode. Figure 2.17 compares the specific gravity and anode thickness of graphite and silicon in the LIBs. Low-grade and unrefined silicon can be used, but competitors use super-refined silicon which is expensive, limited, and C-intensive. Silicon active material is 8.5 times cheaper on a $/kWh basis. And 35% improvement in energy

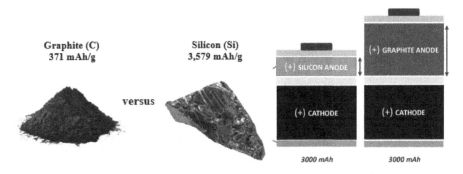

FIGURE 2.17   Comparison of the specific gravity and anode thickness of graphite and silicon in the LIBs.

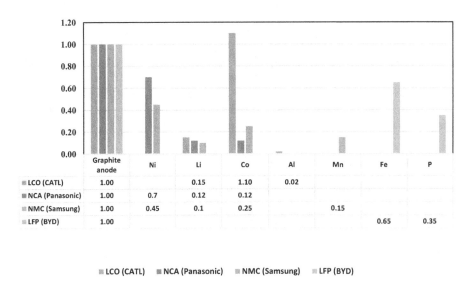

FIGURE 2.18   Graphite components of different LIBs.

capacity is achieved with silicon anodes. Pure silicon anodes are three times thinner than graphite anodes (Wicser, 2023, Energy Next, www.ateotech.com/energy).

Graphite is the main component in the LIB. Figure 2.18 shows the graphite and metal contents of different LIB cathode types (Talbot, 2020). NCA, NMC, and LPF contain more graphite amount than cathode materials (Ni, Co, Li, etc.). Only LCO contains more Co than graphite.

2.8   CATHODE MATERIALS AND MANUFACTURE

Cathode material chemistry chronological cornerstones are summarized below:

Complex Metal Oxides
- **1979:** R&D on LIBs commences using LCO
- **1991:** First LIB is commercialized by Sony (Coke/LCO)

- **1996:** Lithium manganese oxide (LMO) is commercialized
- **1996:** Lithium iron phosphate (LFP) is discovered
- **1999:** Lithium nickel cobalt aluminum oxide (NCA) is discovered
- **2000:** Nickel manganese cobalt chemistries (NMC) appear

Further Development of Chemistry Categories
- Higher Nickel content in NMC to drive up capacity
- Higher Manganese content in NMC to drive higher voltages
- LMFP: Manganese introduction into LFP to drive to higher voltages

Possible Future Cathode Chemistries
- Sulfur and possibly Air (O_2)

2.8.1 Li Demand

Table 2.5 shows the Li requirement for various LIBs per 100 kWh battery capacity which is roughly equivalent to that of the cars most of us use today. LFP requires the least and NCM 1:1:1 requires the maximum amount of Li for battery production. Current global reserves of Li are around 80 Mt. In 2018 and 2019, 95,000 and 77,000 t/y Li were produced, respectively. Thus, the current global Li production from primary and secondary sources should be increased significantly.

2.8.2 Benefits of Using Li Metal as an Electrolyte

The following characteristics make Li metal the perfect anode component for LIBs.

- 0.534 g/cm^3 is the lowest metal density.
- Low possibility of reduction potential: -3.04 V vs SHE.
- Extremely high theoretical specific capacities: 2,061 mAh/cm^3 and 3,861 mAh/g.

The low density of Li helps to reduce overall cell mass and volume, which helps to improve both gravimetric and volumetric capacities and energy densities of Li batteries. Also, the low reduction potential of Li enables the cell to operate at relatively high cell voltage which also increases the energy density of the Li battery.

TABLE 2.5
Li Requirements per 100 kWh Battery

Type	Li (kg)
LFP	7.8
NCA	11.2
NCM 1:1:1	13.9
NCM 6:2:2	12.6
NCM 8:1:1	11.1

The theoretical gravimetric and volumetric capacities of Li metal anode are

$$Q_g = n*F*M_w = 3,861.328 \text{ mAh/g} \tag{2.1}$$

$$Q_v = Q_g*\rho = 2,061.949 \text{ mAh/cm}^3 \tag{2.2}$$

where

Q_g = Gravimetric capacity of Li [mAh/g]
n = Transferred electrons number = 1
F = Faraday's constant = 26.8014814 [Ah/mol]
M_w = Molecular weight of Li = 6.941 [g/mol]
Q_v = Volumetric capacity of Li [mAh/cm³]
ρ = Density of Li = 0.534 [g/cm³]

Li is inherently the most energy-dense anode material for Li batteries since the capacity values given above represent the maximum limit of all possible anode materials that might be used as the anode for Li batteries. However, due to the high chemical and electrochemical reactivities of Li, which cause side reactions during battery cycling, it is challenging to obtain these theoretical values in practice.

2.8.3 ENHANCING BATTERY ENERGY DENSITY BY REPLACING GRAPHITE WITH LI METAL ANODE

In general, there are two representative energy density metrics for batteries: (1) gravimetric energy density (energy stored per unit weight of a battery) and (2) volumetric energy density (energy stored per unit volume of a battery). The low density of Li helps to improve the gravimetric and volumetric energy densities by reducing the anode weight and volume in batteries. Its low reduction potential is necessary to increase an operating cell voltage (V_{cell}). A graphite anode is widely used in commercial LIBs. The graphite anode exhibits a theoretical specific capacity of 372 mAh/g. Comparing the calculated theoretical capacity of Li (3,861 mAh/g), Li metal anode holds about tenfold higher specific capacity than that of graphite. However, the major capacity that dictates the energy density of the battery is the discharge capacity that depends on the cathode. This is because the electrical energy is obtained from the battery during the discharge process. On the other hand, the anode capacity dictates the total storage amount of Li-ions during the charging process. In general, an unequal capacity ratio between the anode and cathode is used when constructing Li batteries. The capacity ratio between the anode (the negative electrode) and cathode (the positive electrode), known as the N/P ratio, is an important cell designing parameter to determine a practical battery performance and energy density. The below equations illustrate how the energy densities of the battery are calculated.

$$V_{cell} = C_{cathode} - V_{anode} \tag{2.3}$$

$$E_g = Q_{dis} * V_{cell} * M_{cell}^{-1} \qquad\qquad (2.4)$$

$$E_v = Q_{dis} * V_{cell} * N_{cell}^{-1} \qquad\qquad (2.5)$$

V_{cell} = Operating voltage of the cell [V vs Li/Li⁺]
$V_{cathode}$ = Operating voltage of the cathode [V vs Li/Li⁺]
V_{anode} = Operating voltage of the anode [V vs Li/Li⁺]
Q_{dis} = Cell discharge capacity [Ah]
E_g = Gravimetric energy density [Wh/kg]
M_{cell} = Total weight of the cell [kg]
E_v = Volumetric energy density [Wh/L]
N_{cell} = Total volume of the cell [L]

The battery's energy densities depend on its capacity, working cell voltage, weight, and volume. The battery energy density is computed using the discharge capacity. Another advantage of having a Li metal anode rather than a graphite anode for Li batteries is that the voltage reference for the operating cell voltage is always with respect to Li/Li⁺. Since the cathode potential is the only function of the cell voltage, which is Li/Li⁺ as the reference, 0 V serves as the working voltage for the Li metal anode. The operational voltage of the cell is reduced by around 0.1 V due to the average working voltage of the graphite anode, which is about 0.1 V. This decrease in working cell voltage translates to about a 3% decrease in energy density assuming the discharge capacity is the same, and state-of-the-art cathode materials exhibit an average operating voltage of 3.9 V vs Li/Li⁺ (http://large.stanford.edu/courses/2020/ph240/kim1/).

2.8.4 CALCULATING THE N/P RATIO FOR THE LI-METAL BATTERY

Areal capacities in mAh/cm² for the Li metal anode and cathode material are frequently used to simplify the calculation of the N/P ratio for Li-metal batteries. It is important to remember that the N/P ratio is calculated using the theoretical and practical capacities of the cathode and Li, respectively. It is thus because Li metal is frequently manufactured industrially as a thin film, with thicknesses varying from 500 to 20 μm, and cycle circumstances have little of an impact on Li's initial capacity. Although the real cathode capacity is susceptible to operating variables such as the operating voltage window, C rate, temperature, etc., the main rationale for utilizing practical capacity, or equivalently measured capacity, for the cathode. As a result, while determining the N/P ratio in the field, it is typical to take into account the observed cathode capacity from the specified operating circumstances. The capacity measurements are normalized by the electrode size to derive the areal capacity of the anode and cathode because the battery electrodes are rather thin. Although areal capacities are now frequently used in the field to determine the N/P ratio, gravimetric capacities are still reported. Because Li metal anode is a thin film, comparing Li thickness to areal capacity provides a useful equation to determine the N/P ratio with Li metal foil. The relationship between the theoretical areal capacity of Li metal foil and its thickness is seen in the equation below.

For ease of calculating the N/P ratio for Li-metal batteries, often areal capacities in units of mAh/cm² for Li metal anode and cathode material are used. It is worth noting that the often theoretical capacity of Li and the practical capacity of the cathode are used for calculating the N/P ratio. This is because Li metal is often industrially processed as a thin film (ranging from 500 to 20 μm in thickness), and the initial capacity of Li is not dramatically affected by cycling conditions. However, the major reason for using practical capacity, equivalently measured capacity, for the cathode is that the actual cathode capacity is sensitive to operating conditions such as operating voltage window, C rate, temperature, etc. Therefore, it is common in the field to consider measured cathode capacity from the defined operating conditions to calculate the N/P ratio. Since the battery electrodes are relatively thin, the capacity values are normalized by the electrode size to derive the areal capacity of the anode and cathode. Gravimetric capacities are reported as well, but it became common in the field to use areal capacities to find the N/P ratio. Since Li metal anode comes as a thin film, a useful equation to calculate the N/P ratio with Li metal foil is relating Li thickness to areal capacity. The below equation relates to how the thickness of Li metal foil is related to its theoretical areal capacity.

$$T_{Li} = (10,000*Q_A*M_w)/(n*F*\rho) \tag{2.6}$$

where

$T_{Li} \equiv$ Li thickness [μm]
$Q_A \equiv$ Li areal capacity [mAh/cm²]
$M_w \equiv$ Li molecular weight $= 6.941$ [g/mol]
$n \equiv$ Transferred number of electrons $= 1$
$F \equiv$ Faraday's constant $= 26,801.4814$ [mAh/mol]
$\rho \equiv$ Li density $= 0.534$ [g/cm³]

For LIB $1.2 \geq N/P \geq 1$ Amount of Li-ion host available in
the anode to that of the cathode

Table 2.6 displays Li's areal capacity concerning its thickness. Keep in mind that these are only theoretical figures. Assuming a completely flat and smooth coating of Li, the thickness is determined. Li areal capacity rises as Li thickness rises.

TABLE 2.6
Li Areal Capacity with Respect to Its Thickness

Li Areal Capacity (mAh/cm²)	Li Thickness (μm)
1	4.85
4	19.40
7	33.95

2.8.5 Intercalation Cathode Materials

A stable host network that can hold guest ions is an intercalation cathode. The host network allows for the reversible insertion and removal of the guest ions. In a LIB, the host network compounds are metal transition metal oxides, chalcogenides, and polyanion compounds, and Li^+ is the guest ion. The crystal structures of these intercalation compounds include layered, spinel, olivine, and tavorite (Figure 2.19). For the cathode materials in LIBs, the layered structure is the earliest type of intercalation compound (Nitta et al., 2015). Due to their greater working voltage and accompanying higher energy storage capabilities, the transition metal oxide and polyanion compounds are the subject of the majority of current intercalation cathode research. Intercalation cathodes typically have 100–200 mAh/g specific capacity and 3–5 V average voltage vs Li/Li$^+$.

The first and most widely used type of layered transition metal oxide cathode is $LiCoO_2$ (LCO). The bulk of commercial LIBs continue to utilize this material, which was first made commercially available by SONY. In octahedral sites, the Co and Li occupy alternating layers to create a hexagonal symmetry (Figure 2.19a). Due to its comparatively high theoretical specific capacity of 274 mAh/g, high theoretical

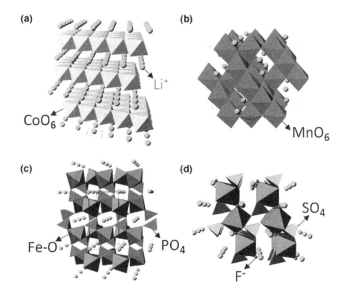

Structure	a) layered	b) spinel	c) olivine	d) tavorite
Formula	($LiCoO_2$)	($LiMn_2O_4$)	($LiFePO_4$)	($LiFeSO_4F$)
Spe. Capacity mAh/g	274	148	170	151
Vol. Capacity, mAh/cm³	1363	596	589	487
Ave. Voltage (V)	3.8	4.1	3.4	3.7
Industrial Level	commercialized	Commercialized	commercialized	research

FIGURE 2.19 Crystal structures, volumetric and specific capacities; average voltage, and level of development of intercalation cathodes (compiled from Nitta et al., 2015).

volumetric capacity of 1,363 mAh/cm³, minimal self-discharge, high discharge voltage, and strong cycling performance, LCO is a particularly alluring cathode material.

Cathode active material manufacturing (co-precipitation flowsheet is given in Figure 2.20. 99.98% $CoSO_4$; $NiSO_4$ and/or $MnSO_4$ and 99% LiOH are used for CAM production. Chemical stirring, co-precipitation, dry solids recovery, lithiation doping, and calcination/coating are the steps. Figure 2.21 presents different cathode choices (such as NMC, NCA, LCO, LMO, and LFP).

NCM composition phase diagram is shown in Figure 2.22. Early NCM variants have discharge capacity between 150 and 165 mAh/g, current Ni-rich variants have between 175 and 200 mAh/g, and future Mn-rich variants will have between 140 and 250 mAh/g.

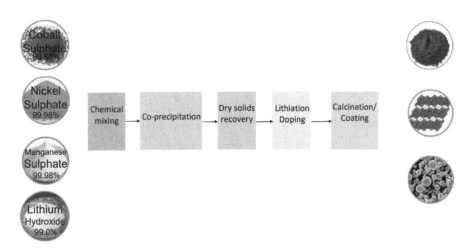

FIGURE 2.20 CAM manufacturing process.

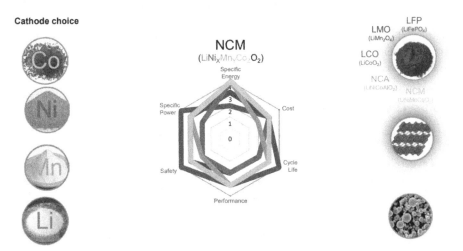

FIGURE 2.21 Cathode material choices comparison.

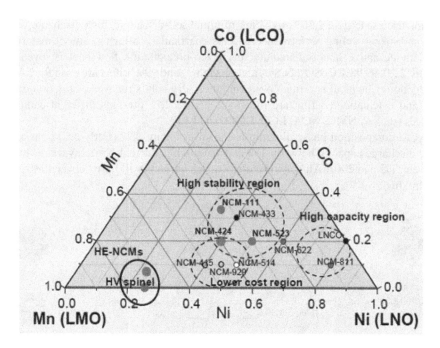

FIGURE 2.22   NCM composition phase diagram.

Early NCM variants (past): NCM 111 – Discharge capacity: ~150 mAh/g and
NCM 523 – Discharge capacity: ~165 mAh/g
Ni-rich variants (now): NCM 622 – Discharge capacity: ~175 mAh/g and
NCM 811 – Discharge capacity: ~200 mAh/g
Mn-rich variants (future): HE-NCM – Discharge capacity: >250 mAh/g and
HV-Spinel – Discharge capacity: ~140 mAh/g.

Figure 2.23 shows the comparison of Li with NCM metal oxide cathode-specific energies. Dendrite formation and controlled Li deposition are key challenges for Li and metal oxide cathodes.

2.8.6   Li vs S as Cathode

Figure 2.24 compares the specific energy capacity of $_3$Li and $_{16}$S.

Targeted Benefits of S
- Theoretic high specific capacity: 1,675 mAh/g,
- Very lightweight cells (promises cells with >400 Wh/kg), and
- No heavy metals in the cathode.

³Li - Lithium
3,860 mAh/g

Metal oxides (NCM)
200+ mAh/g

FIGURE 2.23 Comparison of Li with NCM metal oxide cathode-specific energies.

³Li - Lithium
3,860 mAh/g

¹⁶S - Sulfur
1,675 mAh/g

FIGURE 2.24 Comparison of the specific energy capacity of $_3$Li and $_{16}$S.

Challenges
- Complex working mechanism (polysulfide formation)
- Sulfur is an electrical insulator (requires C)
- Sulfur has low density (impacts Wh/L)
- Average cell voltage is 2.1 V (1.7 V lower compared to LIB)
- Low published energy density values (50% lower than LIB)
- Poor cycle life at full depth of discharge or high rate

Lower Predicted Cost
- Sulfur costs < $150/t*
- Cobalt costs > $33,000/t*
- Nickel costs > $20,500/t*

* Prices: July 2023.

Figure 2.25 presents the Li-S battery cross-section. The anode is made of Li metal foil attached to the Cu foil, and the cathode is made of S/C composite attached to the Al foil. There is a separator between the anode and cathode. A schematic diagram of Li-S cell electrochemistry during charging and discharging is given in Figure 2.26. While discharging S_8 and Li_2S_8 convert to $L_{i2}S$ and the battery voltage drops from 2.8 to 1.7 V at a capacity of 1,000 mAh/g. During discharging the following reaction occurs:

$$S_8 + 16Li^+ + 16e^- \rightarrow 8Li_2S \tag{2.7}$$

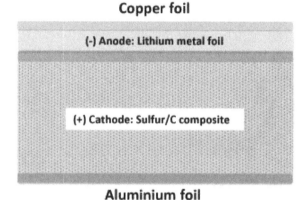

FIGURE 2.25   Li-S battery cross-section.

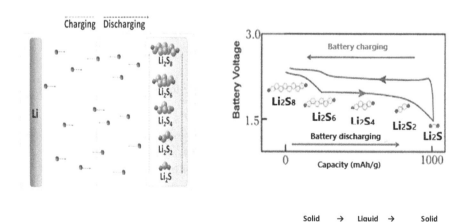

FIGURE 2.26   Schematic diagram of Li-S cell electrochemistry during charging and discharging.

FIGURE 2.27   Li-S battery by Oxis Energy.

Next-generation lithium-sulfur (Li-S) cells are being developed and produced by the UK-based cell company Oxis Energy. In comparison to more traditional cell chemistry, Li-S technology has a variety of advantages, including increased energy density, increased safety, and little environmental impact. Figure 2.27 shows the Li-S battery pouch by Oxis Energy (https://steatite-batteries.co.uk/partners/oxis-energy-batteries/).

OXIS always emphasized that its batteries are up to 60 times lighter than NMC cells and do not require materials such as Co, Mn, Ni, or Cu. OXIS batteries were promised to offer energy densities of 450 Wh/kg and 550 Wh/L (*aircraft*). (https://www.electrive.com/2021/05/21/oxis-energy-is-facing-bankruptcy/).

OXIS Energy's energy-dense Li-S cells were primarily aimed at the aviation industry. With a cathode plant in Port Talbot, Wales, and real cell manufacture in Brazil, the company has been preparing for series production since 2019. In 2020, the business leased a production site in Juiz de Fora from Mercedes-Benz Brazil for this reason. OXIS Energy said in April 2021 that it will send the first Li-S cells to clients and partners for testing in the fall of 2021. A few days later, Bye Aerospace, a US company that has been working with OXIS for some time, announced a nine-seat electric aircraft with a 1,000 km range that is powered by Li-S cells. OXIS has consistently emphasized that its batteries are 60 times lighter and less expensive its batteries than NMC cells and do not need components such as Co, Mn, Ni, or Cu. OXIS batteries were advertised as having 450 Wh/kg and 550 Wh/L (aircraft) energy densities. (https://www.electrive.com/2021/05/21/oxis-energy-is-facing-bankruptcy/).

2.8.7   LI-METAL LICERION TECHNOLOGY

Sion Power (Tuscon, Arizona) is one of the top manufacturers and developers of high-energy Li-metal rechargeable battery technology, with established dendrite-resistant technology (Figure 2.28). Licerion technology is claimed to be the future of batteries. High performance provides big energy in small packages. The patented technology uses a proprietary Li-metal anode and electrolyte coupled with a high-energy cathode (https://sionpower.com/).

Licerion batteries offer the best possible mix of specific energy, energy density, cycle life, and safety, enhancing the performance of e-mobility applications. Sion Power uses vapor-deposited Li (VDLi) in a redundant protection mechanism,

FIGURE 2.28 Sion Power Licerion technology batteries.

which is exclusive to Licerion. The patented method employs three levels of cells, packs safety, and offers an incredibly high level of energy per cell.

2.8.8 Solid-State Electrolytes and Batteries

Conventional LIB electrolytes consist of a mixture of flammable and toxic solvents. The electrolyte fills pores in the anode, cathode, and separator. Electrolytes transport Li ions between the two electrodes. Table 2.7 presents safety, energy density, power capability, manufacturing, and battery cost claims and practical considerations of solid-state electrolytes.

Solid-state electrolyte systems include polymers, oxides, and sulfides. The advantages and disadvantages of each system are presented in Table 2.8 (Wicser, 2023). Polymers have the advantages of ease of scale, cost-effectiveness, good interfacial compatibility, and high flexibility. But high-temperature requirements, lowest ionic conductivity, and limited energy density are disadvantages. Oxides have good ionic conductivity, high strength, and thermal stability advantages. However, poor interfacial compatibility, difficulty in mass production, and the requirement of high sintering temperature are disadvantages. Sulfides have the highest ionic conductivity, good interfacial compatibility, and reasonable scalability are advantages. However, reactivity with water and air, high cell pressure requirement, and toxic gas emissions (i.e. H_2S) are disadvantages.

The chronological direction of LIB with specific energy increase for cathode development is summarized in Figure 2.29. The current state-of-the-art is NCM_{811}/graphite and LFP/graphite cathodes. Silicon anodes will drive to next by a leap in Wh/kg and Wh/L. Solid-state HE or HV cathode/Li metal and silicon and Li-S will have the maximum specific energy (300–600 Wh/kg). Any significant commercialization is expected post-2030 with final practical Wh/kg and Wh/L levels still under development (www.anteotech.com/energy).

TABLE 2.7

Safety, Energy Density, Power Capability, Manufacturing, and Battery Cost of Solid-State Electrolytes and Batteries

	Claim	Practical Consideration
Safety	No flammable electrolyte leads to better safety and thermal stability	Higher energy density paired with pure lithium metal as the anode does not necessarily mean better safety
Energy density	Solid-state electrolytes enable higher energy densities	Without changing to a different anode and cathode chemistry gravimetric energy density would be reduced
Power capability (charge)	Fast charging is advertised by some companies as a feature of solid-state batteries	Most solid-state chemistries struggle to deliver fast charge performance based on interface issues and lower ionic conductivities at room temperature
Manufacturing at scale	Fewer manufacturing steps lead to a reduction in cost	Processing steps are generally more complex and may require additional high-temperature sintering steps and inert gas conditions paired with high capital investment
Battery cost	Simplified battery pack cooling (heating as opposed to cooling) and protection packaging	Battery pack heating requires re-engineering of end application while solid-state batteries generally require higher pressure to work well

TABLE 2.8

Comparison of Solid-State Systems

Polymers PEO + Additives	Oxides Perovskite, NASICON, Garnet	Sulfides Sulfide Glasses & Ceramics, Agryodite
Advantages		
Easy to scale & cost-effective processing,	Good ionic conductivity,	The highest ionic conductivity,
Good interfacial compatibility,	High strength but brittle,	Good interfacial compatibility,
Highly flexible.	Good safety (thermal stability).	Reasonably scalable.
Disadvantages		
High operating temperatures are required,	Poor interfacial compatibility (resistance),	High reactivity with water and air,
The lowest ionic conductivity,	Difficult to scale for mass manufacturing,	High cell pressure is required for performance,
Limited energy density improvement.	High sintering temperatures are required.	Can generate toxic by-products (H_2S).

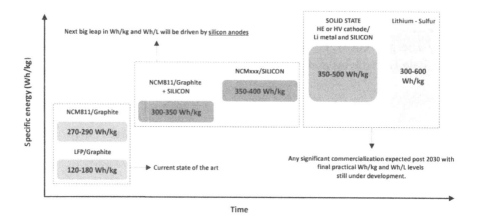

FIGURE 2.29 The future direction of LIB cathode chemistry.

2.9 TESTING LIBs

Each country producing and recycling LIBs should establish an electrochemical testing facility for batteries and material validation and certification. Firstly, the battery module, then battery system integration, and finally battery pack testing should be performed for safety testing and standards (Figure 2.30) (Talbot, 2020).

Battery cell testing can be performed up to 10 V; module testing up to 60 V (up to 2,500 A); and pack testing up to 1,000 V (Figure 2.31) (Talbot, 2020).

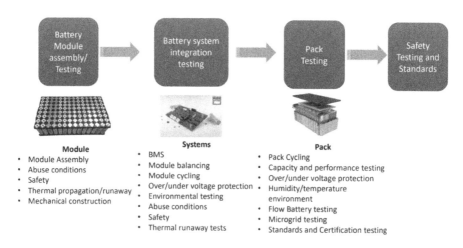

FIGURE 2.30 National battery testing center.

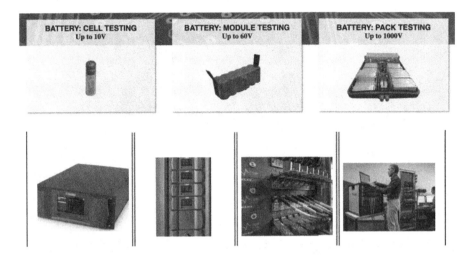

FIGURE 2.31   Battery testing limits for cells, module, and pack; battery testing equipment up to 1,000 V.

2.10   CASE STUDY 1: AUSTRALIA'S BATTERY INDUSTRY SUPPLY CHAIN

Australia is rich in battery mineral resources, but it has not developed LIB manufacturing and recycling capacities. Australia presently produces nine of ten mineral elements required to produce LIB anode and cathode and commercial resources of graphite. In Australia, the University of Melbourne (Chemical Eng. Dept.); CSIRO (Manufacturing Division); and Queensland University of Technology (LIB Pilot plant) work on anode research. Syrah Resources is one the world's largest graphite resources and mine. Talga is Europe's largest and purest graphite mine, and EcoGraf is planning a graphite purification and processing plant in Australia (Talbot, 2020).

Currently, Australia only collects 7% of generated waste. Li-ion and Li primary battery amount was 1,900 and 5,380t in 2012–2013 and 2017–2018, respectively. Average Li-containing battery collection rates were 3.5% and 13% at the same date intervals, respectively. Currently, 6,000t Li batteries are used.

2.10.1   ENVIROSTREAM AUSTRALIA PTY LTD.

Envirostream was founded in Melbourne in 2017. The company focuses on processing EoL alkaline, NiMH, and LIBs. Envirostream estimates every year, more than 300 million batteries are thrown away with ordinary household waste, meaning a staggering 8,000 end up in landfills. It was estimated that there will be 300% annual growth in S-LIB entering the waste stream by 2036. Envirostream developed modular onshore battery processing technology (Figure 2.32). Envirostream aims to recover more than 95% from all types of spent batteries (Mackenzie, 2020; https://envirostream.com.au/).

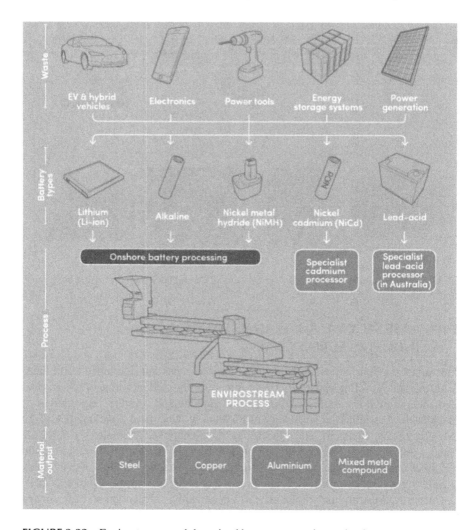

FIGURE 2.32   Envirostream modular mixed battery processing technology.

2.11   CASE STUDY 2: AMERICAN BATTERIES

2.11.1   PolyPlus Battery Company Inc.-Glass-Protected Lithium Battery

The Li-S battery industry is led by PolyPlus. The proprietary non-aqueous electrolyte created and patented by PolyPlus in the 1990s is used in nearly all Li-S batteries, whether they are being produced for sale or are still being developed. Additionally, PolyPlus created a low-cost water-based Li-S chemistry, significantly enhancing the capacity and cycle life of S electrodes. In 1991, PolyPlus was founded upon the invention of the Li/organosulfur battery at the Lawrence Berkley National Laboratory in California, USA. In 2000, Polyplus invented and patented the protected Li electrode (PLE™). In 2011, PolyPlus developed the first water-activated Li battery.

TIME Magazine named it the 50 Best Inventions of 2011. In 2017, PolyPlus introduced glass-protected Li metal batteries. A major invention in Li anodes (Shaibani et al., 2020).

The first conductive glass separator for rechargeable Li metal batteries has been developed by PolyPlus. Thin monolithic glass sheets with suitable flexibility and conductance for the application were identified by PolyPlus scientists, who then created a patented technique to attach Li metal to the monolithic sulfide glass electrolyte. Dendrites are blocked by the monolithic glass sheet, which also promotes effective Li cycling. Continuous thin glass may be made in large quantities and for a reasonable price.

By combining Li metal anodes with traditional Li-ion cathodes, glass-protected Li batteries can double the energy density of Li-ion batteries or offer the same amount of energy in a smaller compact. Solid-state electrolytes may allow the bulky C electrode in Li-ion cells to be replaced with Li metal, according to a long-standing theory among battery developers. By creating a system that is scalable by nature, PolyPlus took this development a step further. The abundant low-cost raw ingredients and well-known high-volume manufacturing are the foundation of the PolyPlus Glass Protected Li Metal Battery technology. We anticipate solid-state Li metal anode pricing to be competitive with Li-ion technology as the technology scales up while providing a step-change in performance increase (https://polyplus.com/product-pipeline/). Figure 2.33 shows the molten conductive glass covering of thin Li metal for solid-state anode laminate production and Figure 2.34 shows melt infusion of Li into lithiophilic scaffold.

2.11.2   SION POWER CORP.

Sion Power was started as a spin-off of Brookhaven National Laboratory in 1989 as Moltech and was founded in Tuscon-Arizona-USA. Sion Power made a strategic decision to transition from Li-S to Li-metal Licerion technology. Applications benefit from Licerion-High Energy (HE) rechargeable batteries' significant performance advantages over competing technologies. designed especially for unscrewed

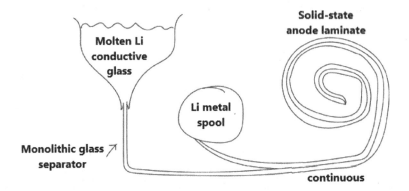

FIGURE 2.33   Molten conductive glass covering of thin Li metal for solid-state anode laminate production.

FIGURE 2.34   Melt infusion of Li into lithiophilic scaffold.

applications in the aircraft industry. Despite being lighter and smaller than a Li-ion cell, Licerion-HE has a greater capacity for energy storage. Longer trips and heavier payloads are possible, thanks to Licerion technology. The answer for next-generation EVs is Licerion EV, which meets or exceeds automotive battery standards. In a 400 Wh/kg; 780 Wh/L; and 17.4 Ah pouch cell, important characteristics include fast charge capability, extended cycle life, and broad temperature range capabilities. Protected Li Anode (PLA) was also created by Sion Power. The PLA incorporates thin, chemically inert ceramic barriers that minimize weight while increasing energy, cycle life, and safety by reducing parasitic reactions. Additionally, Sion Power tests the safety of batteries and cells which include nail penetration, short circuit, overcharge, and heat application tests (https://sionpower.com/products/).

2.12   CHINA'S BATTERY FIRMS ACCELERATE OVERSEAS EXPANSIONS

Major Chinese LIB manufacturer Sunwoda unveiled a plan on July 26, 2023 to build a new energy vehicle (NEV) power battery plant in Hungary, with an investment of no more than $273.57 M for the first phase of this project. More details, including the construction schedules and launch dates, were undisclosed. Sunwoda is not the first Chinese battery company that has invested in battery production in the central European country. China's largest producer Contemporary Amperex Technology (CATL) is building its second European battery production facility in east Hungary's Debrecen, which is expected to start pilot production next year. China's largest EV producer BYD, which is already producing electric buses in Hungary, is also planning to build a battery assembly plant in the country's Fót town. Sunwoda has been expanding its output capacity in recent years, with a combined planned capacity of 500 GWh/y in the coming years. On April 19, 2023, it released a fast-charge battery to be used in EVs that has a driving range of 1,000 km and takes only 10 minutes to charge from 20% to 80%.

Major Chinese LIB manufacturer EVE Energy plans to build a battery production facility in Thailand. It will establish a joint venture with Thailand's renewable energy producer Energy Absolute (EA) which will hold 51% of the venture, with EVE taking 49%. The battery factory will have a nameplate of capacity of more than 6 GWh/y, with an undisclosed launch date. EVE is one of China's ten biggest power battery manufacturers. It installed 6.61 GWh of Li battery power batteries in the first half of this year, accounting for 4.4% of China's total volume.

Other Chinese companies that have accelerated their plans to invest in overseas production recently include Huayou Cobalt. The firm finished building its first overseas battery recycling plant, with a processing capacity of 12,000 t/y for black mass in South Korea's Gwangyang City, and plans to build a facility for $NiSO_4$ and battery precursors in North Gyeongsang province's Pohang City with South Korean battery materials producer Posco Future M.

Chinese battery cathode producer GEM is also on track to build a battery precursor plant in Saemangeum, South Korea in partnership with South Korean battery material firms SK On and Ecopro Materials. Major Chinese LIB manufacturer Gotion High-Tech has begun building a LIB facility in Germany with a design capacity of 20 GWh/y to meet local demand.

Chinese battery feedstock firms Huayou, Tsingshan, GEM, and Lygend have also invested in production projects in Indonesia, which sits on the world's most abundant Ni resources.

The IRA has also spurred greater EV battery investments in northeast Asia, particularly Japan and South Korea this year.

Argus forecasts that global demand for LIBs will rise to 4.5 TWh by this year as a result of the expanding EV sector.

Sichuan Yahua Industrial, a significant Chinese producer of Li-salts, has extended the time frame for which it would provide LiOH to Tesla, a US manufacturer of electric vehicles, from 2025 to 2030.

The two companies have decided to extend their supply deal, which would see Yahua provide Tesla with a total of 207,000–311,000 t of battery-grade LiOH from August 1, 2023 to December 31, 2030. In December 2020, Yahua and Tesla agreed to a supply deal for the procurement of LiOH totaling $630–$880 million during the years 2021–2025. In the Sichuan region of southwest China, Yahua has a total production capacity of 73,000 t/y for Li salts, including Li_2CO_3 and LiOH, with a goal to more than quadruple that capacity to 173,000 t/y by the end of 2025.

It has also agreed to supply 20,000–30,000 t of battery-grade LiOH from its fully owned subsidiary Yahua (Ya'an) to South Korea's SK New Energy (Shanghai) between 2023 and 2025. Additionally, over the years 2023–2025, Yahua (Ya'an) will sell no more than 30,000 t to South Korea's LG Energy Solution.

As a result of increased production capacity and a ramp-up at numerous locations, Tesla raised its global EV production by 86% to 479,700 units in the second quarter of this year. In the most recent few months, the Shanghai EV gigafactory has operating at a full 750,000 units/y capacity. Additionally, it has begun construction on an internal Li refinery in Corpus Christi, Texas, to provide EV battery-grade Li to around 1 million EVs.

The two main Li salt feedstocks utilized in the manufacture of EV power batteries are Li_2CO_3 and LiOH, with the hydroxide used primarily to produce ternary battery cathode active materials and the carbonate used mostly for $LiFePO_4$.

LiOH prices reached record highs of 556,000–571,000 yuan/t ex-works in November of last year due to soaring demand from the EV segment, but have since fallen as a result of battery companies' destocking efforts and China's government's decision to end its subsidies for EV sales. Prices for 56.5% grade hydroxide were last estimated by Argus to be between 250,000 and 275,000 yuan/t ex-works on August 1 as opposed to 255,000–280,000 yuan/t ex-works on July 27 (https://www.argus-media.com/en//news/2475368-chinas-yahua-extends-li-hydroxide-supplies-to-tesla).

2.13 SUMMARY

In household and EV batteries, LIBs have clear fundamental advantages (such as high energy density, high cycle life, low overcharge, less toxicity, high voltage, and high efficiency) over other types of batteries. Li, Co, Ni, Mn, Fe, Al, Cu, F, P, and graphite can be extracted from S-LIBs. LIBs demand is significantly increasing for EVs and consumer electronics sectors.

New electrode materials are still being researched for LIBs to push the limits of cost, energy density, power density, cycle life, and safety. Although there are many potential anode and cathode materials, many of them have poor electrical conductivity, slow Li transport, dissolution or other undesirable interactions with electrolytes, poor thermal stability, rapid volume expansion, and mechanical brittleness. To solve these difficulties, several strategies have been tried. Conversion material technology is gradually getting closer to being widely commercialized as a result of numerous intercalation cathodes that have been released to the market. Recent decades have seen remarkable advancements in the study of LIB electrode materials. In the years to future, LIBs will undoubtedly have an ever-growing impact on our digitalizing lives as new materials and techniques are discovered.

Graphite, silicon, and Li metal can be used as anode material. Mainly graphite is preferred in LIBs production. Expensive synthetic and nano silicon-added graphite is more expensive than natural graphite. Silicon anodes can make batteries smaller, lighter, and cheaper with more charge/discharge cycles. Cathode materials changed from LCO to Li-S and Li-air. Solid-state cathode/Li-metal and silicon and Li-S are the future directions of cathode chemistry.

Introduced three case studies showed that Australia is rich in battery mineral resources, but it has not developed LIB manufacturing and recycling capacities. LIB collection rate was only 13% in Australia. Envirostream aims to recover 95% of all types of spent batteries. In the USA, PolyPlus Battery Com. developed water-based Li-S chemistry. PolyPlus invented the first, patented glass-protected Li metal battery technology for preventing Li dendrites. Sion Power also developed Li-S and Li metal Licerion technology for better performance. The leader in LIBs production and recycling, China battery firms accelerated their overseas expansions in Europe, Thailand, South Korea, Japan, and China for both raw and recycled LIB material production.

3 Lithium-Ion Batteries Structure, Assembly, and Functioning

ABBREVIATIONS

ASSB	All-solid-state-battery
LCO	Lithium cobalt oxide
LFP	Lithium iron phosphate
NCA	Lithium nickel cobalt aluminum oxide
NMC	Lithium nickel manganese oxide
Kts	Kilo tons
LFP	$LiFePO_4$
LIB	Lithium-ion battery
LMO	Lithium manganese oxide
LTO	Lithium titanate oxide
LMN	Lithium manganese nickel
NiCd	Nickel-cadmium
NiMH	Nickel metal hydride
NG	Natural graphite
NMP	N-methyl pyrrolidone
MCMB	Mesocarbon microbead
SG	Synthetic graphite
SEI	Solid electrolyte interface
SBR	Styrene-butadiene rubber
SEI	Solid electrode interphase
SSE	Solid-state electrode
PP	Polypropylene
PE	Polyethylene
PVDF	Polyvinylidene fluoride
MEO	Metal oxide

3.1 HISTORY OF LITHIUM-ION BATTERIES

Lithium (Li) has an atomic weight of three. It has three electrons in the first two orbits around the nucleus. The first orbit is full and the second orbit has one electron. Li tends to lose the last electron to become Li-ion (Li^+) (Figure 3.1).

The oil crises of the 1970s established the groundwork for the lithium-ion battery (LIB). Stanley Whittingham focused on techniques that might result in energy

DOI: 10.1201/9781003384557-3

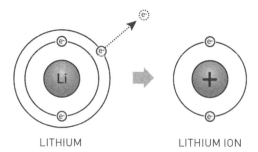

LITHIUM   LITHIUM ION

FIGURE 3.1   Atom structure of Li element and Li-ion.

technology without the need for fossil fuels. He started looking into superconductors and found a very energetic substance, which he used to develop a novel cathode for a Li battery. This was created using titanium disulfide (TiS_2), a material with molecular gaps that can intercalate Li-ions. Metallic Li, which has a powerful drive to release electrons, was used in part to create the anode of the battery. The structure of Whittingham's battery with TiS_2 cathode and Li metal anode is shown in Figure 3.2. This produced a battery with slightly over two volts and extremely high potential. The battery, however, proved too explosive to be practical since metallic Li is reactive (https://www.nobelprize.org/prizes/chemistry/2019/press-release/).

Pure TiS_2 is a semi-metal, and conductivity increases upon insertion of high-conductive Li. Li insertion varies from $1 \geq X \geq 0$ (in Li_xTiS_2), and there is a 10% expansion $TiS_2 \rightarrow LiTiS_2$. Capacity is about 250 Ah/g and voltage is around 1.9 V. This is the major limitation of the TiS_2 cathodes. The energy density is about 480 Wh/kg.

John Goodenough projected that utilizing a metal oxide as opposed to a metal sulfide would increase the cathode's potential even further. He showed in 1980 that cobalt oxide with intercalated Li-ions can yield up to four volts after conducting a thorough search. This was a significant development that would result in batteries with substantially greater power. Figure 3.3 shows the structural illustration of Goodenough's battery with cobalt oxide cathode (https://www.nobelprize.org/prizes/chemistry/2019/press-release/)

In 1985, Akira Yoshino developed the first commercially successful LIB using Goodenough's cathode as a foundation. He employed petroleum coke, a carbon substance that, like the cobalt oxide in the cathode, can intercalate Li-ions, in place of reactive Li in the anode. Figure 3.4 shows the structural illustration of Yoshino's battery with a cobalt oxide cathode and petroleum coke anode (https://www.nobelprize.org/prizes/chemistry/2019/press-release/).

The result was a thin, durable battery that could be recharged numerous times before losing performance. The advantage of LIBs is that they are based on Li-ions flowing back and forth between the anode and cathode rather than chemical processes that degrade the electrodes.

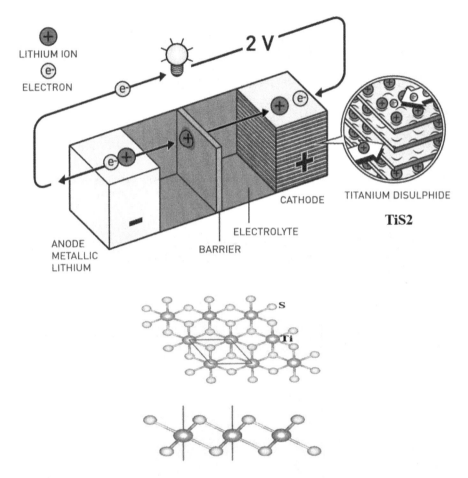

FIGURE 3.2 The structure of Whittingham's battery with TiS_2 cathode and Li metal anode.

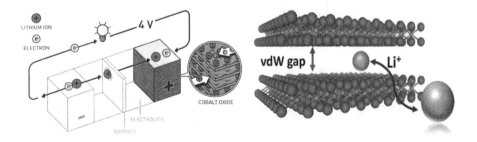

FIGURE 3.3 The structure of Goodenough's battery with cobalt oxide cathode.

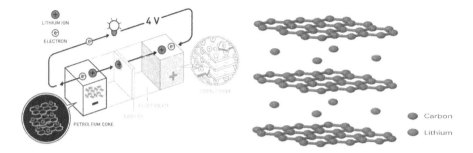

FIGURE 3.4 The structural illustration of Yoshino's battery with cobalt oxide cathode and petroleum coke anode.

2019 NOBEL PRIZE WINNERS IN CHEMISTRY "FOR THE DEVELOPMENT OF LITHIUM-ION BATTERIES"

John B. Goodenough, born 1922 in Jena, Germany. Ph.D. 1952 from the University of Chicago, USA. Virginia H. Cockrell Chair in Engineering at The University of Texas at Austin, USA.

M. Stanley Whittingham, born 1941 in the UK. Ph.D. 1968 from Oxford University, UK. Distinguished Professor at Binghamton University, State University of New York, USA.

Akira Yoshino, born 1948 in Suita, Japan. Ph.D. 2005 from Osaka University, Japan. Honorary Fellow at Asahi Kasei Corporation, Tokyo, Japan and professor at Meijo University, Nagoya, Japan.

© Nobel Media. Photo: A. Mahmoud
John B. Goodenough
Prize share: 1/3

© Nobel Media. Photo: A. Mahmoud
M. Stanley Whittingham
Prize share: 1/3

© Nobel Media. Photo: A. Mahmoud
Akira Yoshino
Prize share: 1/3

(www.nobelprize.org/uploads/2019/10/popular-chemistryprize2019.pdf)

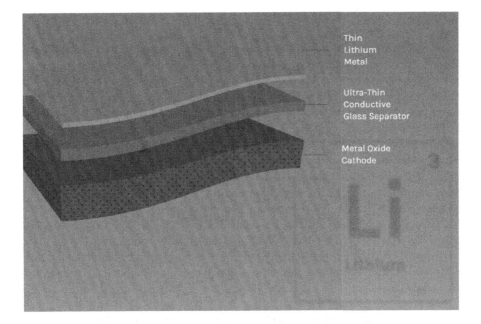

FIGURE 3.5 The first glass-protected Li metal battery in the world made by PolyPlus.

Since Sony first made LIBs available for purchase in 1991, battery makers have steadily made little advancements in the technology. However, there is little chance for progress in the energy density of Li-ion cells, which has plateaued at around 600 Wh/L and 220 Wh/kg, and the products have become commodities. Li-ion has had significant catastrophes like battery fires when attempts have been made to push it above its energy density boundaries.

The only way to meet the need for smaller, lighter batteries is with a novel strategy and a technological revolution. PolyPlus introduced Li metal + conductive glass separators. In comparison to the carbon electrode in Li-ion cells, Li metal has a volumetric capacity density (mAh/cc) that is four times higher and a gravimetric capacity density (mAh/g) that is ten times higher (Figure 3.5) (https://polyplus.com/glass-protected-lithium-battery/#future).

The energy density of rechargeable batteries will increase significantly when thin, lightweight Li metal electrodes with solid-state monolithic glass separators replace bulky, heavy C anodes, while cell safety will also increase. For bonding Li metal to continuous monolithic glass electrolytes, PolyPlus scientists created a patented technique. Continuous monolithic glass avoids dendrites and facilitates the effective cycling of Li metal, as predicted by laboratory cells.

3.2 HOW LITHIUM BATTERY WORK?

The anode, cathode, separator, electrolyte, and current collectors make up a Li battery. As the battery charges, the positive electrode releases part of its Li-ions, which go to the negative electrode through the electrolyte. This energy is captured and stored by the battery. As the Li-ions move, they liberate free electrons, which then flow as a charge to the object being powered (Figure 3.6).

Two electrodes – one positive and one negative – and an electrolyte make up a battery. The positive electrode, also known as the cathode, is composed primarily of Li and a variety of metals, including Co, Ni, Al, Mn, and others. Because of the particle characteristics of graphite, such as its chemical neutrality, thermal resistance, and thermal and electrical conductivity, the negative electrode is an anode. The S-LIBs' chemical composition is listed in Table 3.1. Numerous Li-ions combine to form the electrolyte, where appropriate circulation ensures the battery's operation. Comparatively speaking to the cathode, which can have many chemistries, the anode, and electrolytes share a similar makeup. LIBs preceded NiCd and NiMH batteries, which are still used, but with decreasing market share. Since Cd is a toxic element, the EU forbids NiCd batteries. The market share of NiCd is about 0.5% and declining. NiCd batteries contain about 40% Fe, 22% Ni, 15% Cd, 1% Co, etc. NiMH batteries are used for hybrid EVs (64%) and portable applications (34%). They contain 33% Ni, 30% Fe, 10% REE, 3% Co, 1% Zn, 1% Mn, and 22% other elements along with plastics. Metal concentrations in S-LIBs are often higher than those found in natural ores.

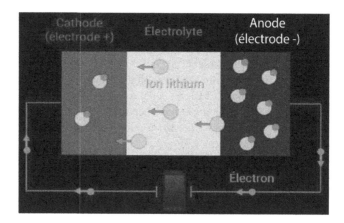

FIGURE 3.6   Functioning of an S-LIB.

TABLE 3.1
A Rough Chemical Composition of S-LIBs

Positive Electrode/Cathode	Electrolyte: Li-Ions	Negative Electrode/Anode
Al plate/foil (mass: 5%–8%)	Li salts	Cu plate/foil (mass: 8%–10%)
Active material: Salts of chosen metals Ni, Co, Mn	$LiPF_6$	Carbon-bearing material, mainly Graphite
Cathode Material Mass: 25%–30%	Electrolyte Mass: 10%–15%	Anode Material Mass: 15%–17%
Binder: PVDF	Solvent: NMP	Additive: Carbon black

3.3 LIB CATHODE TYPES

Existing technologies for LIBs with various cathode chemical compositions offer various levels of power and energy densities for various purposes. LCO, LFP, NCA, NMC, LMO, and LTO batteries. LIBs will dominate the recycling industry. Figure 3.7 shows the summary of the crystal structure and properties of different LIB cathodes. LCO and NMC have a layered structure.

Although the best cathode possibilities are Ni and Co-bearing materials because of their great stability and endurance, Co is limited. Mn is relatively less expensive, and has good rate capabilities, and a high heat threshold, but has poor cycle properties. Therefore, a combination of those three metals is chosen for LIBs that perform well and are reasonably priced (Traore and Kelebek 2023).

3.3.1 LCO (LITHIUM COBALT OXIDE ($LiCoO_2$)) BATTERIES

LCO has a high energy density but limited power and lifespan due to the lower cyclability. They are used for portable small electronic applications (such as smartphones, laptops, digital cameras, and e-bikes). But they are not adapted to transportation. Li-ions move from the anode to the cathode during the discharge mode and reverse flow from the cathode to the anode during the charging mode. The cathode has a layered metal oxide structure. LCO batteries are prone to thermal instabilities, which may cause accidents. LCO batteries may be made up of 22.8% Co, 2.7% Li, 0.2% Ni, and 8% Cu. For LCO, safety is moderate, energy density is very good, power density and cycle lifespan are good, performance is very good, and the cost is poor. The specific capacity is 140 mAh/g.

3.3.2 LFP (LITHIUM IRON PHOSPHATE ($LiFePO_4$)) BATTERIES

LFPs are distinguished by having a high power but low energy density. They are perfect for large vehicles and E-buses because of their low discharge rate. Long cycle life, superior thermal stability, and increased safety with minimal resistance are all characteristics of LFP. It typically takes 3 hours to charge to 3.65 V. LFP was discovered in 1996. About 0.5 million E-buses are in circulation globally, mostly in China. Their cathodes are composed of 7.6 g of Li, 61 g of Fe, 34 g of P, 38 g of Cu, and 13 g of stainless steel in the battery and module housing (g/kg of battery). LFP batteries have excellent safety and very good power density, cycle lifespan, and performance. The cost of LFP batteries is very good. The specific capacity is 170 mAh/g. LFP batteries are used for portable and stationary and need high load current.

3.3.3 NCA (LITHIUM NICKEL COBALT ALUMINUM OXIDE ($LiNiCoAlO_2$)) BATTERIES

NCA offers high specific energy, good specific power, and a long life span. This makes them perfectly adequate for EVs and E-bikes. The cathode contains 86 g of Ni, 16 g of Co, 13 g of Li, and 2.5 g of Al per kg of battery. Module housing contains 50 g Cu, 30 g Al, and 271 g stainless steel per kg of battery. NCA batteries are employed for medical devices, industrial, electrical powertrains, etc.

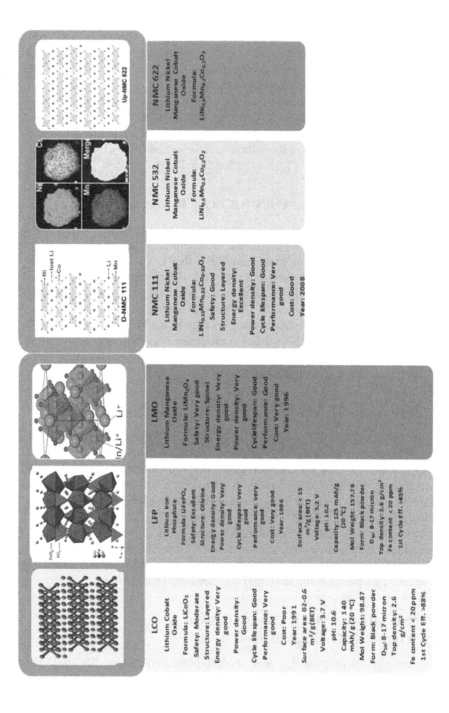

FIGURE 3.7   Comparison of crystal structure and properties of different LIB cathodes.

3.3.4 NMC (LITHIUM NICKEL MANGANESE COBALT OXIDE) BATTERIES

NMCs have high power and energy density. They are used mostly in automotive applications. Their cathode contains 30 g of Co and Ni, 3.6 g of Mn, and 14 g of Li per kg of battery. In the battery housing, there is 15 g Cu, 1.4 kg Al, and 213 g stainless steel per kg of battery (Buchert et al. 2011). NCM batteries have excellent energy density, good safety, power density, cycle lifespan, performance, and cost. The specific capacity is between 180 and 200 mAh/g. NMC batteries are used for e-bikes, medical devices, EVs, etc.

3.3.5 LMO (LITHIUM MANGANESE OXIDE ($LiMn_2O_4$)) BATTERIES

Lithium manganese oxide, or LMO, is a three-dimensional spinel structure (cubic close-packed oxides with eight tetrahedral and four octahedral sites per unit), which enhances ion flow on the electrode and lowers internal resistance while increasing current handling. Most Li-manganese batteries incorporate lithium nickel manganese cobalt (NMC) oxide to increase the specific energy and lengthen the life. The majority of electric vehicle manufacturers used this LMO and NMC combination while creating their next-generation vehicles, including the Nissan Leaf, Chevrolet Volt, and BMW i3. The specific capacity is between 100 and 120 mAh/g. LMO batteries are used for power tools, medical devices, etc.

3.3.6 LTO (LITHIUM TITANATE (Li_2TiO_3))

Li titanate replaces graphite in the anode in the typical LIB. The LTO layer forms a spinel structure. LMO or NMC is used in the cathode, and Li titanate is used in the anode. LTO (commonly $Li_4Ti_5O_{12}$) has advantages over the conventional cobalt-blended Li-ion with the graphite anode by attaining zero-strain property, no solid oxide interface film formation, and no Li plating when fast charging and charging at low temperatures. Li titanate has excellent safety, a long life span, and good performance at low temperatures. The specific capacity is between 50 and 80 Wh/kg.

LFP sales (36%) topped the cathode material market in 2016, followed by LCO (25%), NMC (26%), NCA (9%), and LMO (8%). For all battery producers, using Co presents challenges related to cost and ethical sourcing. Every manufacturer makes an effort to lower the Co content of the batteries.

3.4 ANODE MATERIALS

Mesocarbon microbeads (MCMB) and synthetic (45%) and natural (55%) graphite are two examples of carbon-bearing materials that are utilized to produce the anode material for LIBs. These materials cost between 25% and 28% of the overall cost of each battery (Pan 2020; Traore and Kelebek 2023). About 15%–17% of the weight of the battery is made up of anode materials. Other anodes include tin, transition metal oxides, sulfides, phosphides, and nitrides, surface-functionalized silicon, and high-performance powdered graphene (Arshad et al. 2020; Fan et al. 2012) and carbonaceous materials (carbon nanotubes, nanofibers, porous carbon, carbon black germanium, artificial graphite scraps from carbon materials plants (i.e. petroleum coke graphite, and needle coke graphite).

According to Chehreh Chelgani et al. (2015) and Shaw (2013), the new generation of batteries consumes more than 25% of the world's graphite production. High charge and discharge capacity (per unit weight or volume), little early irreversible capacity losses, excellent charge and discharge cycle characteristics, high electrical conductivity, eco-friendliness, and affordability are expected from a suitable LIB anode. Due to its cycling performance, synthetic graphite (SG) is the most sought-after material, but producing it requires heating carbon precursors to temperatures above 2,800°C, which uses more energy than recovering natural graphite (NG) from sources. It is also important to remember that although though NG has a low price and great performance, its manufacture requires additional processes to make it battery-grade graphite known as spheroidal (rounded particles with a lower surface area) and of purity of 99.95% (Pan 2020; Schulz et al. 2018).

3.4.1 Natural Graphite Structure

A non-metallic material called graphite is increasingly used in the latest generation of LIBs, as well as high-temperature lubricants and electrode materials in industrial electric smelting furnace operations (Jara et al. 2019; Schulz et al. 2018). Inherent hydrophobicity is related to its layer crystal structure (Ubbelohde and Lewis 1960). The layer is held together by a weak van der Waals bond, which determines the natural hydrophobicity and thus floatability. Contact angles can be as high as 80°.

3.4.2 Purification and Surface Modification of Graphite Electrode

The LIB's reversible capacity increases with increasing graphite purity (Suzuki et al. 1999). Therefore, further purification of either natural or SG is required to first get rid of some impurities to make the graphite suitable for an anode-grade battery, which is what it needs to be. Some of these contaminants are even intercalated between graphene layers, mechanically bonded to the graphite surface. Impurity interactions with electrolytes may cause unintended electrolyte degradation and anode aging in LIBs (Traore and Kelebek 2023).

Surface modification methods such as mild oxidation, metal and metal oxide deposition, coating with polymers, coating with other types of carbons, a combination of high-temperature gas treatment and sialylation, and coating polymers via blending can all be used to improve the electrochemical properties of graphite. Purification of the graphite surface improves the electrochemical performance of the electrode materials, including their capacity for reversible reactions, their efficiency during the first cycle of the Columbic reaction, their cycling behavior, and their capacity for high rates (Fu et al. 2006). The many forms of changes and their impact on the surfaces of graphite and its electrochemical properties were reviewed by Traore and Kelebek (2023).

The sharp edges and chips of NGs are smoothened/rounded and become spherical after oxidation followed by calcination and polymerization (Figure 3.8) (Zhao et al. 2008). These modifications have a distinct effect on graphite. The objective, which is

FIGURE 3.8   SEM image of native graphite (a) and modified by thermally and oxidatively treated graphite (b) (Zhao et al. 2008).

to enhance the surface structure of graphite and electrochemical capabilities, has not changed. Smoothing the active edge surfaces (removing impurities), creating a dense oxide layer on the graphite surface, covering active edge structures on the graphite surface with the creation of nanochannels/micropores, increasing electronic conductivity, and inhibiting structural changes during cycling are some improvements in a decrease in the thickness of the solid electrolyte interface (SEI) layer and an increase in the number of host sites for lithium storage.

3.4.3 SURFACE PROPERTIES OF THE SPENT ANODE

Passivation and exfoliation have an impact on the anode's surface during the battery life. The surface of the anode becomes coated with both inorganic and organic material as a result of passivation and exfoliation, which may affect how well S-LIB electrode components separate during flotation. Solvent reductive breakdown and the development of a SEI coating result in anode passivation. This SEI extends battery life by preventing compounds from coming into direct contact with electrolytes through Li intercalation. It is made of electrolyte decomposition products. Because SEI has a thin layer (10–50 nm), analyzing its composition can be difficult. However, several investigations discovered that the SEI is constituted of an inner SEI that is primarily composed of lithium-bearing inorganic species such as LiF or Li_2CO_3, Li_2O, and an outside SEI layer with solvent reduction products as main components (Heiskanen et al. 2019).

The second concept results from an exfoliation process, the gaseous decomposition of solvent molecules co-intercalated with Li-ions between graphene sheets (Balasooriya et al. 2007). The C atoms are not saturated on the margins of graphite, which increases their reactivity to the electrolytes. This is likely to cause the electrolyte solution to break down during the first cycle's charging process, which will lower the Columbic efficiency. The primary causes of exfoliation are the type of electrolyte used and the surface chemistry of the graphite edges.

Due to the presence of a binder, the spent graphite's SEM study (Figure 3.9) reveals a compact structure (Li et al. 2021). A detailed examination of the magnified figure reveals that the particles have a relatively round surface but poor smoothness.

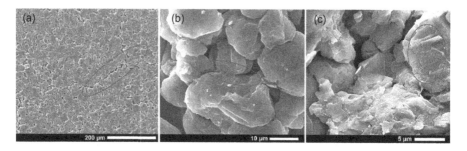

FIGURE 3.9 SEM images of spent graphite from LIBs with different magnifications (Li et al. 2021).

In contrast to surface smoothness, the original layer structure of the spent anode is not as significantly impacted. However, the charge and discharge cycles that result in the anode's expansion and shrinking put certain strains on the lattice layer. Excessive expansion may cause permanent structural damage and cause the anode lattice to collapse (Yu et al. 2021). Additionally, organic compounds from plasticizers, PVDF, and electrolytes were discovered on the surface. As mentioned in this section, graphite is altered before being used in batteries to enhance its electrochemical properties. When a battery is applied, its surface characteristics are also altered. Graphite's intrinsic hydrophobicity makes flotation a practical method for separating metal oxides from graphite. Therefore, all the modification techniques, the coating of SEI, and the induced stresses on the crystal structure may have an important impact on the flotation of this material from the cathodes.

3.5 ELECTROLYTE

As a medium, electrolytes enable the transfer of ions between electrodes and the transformation of chemical energy into electrical energy. Electrolytes come in a variety of forms, including liquid, polymer, and solid-state. Li salts, such as $LiPF_6$, $LIBF_4$, $LiCF_3SO_3$, or $Li(SO_2CF_3)_2$, are liquid electrolytes. $LiPF_6$ is by far the most often utilized. These solvents all contain fluorine compounds, which accounts for their flammability and volatility with flashpoints ranging from 16°C to 33°C (Claus 2011). Solid-state electrolytes (SSEs) have gained popularity recently.

3.6 SEPARATOR

A separator maintains a consistent distance between the anode and the cathode and guards against short-circuiting due to electrode contact. Polymers like polyethylene (PE) or polypropylene (PP) are the most often utilized separator materials. The separator serves as a safety feature by locking the electrodes apart if the cell overheats. The separator must be easily wettable by electrolytes for a high-performance battery, which is typically not the case with standard electrolytes. Separators face additional difficulties relating to their thermal stability and width, both of which are crucial for the developing electric vehicle (EV) sectors. Separators must be updated by methods such as hydrophilic monomer to handle the electrolyte compatibility issue, ceramic coating for thermal performance, and polymer coating to fulfill modern criteria (DeMeuse 2020).

3.7 ADDITIVES

The LIBs also contain additives. $LiPF_6$ salt stabilizer, safety protection agent, SEI forming improver, cathode protection agent, SEI forming improver, and other agents such as solvation enhancer, Al corrosion inhibitor, as well as wetting agents are useful additives that perform the following tasks (Zhang 2006).

3.8 BATTERY ASSEMBLY

The cathode and anode active materials or powders must adhere to their corresponding Al or Cu collector foils during the manufacturing of cylindrical cell batteries. Active powders, a solvent (n-methyl pyrrolidone (NMP)), additives, and binders – polymers that can withstand both heat and electricity – are combined to create a paste for this purpose. Additionally, carbon black can be used to increase conductivity. The most common binders used are Styrene-Butadiene Rubber (SBR) and Polyvinylidene Fluoride (PVDF), which have the large molecular weight and suitable surface chemistry needed for binders of the battery grade. The most widely used material, PVDF, has several benefits, including quick dissolution, simple processing, high throughput, consistent slurry viscosity, and extremely high adhesion during numerous cycles and extensive temperature changes (Claus 2011; Traore and Kelebek 2023). The goal of the NMP solvent is to form a uniform slurry to facilitate coating on the collector plates. NMP is then dried and removed. Because of its high cost and environmental risks (resulting from its flammability and toxicity), alternative solvents are being investigated. Figure 3.10 shows the materials used in battery, electrode manufacturing, and the battery assembly process.

3.9 PERFORMANCE OF VARIOUS COMMERCIAL CATHODE MATERIALS

Table 3.2 presents the average potential difference, specific capacity, and specific energy of various commercial cathode materials (Zou et al. 2013). Average potential difference changes between 3.3 and 4.0 V, specific capacity between 140 and 180 mAh/g· and specific energy between 0.518 and 0.630 kWh/kg. LCO and LMN have toxic metals. LFP has the highest cycle life and LMO has the lowest cycle life. NMC has the highest thermal stability. LCO has moderate thermal stability.

3.10 INTRINSIC VALUE OF LIB CATHODE MATERIALS

Figure 3.11 compares the intrinsic value of the most popular used cathode materials in one ton of S-LIB. LCO is the most expensive while LFP is the most affordable cathode material. The value of one ton of S-LIBs is \$7,708, and the cathode materials comprise \$6,101 of the total (Gratz et al. 2014). The improperly managed S-LIBs threaten biological health and the environment, and wastes resources.

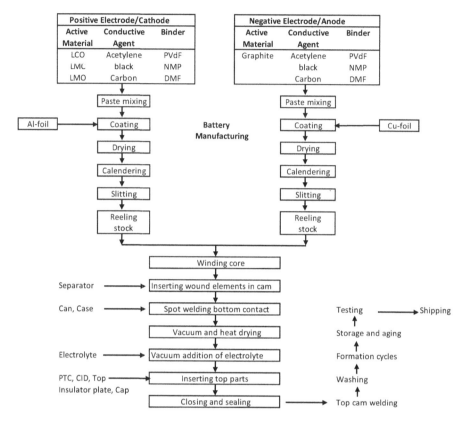

FIGURE 3.10 Schematic of battery assembly processes. (Life-cycle analysis for LIB production and recycling (Gaines 2021).)

TABLE 3.2
Performance of Various Commercial Cathode Materials (Compiled from He et al., 2017).

Cathode Material	Ave. Pot. Difference (V)	Specific Capacity (mAh/g)	Specific Energy (kwh/kg)	Cycle (Time)	Thermal Stability
$LiCoO_2$[a]	3.7	140	0.518	500–1,000	Moderate
$LiMn_2O_4$[c]	4.0	148	0.592	300–700	Low
$LiFePO_4$[c]	3.3	170	0.495	1,000–2,000	Low
$LiNiO_2$	3.5	180	0.630		
$LiNi_{0.33}Mn_{0.33}Co_{0.33}O_2$[b]	3.6	160	0.576	1,000–2,000	High

Source: Compiled from He et al. (2017).
[a] Co is toxic.
[b] Co and Ni are toxic.
[c] Environmentally friendly.

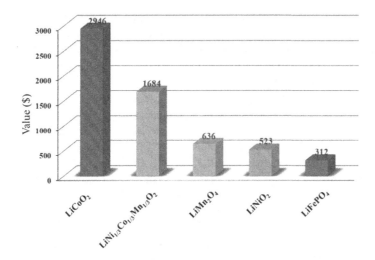

FIGURE 3.11 Intrinsic value of CAM in 1 ton of LIB (Georgi-Maschler et al. 2012; Gratz et al. 2014).

3.11 CHALLENGES IN THE BATTERY STOCK COMPOSITION

It is difficult to make changes to the battery stock composition. The market for LIBs will continue to improve, as shown in Table 3.3. New technologies should start to show up in battery manufacture during the next few years. All of these methods are, however, plagued by technical issues that limit their applicability to specific markets.

The All-Solid-State-Battery (ASSB) employing Li metal and solid electrolyte with a Li anode is the most promising large-scale development. New active materials should raise the battery's capacity, voltage, and density, just as the electrolyte's inorganic condition should increase its safety and minimize the need for a fire protection system, which would decrease the mass and price of the batteries. The most likely time for the commercialization of this technology is between 2025 and 2030. NiCd and NiMH batteries from hybrid EVs have already been recycled, according to LIBs' reevaluation of

TABLE 3.3
Developments in LIBs

New Technology	Use Area	Producer	References
Lithium titanate oxide (LTO)	EVs, e-bikes, mobile phones, laptops, cameras	Mitsubishi, Honda	Dong Joon and Xing Chang (2015)
Nickel-zinc batteries	Heavy vehicles		Parker et al. (2017)
Redox flows, vanadium technologies, and Li-silicon technologies	Stationary applications		Diez et al. (2021)
Titanium-niobium-oxide	Next-generation LIBs	Toshiba	Toshiba (2017)
Lithium-sulfur	Space applications		Nestoridi and Barde (2017)

ASSB, all-solid-state-battery.

the recycling sector. However, the stock's makeup is transitioning to LIBs as a result of rising EV sales. But their different chemical composition will be a key element in their recyclability. Currently, recent trends in metal prices, recycling a battery without Co and even Ni, like an LMO, will be far less profitable. Furthermore, changing chemical compositions require different hydrometallurgical processes and increase processing costs. As a result, the composition of a battery leads to different uses, directly linked to their recycling. Indeed, electrical and electronic equipment (EEEs), which still represent the majority of the LIBs in circulation, are less easily collected and recycled.

A major challenge arises from the optimization of LIB standards and sourcing to allow for the best collection techniques, knowledge of LIB chemistry, and an affordable recycling process, which will require automation and a large-scale process to minimize cost.

3.12   THE NEED TO RECYCLE S-LIBs

The main drivers of S-LIBs recycling are the energy consumption and Green House Gas Emission (GHGE) during battery production, the environmental and financial risks associated with landfill approaches, and the scarcity of some essential natural resources (Harper et al. 2019; Pinegar and Smith 2019) (Table 3.4). Discarded LIBs

TABLE 3.4
Summary of the Need for S-LIB Recycling

Landfill, Incineration, and Stabilization Process	Scarcity of Critical Raw Materials (CRMs)	Depletion of Natural Resources
Used for irreversibly damaged LIBs **Disadvantages:** • Groundwater pollution, which poses a risk to public safety. These are brought on by microorganisms that corrode the battery shell and allow water to seep out heavy metals. • The electrolyte can also react with water to create hazardous HF fumes, while the current Li can react with water to set things ablaze and explode. • Land loss.	The method of recycling will be required due to the rising demand for LIBs and the limited availability of the natural resources needed for battery production. In 2017, the global demand for this metal, around 38 Kts of refined metal, was expected to reach some 117 Kts by 2025, which is three times its value in 2017. The revenue from the LIBs market was approximately 35.3 B$ in 2020 and is expected to double by 2025.	When using the landfill strategy, recycling is anticipated to lessen the exploitation of natural resources, minimize energy use, and eliminate the harmful environmental effects of hazardous components. According to Ordonez et al. (2016), the recovery of Co and Ni from S-LIBs can reduce the demand for fossil fuels by 45.3%, natural resources by 51.3%, and nuclear energy by 57.2%. However, it was noted that recycling would not be able to stop the eventual depletion of resources even if all S-LIBs were recycled at this time (Zeng et al. 2014).

Incineration causes material loss, toxic gas emissions, fires, explosions, and water pollution.

Source: Adapted from Traore and Kelebek (2023).

can be employed in stationary storage or low-speed transportation where less energy is needed. They must be dealt with using techniques such as landfilling, incineration, stabilization, or recycling after their second life is up (Gaines et al. 2021). The only realistic option for a LIB market solution that is also economically, environmentally, and socially sustainable is recycling. Despite the vast majority of techniques and for financial reasons, LIBs recycling is still in its early stages.

Closed-cycle LIBs recycling is urgently required, taking into account safety, environmental effects, and financial affordability. Other battery parts that are lost to the slag in pyrometallurgy or related to contaminants after Co leaching should also be recovered, such as wasted anode materials, binders, additives, and solvents (Sommerville et al. 2021). Recycled anodes and electrolytes were the subject of a recent review by Arshad et al. (2020). The recycling strategy that separates the cathode and anode without changing their morphology, chemical composition, or structural makeup will be beneficial.

CAMs are mostly made up of hydrophilic metal oxides, in contrast to graphite's naturally hydrophobic composition, which is advantageous for flotation separations. Typically, flotation comes before pyro or hydrometallurgy as an important and economical Mineral Processing phase of upgrading (Crundwell et al. 2011). The possibility of flotation for the separation of the cathode and anode materials has been investigated by numerous researchers. As will be mentioned, graphite is the material that floats in all chapters that examine recycling S-LIBs while $LiCoO_2$ is left in the pulp. It is a reverse flotation technique for the metal oxides (MEOs), which are the most desired components. Therefore, selective flotation of graphite is necessary for this process to attain acceptable efficiency.

4 Lithium-Ion Battery Recycling Technologies

ABBREVIATIONS

ANL	Argonne national laboratory
BSM	Battery management system
C	Carbon
CAM	Cathode active material
DEC	Diethyl carbonate
DMAC	N,N-Dimethylacetamide
DMF	N-Dimethylformamide
DMSO	Dimethyl sulfoxide
DR2	Direct-Recycle-Reuse
EC	Ethylene carbonate
EMC	Ethyl methyl carbonate
ERP	Extended producer responsibility
EU	European Union
EW	Electrowinning
HC	Hydrocarbon
IEA	International Energy Agency
LAB	Lead-acid battery
LCO	Lithium cobalt oxide
LFP	Lithium iron phosphate
MC	Mechanochemical
MIBC	Metyl isobutyl carbinol
NMC	Nickel manganese cobalt
NMP	N-Methyl-2-pyrrolidine
LIB	Lithium-ion battery
PC	Polycarbonate
PE	Polyethylene
PVC	Polyvinyl chloride
PVDF	Polyvinylidene fluoride
PTFE	Polytetrafluoroethylene
SAE	Society of automobile engineers
SEI	Solid electrolyte interphase
S-LIB	Spent lithium-ion battery
SX	Solvent extraction
TFA	Trifluoroacetic acid

DOI: 10.1201/9781003384557-4

4.1 CURRENT S-LIB RECYCLING TECHNOLOGIES

Currently, pyrometallurgy, hydrometallurgy, direct recycling, and hybrid procedures are the four main LIB-recycling methods (Figure 4.1). Hydrometallurgical processes use either reagents (leaching) or living organisms (bioleaching). Calcination, roasting, reduction, or chlorine processes can be used in pyrometallurgical processes. Direct recycling can be performed by solid-state synthesis, hydrothermal, electrothermal, or chemical processes. Hybrid processes can use both pyro and hydrometallurgical processes together. Figure 4.2 summarizes the production steps in each process. The pyrometallurgical process is a high-temperature smelting technique, which often involves burning and subsequent hydrometallurgical separation. Utilizing aqueous chemistry, the hydrometallurgical process is carried out by leaching in acids/bases and followed by subsequent concentration and purification. Direct recycling extracts and recovers LIBs' active components while preserving the integrity of the original compound. Direct recycling is being used at a lab and pilot scale, whereas pyrometallurgy and hydrometallurgy are being used on an industrial basis.

Various separation technologies used in mineral processing can be used for S-LIBs recycling (Table 4.1). Particle size, specific gravity, magnetism, and electrostatic conductance were utilized for sorting S-LIB materials. Wet or dry magnetic separators can be used for magnetic steel casing separation. Fine Co and Ni-containing materials can be separated from coarse Cu and Al foils of S-LIBs using sieving. Similar-sized plastic, Al, and Cu components can be separated using density difference by gravity separation. Wet shaking tables and dry air separators can separate plastics from metals (Al and Cu) at +200 µm size. The use of toxic

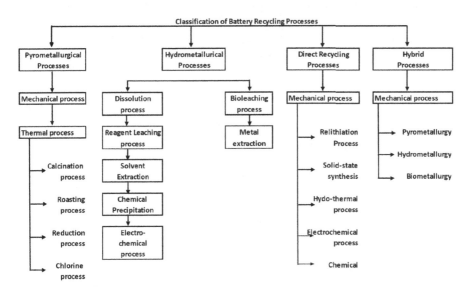

FIGURE 4.1 Classification of LIB recycling processes.

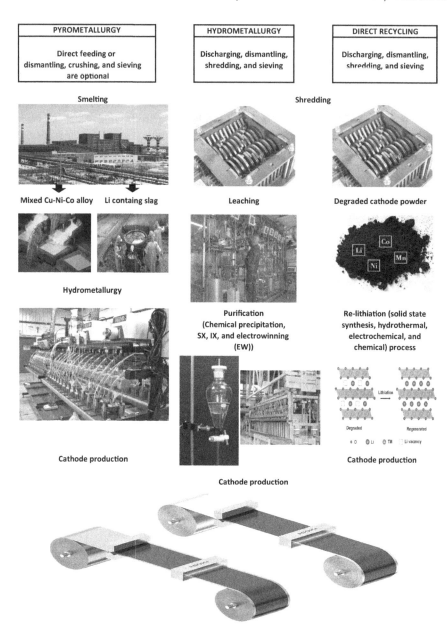

FIGURE 4.2   Different LIB recycling technologies.

TABLE 4.1

Various Separation Methods Used for S-LIBs Separation

Properties	Wet or Dry Process	Separation/Sorting	References
Particle size (200 μm)	Wet/dry sieving	Fines: Co and Li Coarse: Al and Cu Al and Cu can not be separated by size from each other	Bertuol et al. (2015) Wang et al. (2016b) Widijatmoko et al. (2020a, b)
Specific gravity (SG)	Wet shaking table Dry air classifier Heavy medium separation	At +200 μm Plastic and Al-Cu separation Anode & cathode separation	Tedjar and Foudraz (2010) Wills and Finch (2015) Kepler et al. (2016)
Magnetic susceptibility	Wet in fine sizes or dry in coarse sizes	Steel casing sorting	Shin et al. (2005)
Electrostatic conductance	Dry process, $LiCoO_2$ material	Metallic Cu and Al are recovered. Electrode active materials, polymers, mixture (polymers-metal), and the metallic fraction can be separated	Cui and Forssberg (2003) Widijatmoko et al. (2020a, b)
Leaching + precipitation at pH: 4.5 using NaOH (5% Cu is acceptable in NMC)	Wet process	Leachable: $LiCoO_2$, Fe, Al Impurities: Al^{3+}, Cu^{2+}, Fe^{3+} Non-leachable: Graphite, polymeric materials	Widijatmoko et al. (2020a, b)

bromoform as a heavy media was also used. Mechanical, pyrometallurgical, and hydrometallurgical methods can be employed in S-LIBs indirect recycling.

Chen et al. (2016) reused valuable CAMs from waste $LiFePO_4$ batteries. In their method, CAMs were manually removed from Al foils. Cathode materials can be regenerated while preserving functional integrity and reused in new LIBs (Ganter et al., 2014; Li et al., 2013).

Figure 4.3 shows recycling methods of mixed anode/negative and cathode/positive electrode materials from S-LIBs at fine size fractions. After shredding and crushing under −200 μm, expensive cathode material purification can be achieved by thermal treatment, physical separation, and chemical separation. Volatile organic materials (VOCs) (such as binders, carbon black, and graphite) are burned off at temperatures between 500°C and 800°C. Physical (such as heavy medium separation) or physicochemical (froth flotation) separation processes can also be employed for anode and cathode material separation. Chemical leaching, precipitation, solvent extraction (SX), and electrowinning (EW) can also recover cathode material. Physical and physicochemical separation systems separate mixed fine materials more cost-effectively and energy-efficiently.

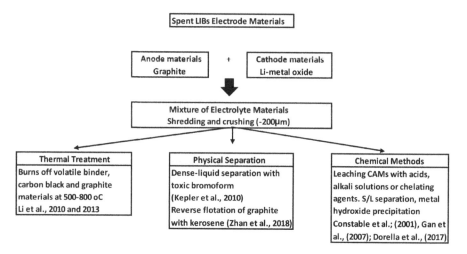

FIGURE 4.3   Recycling technologies for S-LIB electrode materials.

Froth flotation takes advantage of the difference in initial surface wettabilities of the graphite and cathode active materials. According to Zhan et al. (2018), froth flotation is used to extract hydrophilic cathodic $LiCoO_2$ from fine hydrophobic anodic graphite. In flotation, the collector is kerosene (1–2 kg/t)/n-dodecane, while the frother is MIBC/2-octanol. It only takes 2 minutes to condition the frother and 3 minutes to condition the collector. And −212 µm particle size and 1%–2% w/w solid slurry are employed for superior flotation results. To get rid of additional electrolytes before floating, fine materials can be washed with distilled water. More than 95% of the graphite could be recovered in 4 minutes of floating time.

Cathode active materials (CAMs) without graphite should be present in the tailing. At the expense of recovery, kerosene dosage increases CAM contents. Without kerosene, CAM concentration is low, but recovery is strong. Only a few investigations have been conducted on the flotation of LIB material (Kim et al., 2004). Binders were burnt off before floating for an hour at 500°C. After that, flotation is used. Anode, cathode, mixed new, and S-LIBs material's flotation behavior may all be predicted. Flotation has a lot of advantages for recycling LIBs. First off, throughout the recycling process, the physicochemical separation method of flotation successfully maintains the functional integrity of electrode materials. CAMs can therefore be recycled using a lithiation procedure and employed in fresh batteries (Chen et al., 2016). Additionally, flotation divides fine materials based on their chemistry. This greatly reduces secondary waste while improving the energy efficiency of downstream processes. As a high-value feedstock for hydrometallurgical or pyrometallurgical processes, flotation creates metal-reach products. Last but not least, full plant-scale production of LIB materials by flotation beneficiation is commercially viable. For the purpose of recycling LIBs, Zhan et al. (2018) proposed the Direct-Recycle-Reuse (DR2) froth flotation technique, particularly for the inexpensive cathode materials $LiMn_2O_4$ and $LiFePO_4$. For these kinds of LIB chemistries, conventional pyrometallurgy and hydrometallurgy are not cost-effective. Concerns about CAM's purity can arise. Alkaline solutions have the ability to dissolve tiny or ultrafine Al contaminants. By using organic solvents, the binder can be redissolved. $LiCoO_2$ (4.9 g/cm^3) and graphite (2.26 g/cm^3) can be separated by gravity.

4.2 RECYCLING PROCESS

Recycling allows valuable metals/materials to be recovered within a circular economy framework (Figure 4.4). Recycling consumes energy, water, and chemicals, but much less energy than the primary production of Al (95%), Cu (85%), and Fe (74%). A closed-loop process for LIBs would cut out more than 50% of the environmental impact of their manufacturing process.

A wide range of applications including all physical and chemical processes are necessary for an integrated S-LIBs recycling process. It is crucial to choose pretreatment-based mineral processing principles to separate precious materials and eliminate pollutants before chemical treatments are done in the final stage (Cuhadar et al., 2023). Following the collection of S-LIBs, the recycling process may involve the following steps: shredding, disassembly, combustion, or acid leaching for the extraction of elements and their recovery (Figure 4.5). Mechanical, pyrometallurgical, hydrometallurgical, combined pyro and hydrometallurgical, or direct extraction methods can be used. A biometallurgical process uses microorganisms, which allow the recovery of insoluble substances under an aqueous form. The recovered elements can then be purified through other processes. Even though it uses less energy, this process typically takes longer than conventional ones and is only effective with

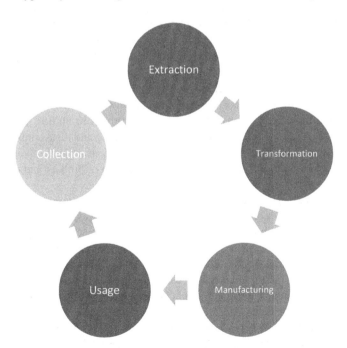

FIGURE 4.4   A closed-loop circular economy framework for LIB recycling.

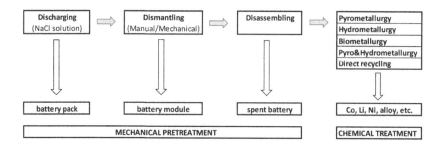

FIGURE 4.5   A comprehensive process of recycling S-LIBs.

very low S/L ratios. On a large industrial scale, these techniques are not yet applicable. Discharging the battery pack, disassembling the battery module, and disassembling the S-LIBs are the three processes of mechanical pretreatment. Following that, pyro- or hydrometallurgical techniques are typically employed. The recovery of important metals is the main focus of pyro- and hydrometallurgy methods. Table 4.2 lists the recovered materials, benefits, and drawbacks of various LIB recycling procedures.

The pyrometallurgical process refers to the incineration or smelting of LIBs at very high temperatures, which eliminates organic elements (the C/graphite anode) and the polymer separator/binders by burning off. Three main processes are involved in pyrometallurgical treatment: pyrolysis (the degradation of organic components); metals reduction for the creation of metal alloys at a temperature of about 1,500°C; and gas incineration and quenching at a temperature of about 1,000°C to prevent the release of dioxin. This leaves a powder of Ni-Co alloys, which are then treated chemically. This process consumes high energy and loses Li and Al in the slags (Meshram et al., 2015; Gaines, 2018; Zhang et al., 2020). Furthermore, mixed Ni-Co alloys need to retreat to obtain Ni-Co sulfates to reuse them in new batteries. Loss of Li and Mn in the slag and high energy consumption are some other difficulties. The benefit of the pyrometallurgical treatment is avoiding the shredding and disassembling phases. Everything burns, and there is some energy recovery from organic materials to fuel furnaces. To handle the HF produced by the breakdown of binders and electrolytes, costly gas treatment equipment is needed.

The hydrometallurgical process requires mechanical pretreatment. But shredding and dismantling phases generate a loss of materials and raise safety issues due to the remaining charge in the battery. They undergo pretreatment before being dissolved in acidic or alkaline solutions for elemental separation. Even though the liquid solution is almost immediately useable for the production of Ni-Co sulfate, and even though it is difficult to separate these two elements due to their similar physical properties, some elements, such as Li and Cu, may be lost in the process. To prevent harmful chemical reactions, each type of battery must also undergo a specialized hydrometallurgical treatment. However, the stages of shredding and dismantling result in material losses and raise safety concerns depending on the remaining charge of the battery. If a battery is still charged, it can explode while being shredded.

TABLE 4.2

Comparison of the Advantages and Disadvantages of Different LIB Recycling Processes along with Recovered Materials (Compiled from https://pubs.acs.org/doi/10.1021/acs.energyfuels.1c02489)

Recycling Methods	Recovered Material	Pros	Cons	Industrial Examples
Mechanical process	Li_2CO_3	1. Applicable to all battery chemistry and configuration 2. Energy consumption is low 3. Enhanced leaching efficiency 4. Cheap process	1. Mechanical process alone not sufficient to recover materials 2. Should be combined with hydrometallurgical process	1. Taxco process (USA) 2. SNAM (France)
Pyrometallurgy (Smelting)	Cathode Co, Ni, Cu, and Fe molten alloy > 90% recovery Li and Al are lost in the slag Anode: not possible to recover, burns, generate heat	1. Effectively recover Ni, Co, and Cu 2. Recycling efficiency is high 3. Remove toxicity 4. Scale up is easy and simplified logistic 5. Already exists at industrial scale and mature technique 6. No need for pretreatment	1. Inherit chemistry of smelter traps, other elements in the slag 2. Recovering Li from slag by hydro- metallurgy is expensive and inefficient 3. Most chemicals in the battery lost 4. Suboptimal from lifecycle perspective 5. Energy intensity, toxic gas emissions 6. High investment for gas cleaning system	1. Umicore (Belgium) 2. Batrec (Switzerland) 3. Accurec (Germany) 4. Nippon Mining and Metals (Japan)

(Continued)

TABLE 4.2 (Continued)

Comparison of the Advantages and Disadvantages of Different LIB Recycling Processes along with Recovered Materials (Compiled from https://pubs.acs.org/doi/10.1021/acs.energyfuels.1c02489).

Recycling Methods	Recovered Material	Pros	Cons	Industrial Examples
Hydrometallurgy (Leaching)	Cathodic Co, Li, Ni, Al, Cu, Li$_2$CO$_3$ recovable as salt > 95% Anode: destroyed	1. Easy recycling and simplified procedure to reuse as battery materials 2. Almost all material recovered at high efficiency 3. Chemistry and procedure are fairly mature 4. Low gas emission 5. Low energy consumption	1. Environmental and revenue issues in industrial scale due to use of hot water, acid/alkaline solutions, solvents 2. Source materials need known (battery chemistry) to tailor efficient process 3. Potential revenue and costs that make hydrometallurgy approaches promising are highly dependent on scale	1. Recupyl (France) 2. G&P Batteries (England) 3. Retriev Technologies (Canada) 4. Eurediez (France)
Direct recycling	Almost all components except separators Cathode Co, Li, Ni, Mn, Al, Cu recoverable as battery materials Anode: Graphite, activated carbon	1. Potential recovery of all battery materials in a reusable form 2. Higher revenues and lower environmental impacts 3. Process not dependent on a scale as pyrometallurgy and hydro metallurgy methods	1. Recently demonstrated at a workable scale and quality 2. Process depends on knowing battery composition and battery chemistry 3. Old cathode chemistries may not be desired to be reused in new generation LIBs 4. Manual approach to breaking the cell and extracting is contents may be difficult to scale	1. On To Tech (USA)

Source: Compiled from https://pubs.acs.org/doi/10.1021/acs.energyfuels.1c02489.

Lastly, solvent costs and their environmental impacts are other additional problems for hydrometallurgical methods.

A combined pyro-hydro method may be preferrable for two reasons. The first benefit of pyro treatment is the avoidance of the disadvantages linked to safety concerns resulting from the various battery types, their makeup, and their level of charge. The further separation and treatment of various components recovered in slags using various lixiviants can subsequently be done using hydrometallurgical treatment.

It is economically advantageous to recycle Li and Co from S-LIBs, and recovered precious metals can be utilized to make new cathodes. Critical metal Co, which is employed in the cathode layer as a metal oxide, is less common and more expensive than Mn and Ni. Due to Co's danger to people (i.e., its carcinogenic, mutagenic, and toxic effects on reproduction), recycling Co from LIBs is crucial. From a health and environmental perspective, recycling Ni from S-LIBs is also crucial. Because of its small amount and low spot price, Mn is not considered to be a high-value commodity. Due to the rising demand for LIBs in portable electronic devices, the significance of Li metal has expanded. Li recycling from S-LIBs has an important benefit and advantage in the development of next-generation EVs since the demand for Li may exceed the supply if not enough money is invested in Li extraction in the near future.

S-LIB recycling is currently a crucial and significant task for commercial recyclers and academic research institutions. This is due to the battery's short lifespan and additional factors like environmental impact and the possibility of recovering expensive critical materials. When is the battery deemed to be at its EoL is a major concern. First and foremost, it is important to understand that the batteries' degradation mechanisms include structural failure and particle cracking, corrosion of the current collector, binder decomposition, dissolution, dendritic formation in both cathode and anode materials, deterioration and aging of the electrolyte, puncture, and clogging of the separators. Many S-LIBs are considered at the EoL when their storage capacities fade by 20%–30% of the original storage capacity.

Four alternative metallurgical techniques (i.e. mechanical process, pyrometallurgy, hydrometallurgy, and direct physical recycling) have been used to recover important CAMs from S-LIBs. These techniques can be employed separately or in combination, as in the case of pyrometallurgy, which requires a hydrometallurgical procedure to recover valuable elements like Co and Ni. The ability to reuse some leaching chemicals after filtration and purification, the ability to reuse some leaching chemicals after filtering and purification, low gas emission, and high concentration metal recovery make hydrometallurgy the most effective recycling process out of these four general recycling methods. The type of the metal and the acid efficiency are the primary elements in the recovery of the high concentration of PLS.

4.3 DIRECT RECYCLING

All the components may be recycled together through direct recycling (Harper et al., 2019). If done in a closed loop (recycling, regeneration, and reuse), it may be more cost-effective than pyrometallurgy or hydrometallurgy (Al-Shammari and Farhad, 2021). The discarded cathodes can be directly lithiated and used as cathode materials for S-LIBs if they can be successfully recovered from direct recycling. According to several investigations, the so-recycled cathode exhibited electrochemical performance and crystal structure that was comparable to that of commercial $LiCoO_2$ (Yang et al., 2020). This procedure uses less energy, which reduces associated expenses and has a less negative effect on the environment.

The manufacturing of recycled cathode via pyrometallurgy, hydrometallurgy, and direct recycling has been found to be more energy-efficient than the creation of new $LiCoO_2$ material, to use $LiCoO_2$ as an example. The least energy-intensive of all three methods and the one that produces the fewest emissions and wastes is direct recycling. Pyrometallurgy, hydrometallurgy, and direct recycling each need 53.8, 41.6, and 6 MJ for a kilogram of $LiCoO_2$, respectively (Yang et al., 2020).

On a lab scale, direct recycling begins with a cautious harvest of used cathodes (no cross-contamination with the anode) and replenishing the Li deficiency to improve the electrode's value (Ganter et al., 2014). Either the solid-state, hydrothermal, or reductive method of lithiation is used. The cathode produced by hydrothermal treatment is purer and has more electrochemical capacity than those produced by the other methods (Wang and Whitacre, 2018). The lithiation processes were evaluated using Li-bearing salts such as LiI (Ganter et al., 2014), Li_2CO_3 (Chen et al., 2016; Li et al., 2017), LiOH (Gao et al., 2020a), and Li_2SO_4 (Shi et al., 2018).

The fresh battery performs poorly because the regenerated cathode particles contain binder (PVDF). The performance of the recycled cathode is better than that of commercial cathodes when the polymer is removed prior to lithiation (Chen et al., 2016; Gao et al., 2020a; Li et al., 2017; Shi et al., 2018). Heat treatment can recover the morphology and structure of the damaged cathode (Yang et al., 2021). Its performance is also enhanced by additional treatments like Al_2O_3 coating (Gao et al., 2020b) and vanadium (V) doping (Liu et al., 2021) on regenerated cathodes. As was previously said, direct recycling is quite alluring since the recovered cathode can reach electrochemical performances that are comparable to those of commercial cathodes. However, S-LIB is generally crushed in its entirety in pilot and plant size testing, producing a combination of electrode materials. For direct recycling, anode and cathode separation is essential (Zhang et al., 2020). To remove the graphite from the metal oxides so that it can be recycled for the production of direct cathodes, a process like flotation can be used. Due to the intrinsic wettability difference between metal oxides and graphite, this separation seems plausible. Similar to the flotation of graphite, this process may only need a small amount of inexpensive reagents, such as petroleum derivatives as collectors, a frother (such as MIBC), and occasionally a depressant when necessary (Peng et al., 2016; Shi et al., 2015).

A case study on direct recycling in which electrolytes are extracted using CO_2 was published by Sloop et al. in 2020. Figure 4.6 illustrates the procedure. Additionally, it is a method of "direct recycling." For testing purposes, the recycled elements in this study are rebuilt in battery cells. According to reports, the finished goods showed usable capability in the first complete cells produced from cathodes and anodes that were directly recycled. However, as previously stated, the direct recycling method may have trouble handling feedstocks with uncertain or poorly described provenance, and there may be financial resistance to material reuse due to the material's dubious product quality.

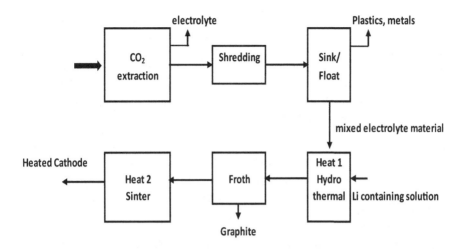

FIGURE 4.6 Direct recycling process with the extraction of the electrolyte using CO_2.

The first direct recycling of cathodes, anodes, and electrolytes from cells sourced from a battery recall was demonstrated by Sloop et al. (2020) on an industrial scale. To reduce process costs and environmental/safety risks, actions were created. These consist of

- electrolyte extraction to mitigate toxicity
- cell disassembly to reduce Fe and Cu contamination
- hydrothermal processing, which decomposes binder quickly and safely, enables easy separation of anode and cathode without dense fluids, restores Li inventory in the cathode, and removes trace metals; aqueous harvesting of electrodes to avoid organic solvents; high pH to prevent irreversible decomposition of the cathodes; and NCM one pot. In the first fully constructed cells created from directly recycled components, the finished products showed functional capabilities.

In conclusion, direct recycling seems to be a viable approach for the recycling of S-LIBs due to benefits including simplicity and affordable recovery of all battery components. As a result, the materials that are recovered can be employed right away either in new batteries or in other sectors for the viability of industrial-scale direct recycling as opposed to manual electrode recovery and disassembly.

4.4 INDIRECT RECYCLING

4.4.1 PRETREATMENT METHODS FOR S-LIBS

Pretreatments often try to segregate S-LIB materials and components according to several physical characteristics, such as form, density, conductivity, magnetic property, etc. (Kaya, 2022). The components, materials, and metallic scraps with comparable physical qualities can be separated and enriched with the aid of pretreatments, improving the recovery rate and lowering the energy need of the ensuing pyrometallurgical or hydrometallurgical processes. The benefits of pretreatments in S-LIB recycling procedures have been shown in some research. A two-stage crushing and sieving technique, for instance, was suggested by Shin et al. (2005) before the recovery of Li and Co by the hydrometallurgical approach. Their pretreatment was thought to be useful in the recycling sector and helpful for boosting the hydrometallurgical process' extraction efficiency peeling off and collecting $LiCoO_2$ from S-LIBs required thermal pretreatment and shredding, according to Lee and Rhee (2002). Li et al. (2009) combined crushing and ultrasonic washing as the pretreatment process in LIBs recycling to further increase the energy efficiency of pretreatment. In their research, they looked into the effects of the crusher screen aperture as well as the temperature and length of ultrasonic washing. The highest selectivity for enriching was determined to be a 12 mm aperture screen. For separating $LiCoO_2$ and crushed current Al-foil collector, containing scraps and 15 minutes of ultrasonic washing with agitation at room temperature were advised. Because ultrasonic washing uses so much energy, Li's approach (2009) is more energy-efficient than those published by Shin et al. (2005) and Lee and Rhee (2003).

Li and other metals that make up these batteries are incredibly expensive. The cost of raw Li is around seven times more expensive than lead (Pb), but unlike Li batteries, almost all lead-acid batteries (LABs) get recycled. Therefore, there's something beyond just pure economics at play. As it turns out there are valid justifications for why LIBs recycling hasn't happened yet taken place. However, some businesses anticipate things will change that, which is fortunate because the transition to renewable energy sources will depend heavily on LIB recycling. There is a huge difference between LIBs and LABs. According to the International Energy Agency (IEA) report in 2021, the average price of battery-grade Li_2CO_3 was \$17,000 per metric ton as compared to \$2,425 for Pb in North American markets and battery raw materials now account for more than half of battery cost. In terms of the supply of new materials, the recycling imbalance is also illogical. A recent USGS analysis estimates that there are 89 million tons of Li sources worldwide, the majority of which come from S. America. The global supply of Pb, however, was 22 times more than that of Li, at 2 billion tons.

Despite the lower availability of Li, a prior study revealed that fewer than 1% of LIBs—which are most frequently used in gas automobiles and power grids— are recycled in the USA and EU as opposed to 99% of LABs. The study found that barriers to recycling include things like rapidly changing battery technology, expensive shipment of hazardous items, and insufficient government oversight (https://arstechnica.com/science/2022/04/lithium-costs-a-lot-of-money-so-why-arent-we-recycling-lithium-batteries/).

However, it is necessary to overcome these recycling obstacles. Comparing LIBs to LABs, more energy can be stored in a smaller space. They are essential for reducing the carbon footprint of transportation and making it possible for more people to switch to renewable energy sources by assisting in ensuring a steady flow of otherwise erratic power from wind and solar. It will take a tremendous amount of effort to make these changes on a global level. As a result, since 2020, the global Li consumption has climbed by 33%. The IEA predicts a 43-fold increase in demand for Li if renewable energy targets necessary to halt climate change are to be met. The use of these batteries may be restricted by components other than Li. Batteries have elements in their anode and cathode that could also experience supply shortages.

Pretreatment is essential to obtain electrode materials (cathode and anode) from S-LIBs and move on to the next hydrometallurgy step or direct recycling. Before continuing with the processing, the S-LIB needs to be cleaned of various metals, plastics, and polymers it contains. It is very difficult to recover the pricey Co metal, which is found in large concentrations in battery cathodes (such as LCO and NMC). As a result, substantial concentrations of important metals may be lost during direct processing without pretreatment. Pretreatment has been used from the beginning of S-LIB recycling research, and during the past 10 years, a lot of studies have surfaced that concentrate on pretreatment techniques for S-LIBs. As a result, it would be appropriate to group different emerging techniques in the pretreatment stage and review the pertinent literature at this point. This section thoroughly examines the pertinent literature while concentrating solely on the pretreatment procedure itself. Discharging, disassembling, comminuting, classifying, mechanical separation, dissolving, ultrasonic washing, and thermal treatment can be used to categorize the pretreatment processes' scope and order (Kim et al., 2021). Here, each category's present condition and technological advancement are covered in great detail.

Pretreatment S-LIBs are intended to release the active components from the current collectors, carbon conductive agents, and polymer binder PVDF. To prevent thermal runaway, when poisonous and corrosive compounds can be released, as well as loss burning of electrolytes and polymers, this pretreatment often begins with S-LIB cell stabilization (Sommerville et al., 2020). In the literature, a variety of stabilizing techniques in various media, including electrical, electrolyte solution, and cryogenic, can be found (Kaya, 2022). Mechanical separation, solution treatment separation, and calcination treatment separation are the three primary categories of pretreatment methods that are used to remove the cathode and/or anode active material from the discarded battery after discharge. Table 4.3 compares the advantages and disadvantages of numerous pretreatment techniques.

4.4.1.1 Discharging

The major goal of this procedure is to lessen or eliminate the possibility of short-circuiting and spontaneous self-ignition of S-LIBs. To release any residual energy in the battery, LIBs need first to be drained. It is challenging to determine the S-LIBs' residual power. The battery may still have some power left after its charge has been completely discharged. Additionally, treating the batteries directly is ineffective due to the variety of elements included in LIBs. Consequently, a pretreatment procedure is crucial. Only when they are completely discharged, S-LIBs can be pretreated and separated safely. Otherwise, a short circuit could cause the battery to blow up or release dangerous fumes. Electronic techniques and conductive liquids/metals can

TABLE 4.3

Comparison of the Advantages and Disadvantages of Different Pretreatment Methods

Pretreatment Method	Advantages	Disadvantages
Alkaline NaOH dissolution	Simple treatment, High separation rate.	Al recovery is difficult, Alkali wastewater discharge.
Ultrasonic-assisted operation	Simple operation, No gas emission.	High noise pollution, High capital investment.
Solvent extraction (SX)	High separation efficiency	Solvents are expensive, Environmental hazards are high.
Thermal treatment (Calcination process)	Simple treatment, High capacity.	Energy consumption is high, High device investment, Toxic gas emissions,
Mechanical methods	Suitable operation, Simple operation.	Toxic gas emissions, Cannot separate mixed and all kinds of electronic components in S-LIBs

be used for discharging batteries. Ku et al. (2016) discharged S-LIBs to less than 0.1 V by using a discharger. The discharging cycle converts transition metals into a deactivated divalent state (Illes and Kekesi, 2023). Divalent metals can be easily solubilized with H_2SO_4. S-LIBs firstly discharged to 1.0–2.0 V by connecting with a load of 10–30 W halogen bulbs in a circuit to prevent explosion and firing during the disassembly process (Zhan et al., 2018). S-LIBs can also be discharged by connection to n 56 Ω resistor until 0.3 V which is safe (Widijatmoko et al., 2020). Under the fume hood, S-LIBs are unfolded to separate the anode, cathode, separator, and other parts to prevent harmful organic solvents (ethylene carbonate (EC) or ethyl methyl carbonate (EMC)).

A typical discharging method is to immerse the S-LIBs in a salt solution, such as NaCl, for electrochemical discharge. Even though NaCl is the most often used conductive solution, research on the corrosion and discharge rates of other conductive liquids is still ongoing. Because of radical oxidation, which is brought on by the mechanical shock of Li metal induced by battery overcharge from exposure to the air, LIBs frequently burst during recycling, which makes them more dangerous than ordinary batteries. He et al. (2015) totally discharged the LIBs using a 5% NaCl solution. Battery discharge time is a crucial element in determining how quickly the recycling process can go on to the next stage. Shaw-Stewart et al. (2019) investigated the discharge properties of fully charged LIBs using various saline and basic solutions. Li et al. (2016a) investigated the discharge efficiency utilizing a range of NaCl solution concentrations and discharge timings. With a 10% by-weight NaCl solution, they saw a discharge efficiency of 72% in 358 minutes. The discharge of brand-new batteries was investigated by Lu et al. (2013) using three different NaCl solution concentrations (1%, 5%, and 10%). The cell voltage dramatically decreased when the NaCl solution's concentration was high. This presents a serious risk given the way that crushing and grinding are currently done. Therefore, S-LIBs should be soaked/immersed in unsaturated salt water (such as NaCl, KCl, $NaNO_3$, Na_2SO_4, $ZnSO_4$, $MgSO_4$, $MnSO_4$, and $FeSO_4$) in order to discharge the energy inside of them prior to the crushing and grinding processes (Kim et al., 2021; Ojanen et al., 2018; Prazanova et al., 2022; Xiao et al., 2020). According to Xiao et al. (2020), the order of NaCl > $FeSO_4$ > $MnSO_4$ resulted in a decrease in total discharge efficiency. This procedure, which is the most common stabilizing technique, lessens the likelihood of material deposits in the anode short-circuiting and undergoing exothermic processes. In the case of NaCl, discharge speed rises with salt content; however, large concentrations of NaCl (approximately 20% wt.) are undesired due to the violent and quick reaction with S-LIBs (Li et al., 2016a). Based on time and economic considerations, a dosage of 10% weight was discovered to be the most effective method for discharging S-LIBs.

For instance, the inflammatory response of Li to oxygen and water causes the highly flammable organic solvents to flare up. To prepare S-LIBs for hydrometallurgical recycling, prior discharge treatment is required (Contestabile et al., 1999). A suitable method for large-scale discharge can be the physical discharge in a solid medium like graphite powder and copper. However, it cannot be utilized extensively due to a contact issue, the readily oxidized metal surface, and the possibility of an explosion caused by graphite dust. Due to the low quantities of the LIB-constituting elements in both the supernate and the sediment, $FeSO_4$ was recommended as the most ecologically benign discharge solution (Kim et al., 2021).

To recycle LIB cells, the Batrec company mostly uses a mechanical process. The batteries are first crushed in an environment of inert CO_2 gas. The resulting evaporation causes the volatile organic electrolyte to become a non-useful condensate, which is then collected. The S-LIBs are then handled using either manual disassembly or mechanical separation. The cathode, anode, and other parts are often separated manually. It might be challenging to separate the cathode material from the foil since the binder PVDF or PTFE often adheres the cathode material to the aluminum foil. A solvent dissolution method, a NaOH dissolution method, an ultrasonic-assisted separation method, a heat treatment approach, and a mechanical method have all been evaluated to successfully separate the cathode material from the foil (Zhang et al., 2018b).

Kar (2023) compared four different LIB pretreatment options including discharging for the safety of the whole process, solvent treatment to dissolve the binder in the LIB structure, and grinding to the optimum size for the different separation and extraction methods. Discharge in 10% NaCl solution was compared with cryogenic discharge, and cryogenic discharge was found to be more effective due to both shorter processing time and reliability. For the next step, it was noticed that usage of the NMP solution caused problems in grinding.

4.4.1.2 Dismantling

S-LIBs are manually disassembled to remove the cathode, anode, organic separator, steel, and plastic from the battery chamber after they have been fully discharged. However, when large-scale recycling operations are taken into account, it is quite challenging to manually separate the various components of S-LIBs. For laboratory experiments, it is common practice to physically disassemble individual battery cells;

however, this method is inefficient and costly for processing small battery cells for industrial use. From a cost and effectiveness standpoint, mechanical techniques such as crushing, sieving, magnetic separation, and classification may be more advantageous for dismantling. Crushing was described as a pretreatment procedure for S-LIBs by Zhang et al. (2013). They claimed that plastics, aluminum, copper, and other metals were removed from the ground-up components as large particles, and a mixture of fine graphite, and $LiCoO_2$ electrode materials. Then, this mixture undergoes several steps to recover the high-value elements such as Co and Li.

Although it is difficult due to the wide range of battery cell designs, automatic or robotic disassembly is possible. Therefore, before materials can be recovered, the majority of processing flowsheets proposed for small LIBs include shredding. After shredding, the various components must first be sorted. Typically, sieving is used to separate the Al-foils and Cu-foils holding the electrode active components. Part process flowsheets at this point incorporate a step before or after screening to evaporate volatile organics since part of the active substance may still adhere to the foils. By removing the binder, the amount of active material adhered to the foils is decreased, and the electrolyte is also evaporated in this phase. Z-folded electrode-separator compounds were the term Li et al.'s suggested automatic disassembly process for recycling pouch-type LIBs in 2019 (Li et al., 2019). Without using any harmful pressures, the suggested electrode sorting method extracted the cathode and the anode sheets. The cathode and the anode sheets attached to the opposing sides of the Z-folded separators were scraped off using specialized toolsets by automatically extending and feeding the separators.

4.4.1.3   Mechanical Pretreatment

When recycling S-LIBs, a hydrometallurgical or pyrometallurgical procedure will be used, both of which need physical treatment to remove the outer cases and shells and concentrate the metallic fraction. The target metals are recovered more effectively, and the purification process is made easier by mechanical treatment and pre-enrichment prior to leaching. To liberate the electrode materials from the disassembled S-LIBs, which are referred to as "black mass," crushing and grinding are necessary. This comminution stage is crucial for the hydrometallurgical or direct recycling procedures. The physical size reduction (crushing and grinding) and classification (sieving and separation) of the discharged LIBs used in mechanical pretreatment methods are efficient procedures. Because the mechanical characteristics of the battery's many compartments vary, a shredder or mill that can uniformly crush or grind every compartment is required. Typically, air is closed off to produce a grinding chamber. There are many tools that can be employed, including shredders, hammer mills, granulators, and blade crushers. The fine size CAM, which contains crucial and expensive metals (such as Co and Li), is the most valuable component of S-LIBs. The fundamental components of each S-LIB cell are the Cu and Al foils, Fe and Ni casing metals, and polymers that build in the coarse fraction as a result of mechanical processes. Due to the presence of PVDF binder, the majority of the CAMs were still present in the coarse size range ($>850\,\mu m$) despite the comparatively fine size of $LiCoO_2$ (1.50–$7.80\,\mu m$) (Kim et al., 2004; Widijatmoko et al., 2020a). Plastics and metals have highly different specific gravities, making them stand out as excellent candidates for gravity separation. Magnetic separation of diamagnetic Cu & Al particles from ferromagnetic Fe & Ni metals is possible. Hydrophobic plastic particles can be easily separated from hydrophilic metals by the froth flotation process (Cuhadar et al., 2023).

Air classification (Diekmann et al., 2016), ultrasonic washing coupled with agitation (Zeng et al., 2014), eddy current (Bi et al., 2019), high shear mixer or blender (Shin et al., 2020; Wu et al., 2021; Zhan et al., 2018, 2020a), attrition scrubbing in the presence of low Fe silica sand (Widijatmoko et al., 2020a) can help further separate the black mass from the Al and Cu current collector. Application reports for magnetic and density separation are also available (Pinegar and Smith, 2019; Pudas et al., 2015; Shin et al., 2005; Tedjar and Foudraz, 2010). Graphite and metal oxides have different relative densities of 2.2 and 4.4 g/cm^3, hence separation using this technique has also been reported (Kepler et al., 2016; Zhang et al., 2018c).

Parvali et al. (2019) determined total cumulative particle size distribution based on the chemical analyses of LIBs in the size between 125 µm and 7 mm before mechanical and hydromechanical processing using HCl media. Al and Cu foils and Fe were in relatively coarse size fractions and Li, Co, Ni, Mn, Al, and Fe were found in fine size classes. The cumulative distribution of metals increases with particle size increases (Figure 4.7).

A paper shredder was used for the cathode and anode layers to cut into 5*12 mm. The shredded metal foil is mixed with distilled water five times to rinse off electrolytes and other soluble organic compounds. A slurry of shredded metal foils with electrode active materials was blended with a high-speed blender. And 212 µm wet sieving separates fine liberated active battery materials and coarse metal foils. Widijatmoko et al. (2020) used Retsch SM2000 cutting mill with 8 mm screen size for the first stage of liberation. Samples were then washed to remove volatile organic matter and then dried at 80°C until constant weight. Samples were riffled and ferromagnetic materials were removed by permanent magnets. Attrition scrubbing with

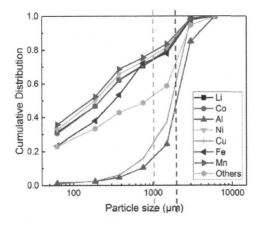

FIGURE 4.7 Chemical composition of different particle sizes varies a lot in LIB waste.

2,360–850 µm low iron silica sand at 1,000 rpm can be performed. The pulp density can be around 70% and the S-LIBs to silica media ratio of 10% is enough for 2.5 minutes attrition scrubbing. Wet sieving from 212 to 38 µm produces fine CAMs in the fine size fraction. After sieving, materials can be dried at 80°C until constant weight is achieved. Coarse size fraction (>212 µm) is dry screened into different size fractions. Widijatmoko et al. (2020a) found that attrition scrubbing selectively increased Co recovery up to 80% with 7% Al and 6.1% Cu impurities at a −38 µm size fraction.

For the first industrial-scale battery recycling facility, the Batenus hydrometallurgical process shreds a variety of different battery types in a gastight unit (Frohlich and Sewing, 1995). The Recupyl method involves two steps of crushing and is carried out in an inert environment (Pinegar and Smith, 2019). A low-speed rotary mill is used for the first stage of crushing, and a high-speed impact mill is used for the second stage. For the simultaneous recycling of the NiMH, LIB, and primary Li batteries, the crushing operations using a two-blade rotor crusher and hammer crushing were carried out (Granata et al., 2012, 2012a). Without any controlling sieve and with a 5 mm sieve, a two-blade rotor and hammer crusher were utilized, respectively. The second step crushing increases the yield of the black mass from 60% to 75% without the addition of the impurities from the Cu and the Al current collectors (Kim et al., 2021).

After being crushed and ground, the active (cathode and anode) and inert (Al and Cu) materials are separated using the differences in density and magnetic susceptibility qualities. Zhang et al. (2018b) produced a battery piece by shear-crushing a spent battery that had been drained via immersion in a NaCl solution. They ground the battery for 20 seconds at 3,000 rpm using impact-crush equipment with blade crushers. By use of dry sifting, the crushed bits were divided into four size groups. The 0.075 mm cathode and anode components were isolated from the bigger plastic fragments, Al, Cu, and polymers. $LiCoO_2$ and graphite were separated using froth flotation following the removal of the HC material from the surface. The most used S-LIB separation technique is mechanical pretreatment because it is straightforward and scalable. The process can generate noise, dust, and hazardous gases, so it's important to put up a reliable separation system. Additionally, utilizing this procedure makes it challenging to guarantee the proper separation of all elements.

An Al-enriched coarse fraction (>2 mm), a Cu-Al-enriched medium fraction (0.25–2 mm), and a Co and a graphite-enriched fine fraction (0.25 mm) make up the three components of the size-reduced LIBs products. The cathode materials derived from the 0.25 mm fine fraction maintained their original crystalline structure and chemical condition in LIBs, according to the study of mineral phase and chemical state. However, a layer of HCs was present on the surface of these powders, making froth flotation operations challenging. Fine portions of the black mass are present, whereas large fractions typically contain less desirable materials including plastics, cell casings, and current collectors. To extract metals from S-LIBs, Shin et al. (2005) presented a combination technique that utilized mechanical separation for the collection of cathode materials, followed by a hydrometallurgical process for metal recovery. Enriched particles of $LiCoO_2$ were obtained through a series of mechanical treatments such as crushing, sieving, and magnetic separation; next, $LiCoO_2$

and small pieces of Al-foil were separated by finely grinding the $LiCoO_2$-enriched particles (Figure 4.8). Before the metal-leaching process, mechanical separation techniques can increase the effectiveness of the targeted metal recovery. The fundamental drawback of mechanical approaches is that $LiPF_6$, DEC, and PC decomposition during mechanical operations represent a threat to the environment as well as the fact that the components of S-LIBs cannot be totally separated from one another. Although several pretreatment strategies have been established by researchers, pretreatment of S-LIBs still faces difficulties. The comminution process utilized affects selective liberation. Impact, shear, abrasion, etc. can offer liberation. Surface abrasion and the production of tiny particles are promoted by the impact and shearing action that attrition scrubbing causes between the particles. An abrasioncam medium can be made of inert silica sand (Widijatmoko et al., 2020a).

Mechanical separation processes have the drawback of not completely separating all types of components in S-LIBs from one another because LIB is made up of numerous metals, organic substances, and inorganic substances that interact with one another, has a small volume, and a very accurate, fine, and complicated structure. This makes it difficult to disentangle these parts from one another.

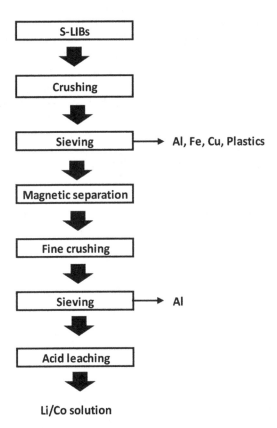

FIGURE 4.8 Metal recovery flowsheet from lithium-ion batteries.

There are two types of LIBs crushing and grinding: dry and wet. Dry crushing is conducted in a gastight unit in an inert atmosphere. After at least 24 hours of soaking in saline water and cutting into small pieces, dry crushing can often be done in two phases (first, shear-crushing for chopping, and second, impact crushing). In comparison to the wet crushing approach, overcrushing occurs less frequently with dry crushing (Kim et al., 2021). Poor liberation results from cutting mill crushing done in one stage. The selective liberation of LCO can be accomplished through second-stage milling. With the addition of water, a blade crusher is capable of doing wet crushing. The plastic-containing shredder slurry's output floated on the water and could be manually removed, but the other parts, such as the electrode materials, could be sieved apart. The LCO and graphite electrode materials could be released from the Al and Cu-foils using the dry crushing approach without overcrushing the other S-LIB components, thanks to the selective crushing capabilities of the S-LIBs (Kim et al., 2021). Laminate is made of active materials that have been separated from its collector but are still attached. For particle sizes 38–850 m, this kind of particle is available as an aggregate. The planetary ball mill's mechanochemical reduction procedure increased how effectively the Li and Co were leached (Guan et al., 2017).

The grinding flotation was suggested for the separation and recovery of LCO and graphite from the S-LIBs (Yu et al., 2018). First, an impact crusher was used to reduce the size of the cathode and anode strips to 0.074 mm. A ball mill was used to grind the resulting combined electrode powder for 5 minutes. Under the horizontal shear force generated by the grinding media, the lamellar graphite structure slips, flakes, and exposes a significant number of new hydrophobic surfaces. On the other hand, the organic film-coated LCO particles have some of their hydrophilic surfaces restored. The adhesion of LCO and graphite may develop and progressively worsen under the influence of the vertical rolling pressure. The large wettability difference produced a satisfactory flotation concentrate grade even though the adhesion behavior in the 5 minutes grinding process would cause some LCO to concentrate to follow graphite into the foam layer and limit the flotation recovery rate.

Tanii et al. (2003) patented a process in which the battery plastic cases were cooled to −50°C or lower temperature and mechanically separated from sealed battery cells. Then the sealed battery cells were heated to over 200°C in a non-oxidizing atmosphere to separate mainly the organic materials, while the targeted valuable materials were separated in one or more separation processes.

Kar (2023) and Kar et al. (2023) used two different grinding methods, i.e. dry and cryogenic, for $LiCoO_2$ batteries. Cryogenic discharging and grinding, without NMP solvent treatment, were found to be the best option to obtain black mass in the <2 mm size fraction. Then magnetic and electrostatic separation can remove impurities and graphite can be separated by reverse flotation. Li and Al extraction can be achieved by bioleaching.

4.4.1.4 Separation Process

The crude separation based on the particle size, magnetic susceptibility, conductivity, and surface chemical properties (such as magnetic, eddy current, electrostatic, gravity separation, and froth flotation) is used after the categorization of the S-LIB comminution products. In the S-LIB recycling process, magnetic separation has frequently been utilized to remove the components that contain iron. The cathode, which housed the LCO active materials and the Al current collector, the anode, the steel casings, and the plastic packaging were separated by magnetic separation following vibration screening (Shin et al., 2005). An eddy current separation device is utilized for the recycling of the LFP-based LIBs. Eddy current separation enables the separation of electrical conductors from non-conductors or less conductive materials. Al and Cu foils can be divided into particles with a size range of 2.0–20 mm since they are both conductive metals that produce a high-intensity eddy current in an alternating magnetic field. Using gravity, it is possible to separate Cu, PE separators, NMC active material, and Al-foil. Shaking tables are used in the Recupyl process on an industrial scale to separate non-magnetic materials based on variations in density (Pinegar and Smith, 2019). Air separation can be used to separate the Al-shell and separators for a +2.0 mm fraction. Electrical conductors can be separated, thanks to eddy current separation. Electrostatic separation can be used to separate Al-Cu and the separator for the −2.0 + 0.5 mm size fraction. Grinding and dry sieving from 0.075 mm is carried out for the size fraction of −0.5 + 0.075 mm. To separate $LiCoO_2$ and graphite, size fractions of +0.075 mm and −0.075 mm are sent to electrostatic separation and froth flotation, respectively (ref addedGratz et al., 2014; Kim et al., 2021). Fenton-assisted flotation (Yu et al., 2017) or flotation modified by roasting can both modify the surface. $LiCoO_2$ grade increased from 55.6% with direct flotation to 60% with Fenton-assisting flotation. Both graphite and CAMs undergo treatment with Fenton chemicals to remove the organic hydrophobic binder. During flotation, hydrophilic LCO sinks while hydrophobic graphite floats. The best selectivity between tailings and concentrate was achieved by the roasting-modified flotation, which produced an 89.6% LCO concentrate grade (Zhang et al., 2018b).

4.4.1.5 Flotation for Separating Active Mass

Hydrophobic materials can be selectively separated from hydrophilic materials using the froth flotation technique. It has long been employed in the mining sector. Anode materials in LIBs, often graphite, are hydrophobic by nature, whereas cathode materials, like $LiCoO_2$, are hydrophilic. So, to separate these two components, researchers aim to use froth flotation.

If properly developed, froth flotation, a physicochemical process, can benefit the LIB recycling sector in a number of ways. According to Yu et al. (2017) and Zhang et al. (2018b), flotation is a good choice for their separation because of the variations in the initial surface wettabilities of the cathode and graphite materials.

Using flotation seems to be a generally sound strategy. However, the flotation process exhibits fairly low selectivities despite the anticipated difference in surface wettability between the electrodes (graphite is naturally hydrophobic, whereas the cathode is hydrophilic). Three basic explanations can be invoked to account for the lack of selectivity, in addition to the fine-particle nature of electrode materials that may encourage entrainment.

- Despite being entirely hydrophilic in their initial state, the surfaces of cathode components like $LiCoO_2$ become hydrophobic throughout the battery-making process, resulting in a co-flotation behavior with the naturally hydrophobic graphite. According to analyses of various S-LIBs powders, an organic layer covering the graphite and cathode surfaces is primarily made up of polyvinylidene fluoride (PVDF) and styrene-butadiene rubber (SBR), which together make about 75% of the organic compounds. This area and the reversal of this impact are the focus of the vast majority of the studies covered in this publication.

- The multiple surface modifications used to electrode materials to improve the electrochemical efficiency of the battery may also make selective flotation separation of the electrode components more difficult. Given that graphite has an anisotropic crystalline structure with faces and edges, this has primarily been the case for graphite. For a better understanding of their function in the surface chemistry of flotation separation as a component of the entire recycling process, more study is required.

- Finally, after being used in batteries, graphite loses its natural hydrophobicity, which corresponds to captive bubble contact angles of 80°. Due to pressures on the crystal structure, the solid electrolyte interface (SEI) film, and the presence of certain metal oxides in the graphite lattice in its anode capacity, graphite loses some of its floatability. Further study is required on the surface chemistry of graphite flotation for a more effective separation from the cathode components.

In an effort to find an effective recycling process employing flotation, many researchers investigated the removal of organic binders using various pretreatment approaches. Numerous novel pretreatments, including the Fenton reaction, solvent washing, mechanical grinding (with and without cryogeny), heat treatment (pyrolysis and roasting), and ultrasonics, have been discovered to reduce or completely eliminate the negative effects of these binding agents, thereby promoting selective flotation recovery. They've made a variety of success claims. However, several of these pretreatments suffer from cost inefficiency or have negative effects on health, safety, and the environment. The heat treatment at 500°C–550°C seems to offer the highest promise of these techniques. For the oxidative elimination of organic coating to permit good flotation separations, roasting was shown to be particularly effective. Such techniques are advantageous because they can change the surface chemistry of a flotation system by increasing the difference between the cathode and anode materials' levels of wettability. However, roasting can also have a detrimental effect on how well graphite floats by producing an excessive amount of functional groups that contain oxygen (thus the requirement for optimization). Given that many reported occurrences involve microscale or small-scale studies, some of the approaches evaluated seem deserving of a second look for larger-scale confirmation under the specified ideal conditions. To find new and more effective ways to recycle used LIBs, more research is required.

The remaining black mass, which was 400 μ in size, was recovered using a flotation process (Ruismäki et al., 2020). To create a high-purity cathode material, another method combined high-speed shearing with flotation for the removal of PVDF and carbon black (Zhan et al., 2020b). De-agglomeration reduces average particle size to 10 μm from a value between 74 and 105 μm in the absence of de-agglomeration treatment, which can break covalent bonds within PVDF molecules as well as intermolecular interactions between PVDF and cathodes (Zhan et al., 2020a).

The collector utilized in the flotation method of recycling is often a petroleum derivative, such as kerosene and n-dodecane. The reported frothers thus far are MIBC and 2-octanol. Depending on the mechanical device being utilized, the impeller speed is typically maintained between 1,500 and 2,500 rpm (He et al., 2017a; Liu et al., 2020; Swoffer, 2013; Yu et al., 2018; Zhan et al., 2018). To avoid the entrapment of metal oxides in the graphite concentrate, high-speed shearing is utilized to shatter any potential agglomerates and disperse particles (Vanderbruggen et al., 2022).

Graphite and LCO have particle sizes of about 30 and 44 μm, respectively, which is a necessary component of the battery-making process for improved electrochemical performances. The hydraulic entrainment phenomena may be problematic in the flotation stage as a result of the tiny particle size of electrode materials, necessitating appropriate cleaning procedures (Zhan et al., 2018). With the release analysis, which includes various phases (roughers, scavengers, and cleanings) on the LIB powders, flotation experiments were carried out (Shin et al., 2020; Zhan et al., 2018). Kerosene and MIBC were used as collectors and frothers, respectively. Graphite was floatable with a recovery of 92% following flotation tests on individual anode and cathode materials (removed from fresh batteries); however, the cathode powder recovery suggested a wider range of 8.1%–31%, depending on the type of cathode.

Zhan et al. (2018) presented an experimental study employing froth flotation to separate anode and cathode material (fine powder following mechanical separation) (Figure 4.9). Over 90% of anode materials were reportedly floated in froth layers, compared to 10%–30% of cathode materials. At 60 and 18 μm grinding sizes, respectively, froth concentrates include 16% and 22% of active cathode materials in terms of the purity of the graphite particles. It is thought to be challenging to

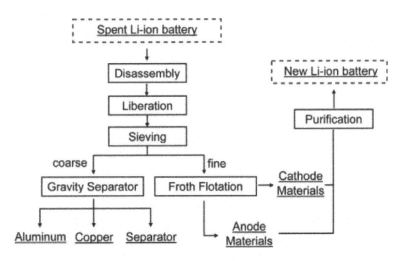

FIGURE 4.9 Froth flotation separation of active mass for the recycling process.

further improve the purity of the graphite components used in the froth products. The authors concluded that froth flotation is not appropriate for producing high-purity graphite material as a result. However, the output cathode material has a greater purity level than the output anode material. Unfortunately, the precise cathode purity is not reported. The scientists noted that the purity of the recycled cathode material may vary depending on the feed samples and grinding settings, raising questions about its suitability for use in fresh cells. The technology needs to demonstrate its feasibility because it is still in the development stage. However, it offers an alternative method for separating the electrode components, boosts the concentration of precious cathode metals (such as Co and Ni), and consequently raises the price of active mass powder.

Graphite consistently exhibits a higher flotation rate than the cathode material in all examined battery samples (LCO, LMO, NCA, or LFP), highlighting the crucial role flotation can play in the recycling process. These investigations' findings also show that wasted $LiCoO_2$ is recovered at a rate that is significantly higher than that of pure $LiCoO_2$, which is evidence of surface modification during battery production that increases hydrophobicity. Material from S-LIBs appears to be less floatable than its fresh counterpart in the anode, which raises the possibility that commercial anodes' hydrophobicity level is negatively impacted over the course of a battery's lifespan. This degradation is due to the presence of hydrophilic sites produced, such as a 20 nm-thick oxygen-rich layer on the surface of the graphite that can be traced to the SEI layer, which reduces the floatability of the graphite (Zhan et al., 2020b). Metal oxides (CAMs) reporting to the graphite concentrate are further impacted by difficulties with flotation selectivity.

Depending on the pre-flotation comminution techniques used, this percentage ranges from 16% to 22%, despite the claimed eradication of hydraulic entrainment in release analysis (Forrest et al., 1994; Zhan et al., 2018). It should be noted that between 10% and 50% of metal oxides report to the graphite concentrate in the flotation investigations published so far (Vanderbruggen et al., 2022). With a 66.1% recovery of cathode concentrate in one instance, a fourfold increase in collector (kerosene) dose was reported to increase the metal oxide loss to the graphite concentrate as evidence of poor performance (Shin et al., 2020). Additionally, it is important to note that some microscopic Cu, Al, and plastic fragments indicate the froth fraction. This unfavorable effect was a result of the leftover polymers' low density and hydrophobicity (Ruismäki et al., 2020).

The selectivity of traditional flotation is weak. The elimination of PVDF and electrolytes was shown to be necessary for efficient anode and cathode separations (Zhan et al., 2020b). With an average value of roughly 56 and 54, respectively, the PVDF was said to lessen the difference in contact angle between the anode and cathode (Wang et al., 2018). Two different forms of organic binders, namely PVDF and SBR, were reported in LIBs from mobile phones (0.2 mm).

As a result, many reasons can be accountable for inefficiency (e.g. the non-selectivity/ incomplete recovery) during the flotation separation of graphite from Li-bearing metal oxides in S-LIBs. The surfaces of both metal oxides and graphite are covered with interfering species (such as organic layers, inorganic sites, etc.) that can reduce the efficiency or even render flotation inefficient. So, to achieve a good flotation recovery and acceptable grade, these substances should be removed. The main challenge is the removal of organic layers. Several pretreatments before flotation were proposed in the literature (Yu et al., 2017; Zhang et al., 2014b).

Despite all the challenges involved in recycling S-LIBs, certain strategies that could help successfully recover valuable elements from batteries have been researched. To physically separate graphite and metal oxides, a variety of innovations have been put forth and tested. These remedies include surface pretreatments like the Fenton reaction, attrition grinding, pyrolysis, and roasting treatments, which primarily aim to remove the organic layer that is causing the selective flotation of graphite from $LiCoO_2$ to be severely impacted. The three most popular pretreatment techniques before flotation are fentom, grinding, and heat treatment.

Fenton Pretreatment and Flotation: For the recovery of LCO and graphite in froth flotation, He et al. (2017a) changed the surface of the crushed electrode materials using the Fenton reagent. Fenton reagent helps with flotation by removing the organic binder layer. First, the organic outer layer was removed from the electrode components from the S-LIBs using the Fenton reagent. The majority of the organic binder layers that were placed on the surface of the electrode materials were removed under ideal conditions comprising a 1:120 Fe^{2+}/H_2O_2 ratio and a 75:1 L/S ratio. The amount of PVDF dropped from 14.28% to 9.98%. The original wettability of LCO and graphite was restored after being altered by the Fenton reagent, allowing the separation of LCO and graphite by froth flotation. Following the Fenton treatment, electrode particles are wrapped with an inorganic hydrophobic $Fe(OH)_3$ layer, leading to poor flotation.

The Fenton pretreatment can remove the binder coating layer and promote $LiCoO_2$ and graphite selectivity. However, compared to traditional flotation without the Fenton reaction, the quality of $LiCoO_2$ was only marginally improved. A close examination of the procedure showed that Fenton pretreatment resulted in the formation of an inorganic layer of an iron-containing material on the particle surfaces as a result of the adhesion of many $Fe(OH)_3$ flocs to the surface of the particles:

$$6Fe^{2+} + 6H_2O_2 + 6H^+ \rightarrow 5Fe^{3+} + 6H_2O + Fe(OH)_3\downarrow + 3OH \qquad (4.1)$$

According to the equation above, the H^+ ions are used to create the hydroxyl radical, which causes the pH of the solution to rise quickly at the start of the Fenton reaction. This is followed by the precipitation of Fe on particle surfaces, which results in the production of an inorganic layer of a substance containing iron at pH 4. It was found that the grade of concentrates improved by 15% when the pH was consistently kept below 3 by adding HCl during the Fenton treatment. However, unless additional developing investigations are undertaken, the mixture of HCl and the remaining H_2O_2 might have a leaching impact on $LiCoO_2$, making this type of method less desirable for recycling by flotation (Yu et. al., 2017).

Yu et al. (2017) examined the impact of semi-solid phase Fenton's secondary product on the floatability of LCO electrode material made from S-LIBs. They conducted a 20-minutes Fenton treatment using 500 mL of 0.8 M H_2O_2, 33.3 mL of 0.1 M $FeSO_4$, and 7 g of LCO sample. Following filtration and drying, flotation was carried out with the addition of 300 g/t kerosene and 150 g/L 2-octanol after a conditioning time of 6 minutes. Ultrasonic cleaning and pyrolysis at 500°C can clean the surface to improve floatability. LCO grade was raised by ultrasonic-assisted flotation from 67.3% to 93.9%, and recovery was increased from 74.6% to 96.9%. At 450°C for 15 minutes, the best roasting treatment was accomplished (Wang et al., 2018). This led to a 97.7% Co recovery rate by removing the outer layer of LCO and protecting the graphite from combustion. The ideal pyrolysis settings were employed by Zhang et al. (2020b) at 550°C with 10°C/minute and a 15-minutes pyrolysis period. The cathode grade only slightly increased from 94.7% to 94.9%, but its recovery increased from 83.8% to 91.4%, supporting the idea that wet ball grinding can effectively remove any pyrolytic carbon residue that remains on the surface of the cathode materials. The flotation procedure was improved by two-stage pyrolysis, which raised the LCO grade to 98%.

Grinding Pretreatment and Flotation: Grinding is effective in promoting flotation and has no secondary pollution effect. But, grinding consumes about 30%–70% of energy input in mineral processing. Therefore, unnecessary fine size reduction should be avoided. The anode material is softer than cathode material when it comes to the crushing characteristics of LIB materials. It was previously recognized that efficient grinding could aid in removing the organic layers from electrode materials like graphite and $LiCoO_2$ in order to reveal their original surfaces. The crystalline surface properties of graphite can be restored using conditions like a little grinding time, little attrition force, and low atmospheric pressure (achievable by eliminating both the oxygen and adsorbed water molecules on graphite particles).

According to the size reduction mechanism, when weak Van der Walls bonds break, graphite particles separate, creating additional hydrophobic sites that are crucial for flotation. The result of grinding is primarily the removal of the organic layer via abrasion of the cathode surface. By activating some Li ions already present in the materials, the grinding action on the cathodes aids in the breakdown of the C-F's original hydrophobic structure by forming (Li-F). While removing roughly 25% of the PVDF, the creation of Li-F increases the difference in hydrophobicity between cathodes and anodes. Grinding-assisted flotation causes abrasion of the organic film covering the LCO particles, which results in their original hydrophilic surface, while destruction of graphite's lamellar structure produces hydrophobic new surfaces (Liu et al., 2020; Yu et al., 2018; Zhan et al., 2018; Zhang et al., 2020b). Thus, there is a significant wettability difference between graphite and LCO, which results in high flotation selectivity. It was discovered that, at the same recovery, fine grinding could result in a purer product than coarse grinding.

Cryogenic Grinding + Flotation: Liquid N_2 was used to chill the cryogenic single-ball milling, which was paired with froth flotation (Liu et al., 2020). After 9 minutes of cryogenic grinding, the LCO concentrates' grade and recovery rate both improved. The organic material is removed from the LCO and graphite surfaces using cryogenic grinding. By adhering to the flotation bubbles for graphite flotation, and keeping LCO in the slurry, cryogenic grinding improves the flotation of LCO and graphite. The best collectors, frothers, regulators, and depressants for flotation kerosene were MIBC, NaOH, and amylum (Shin et al., 2020; Zhao and Zhang, 2020). To recover cathode active material from LCO-based S-LIBs, Wang et al. (2019a) studied cryogenic ball milling. Up to 87.3% more fine cathode material could be peeled off of Al-Foil using a liquid N_2 pretreatment at −196°C K for 5 minutes and grinding for 30 seconds. The transition temperature for PVDF glass is roughly 38°C. For the separation of LCO and graphite, Liu et al. (2020) combined froth flotation with cryogenic ball milling utilizing liquid N_2.

Cryogenic grinding is performed in the presence of liquid nitrogen. Cryogenic grinding prevents the burning and blasting of S-LIBs during size reduction. There is no need to discharge S-LIBs before cryogenic grinding. After cryogenic treatment, flotation performance is significantly increased. Cryogenic grinding effectively dissociates the agglomerated CAMs while inducing changes in the structure of PVDF at around −196°C. The obtained glassy PVDF is easily peeled off from the $LiCoO_2$ surface. Cryogenic grinding increases both the hydrophobicity of LCO and promotes the hydrophobicity of graphite. More than 6 minutes grinding time generates finer which may be lost LCO to graphite floats due to entrainment and higher energy consumption (Traore and Kelebek, 2023).

Single New Material Flotation: Table 4.4 summarizes the flotation of new and S-LIB materials with a 1% solid ratio in 4 minutes flotations (top size 212 µm) results. Pure graphite and $LiCoO_2$ recovery for 4 minutes were 98% and 8%, respectively. Graphite floated due to natural hydrophobicity. Partial flotation of $LiCoO_2$ was because of fine particle entrainment with water. The entrainment was proportional to the water recovery. From new LIBs flotation, anode recovery changed from 90.3% to 98.6% while cathode recovery changed from 8.8% to 30.5%. From S-LIBs flotation, anode recovery changed from 82.1% to 97.3%, whereas cathode recovery changed from 8.2% to 36%. Low anode recovery and high cathode recovery from S-LIBs were related to the coating of CAM with conductive additives and binders on Al-foils. Upon liberation, some CAMs are covered by binders and conductive additives such as C black, rendering a fraction of cathode material hydrophobic.

TGA analysis can be used to determine the decomposition temperatures of hydrophobic binder and C additives under atmospheric conditions. Binder decomposition

TABLE 4.4

Flotation of New and S-LIB Materials with 1% Solid Ratio in 4 Minutes Flotation (Top Size 212 μm) (compiled from Zhan et al., 2018)

	Anode recovery Hydrophobic graphite (contact angle is 80°)	Cathode Recovery Hydrophilic $LiCoO_2$ Easily hydrated			
New single material	>98	8% (due to entrainment)			
	Battery type	LCO	LMO	NCA	LFP
New LIBs	Anode recovery	96.18%	97.56%	98.61%	90.25
	Cathode recovery	30.51%	8.84%	22.56%	18.80%
S-LIBs	Anode recovery	82.05%	85.73%	97.33%	90.77%
	Cathode recovery	20.11%	8.21%	21.44%	35.99%

Source: Compiled from Zhan et al. (2018b).

occurs between 300°C and 500°C. At 800°C, both binder and additives decompose. The total weight loss of the sample is about 4.5%. Liberated cathode material with binder and additives is hydrophobic and floats to the froth phase. Fine grinding (about 18 μm) can produce a purer product than coarse grinding (60 μm) at the same recovery (Zhan et al., 2018).

Froth flotation can be used to separate active anode and cathode materials from LIBs. Ninety percent of anode materials can be recovered with 10%–30% cathode material using 2 kg/t kerosene. New and spent batteries show similar flotabilities with the exception of a slightly lower recovery for anode materials. The partial flotability of CAMs was due to the coverage of binder and additive materials on surfaces. A fraction of graphite is nonfloatable and goes to the tailing. This impurity problem can be solved with fine grinding. Tailing product with 88.2% CAMs was achieved at a total recovery of 74.3%, showing the effectiveness of the flotation process producing pure cathode materials from S-LIBs (Zhan et al., 2018).

The particle size of $LiCoO_2$ and graphite is around 30 and 44 μm, respectively, which are natural characteristics of the battery production process for enhanced electrochemical efficiencies. Due to these fine-sized electrodes, hydraulic entrainment can occur during flotation, which may require some cleaning stages (Traore and Kelebek, 2023). Zhan et al., (2018) floated LCO, NCA, LMO, and LFP anode and cathode electrode materials separately. The cathode powder recovery changed from 8.1% to 31% at above 92% graphite recovery. Maximum selectivity between anode and cathode material was obtained at 4 minutes flotation time for LMO and NCA.

The effective separation between anode and cathode materials depends on the removal of PVDF and electrolyte. The PVDF reduces the difference in contact angle between the anode and cathode. $LiCoO_2$ becomes more hydrophobic in the presence of an organic binder covering. This may cause poor selectivity between anode and cathode material separation and inefficient graphite flotation. Thus, these substances should be removed before flotation. Surface pretreatments (Fenton reaction), attrition grinding, pyrolysis, and roasting treatments are used to separate $LiCoO_2$ and graphite in the recycling of S-LIBs. These procedures primarily aim to remove the binding organic layer that is the cause of the negative selectivity of graphite from LCO.

4.4.1.6 Electrohydraulic Fragmentation Using Shock Waves

The Fraunhofer Project Group for Materials Recycling and Resource Strategies (IWKS) and a number of partners developed an original electrohydraulic fragmentation method using shock waves during the 2016–2019 period of the NEW-BAT project (https://www.r4-innovation.de/de/new-bat.html). The electrohydraulic fragmentation employing shock waves is the cutting-edge foundation of recycling technology. The substance is added to a liquid, like water or another liquid. An electrical discharge produces the shock waves. Uniform shock waves struck the materials. This enables easy and gentle separation of the components by splitting the composites at the material boundaries in a non-contact manner (https://www.r4-innovation.de/de/new-bat.html; https://www.isc.fraunhofer.de/en/press-and-media/press-releases/efficient-recycling-of-lithium-ionbatteries-launch-of-research-project-new-bat.html).

A schematic diagram of the electrohydraulic fragmentation is shown in Figure 4.10. Figure 4.11 shows the flowchart for the processing (https://www.iwks.fraunhofer.de/en/competencies/Separation-and-Sorting-Technologies/Physical-Separation-and-Sorting-Technologies/Electro -hydraulicFragmentation-EHF.html). The batteries are first discharged, and after that, they are broken down to the cell level. The cells are split open, and their parts are separated by mechanical shockwaves in a fluid media (such as water) in the next electrohydraulic fragmentation process.

The mixed components are then mechanically separated (in a manner similar to the previously described classical mechanical separation), yielding fractions that are similar to those obtained from currently used mechanical treatments: mixed active mass, plastic parts and separator, metal foils, and cell housing. Additional processing of the active mass is still required, such as hydrometallurgy, separation into anode and cathode components, and refurbishing (direct recycling). However, since the current processing volumes are still lab scale, it is not yet possible to evaluate factors like energy efficiency and financial viability. Due to the large energy consumption, a low level of energy efficiency is anticipated. Additionally, it is stated that the battery must be completely depleted before electrohydraulic fragmentation to protect the LIBs from thermal runaway. However, the industry today spends a lot of time and effort on this deep discharge. In light of the safety considerations, this procedure may therefore be better suited to treating crushed or shredded materials as opposed to non-deactivated cells.

4.4.1.7 Mechanochemical Process

Using the mechanochemical approach, Saeki et al. (2004) developed a successful procedure for recovering Co and Li from S- LIB. The procedure entails co-grinding $LiCoO_2$ and polyvinyl chloride (PVC) in an air-filled planetary ball mill to create Li and Co-chlorides, and then leaching the ground product with water to extract Co and Li. $LiCoO_2$ and PVC undergo a mechanochemical (MC) reaction

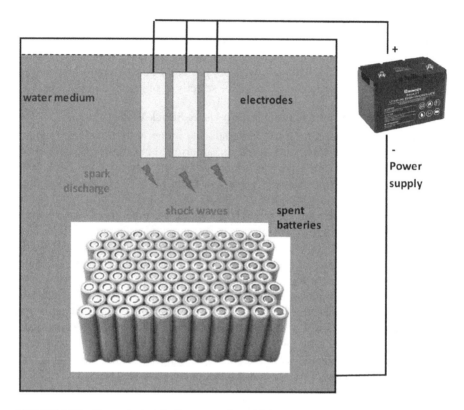

FIGURE 4.10 Electrohydraulic Fragmentation of LIB material (by Fraunhofer IWKS).

during the grinding stage, forming chlorides that are soluble in water. Therefore, the grinding process is crucial to increasing yield. As a source of chloride for the MC reaction, PVC is crucial. Mechanochemical (MC) processes are aided by grinding, which also increases the extraction yields of Co and Li. The recoveries of Co and Li reach over 90% and almost 100%, respectively, after the 30-minute grinding. Over time, the PVC sample's chlorine content has changed to about 90% inorganic chlorides. The idea of this procedure is to recycle usable elements from PVC and battery waste.

It was reported that room temperature extraction of valuable components from $LiCo_{0.2}Ni_{0.8}O_2$ scrap containing the PVDF has been carried out using $1.0\,N\ HNO_3$ solution after mechanochemical (MC) treatment by a planetary mill with and without Al_2O_3 powder. Crystalline $LiCo_{0.2}Ni_{0.8}O_2$ in the scrap was pulverized and became amorphous by MC treatment for 60 and 240 minutes, respectively, with and without the assistance of Al_2O_3 powder. This demonstrates that the addition of Al_2O_3 is very effective for MC treatment. As a result, a high yield of more than 90% of Co, Ni, and Li was recovered from the amorphous scrap sample. When the Al_2O_3 powder was introduced to the scrap during the MC treatment, about 1% of the fluorine in PVDF was dissolved in the filtrate, but no fluorine was detected in the filtrate obtained from the ground scrap sample without Al_2O_3 powder.

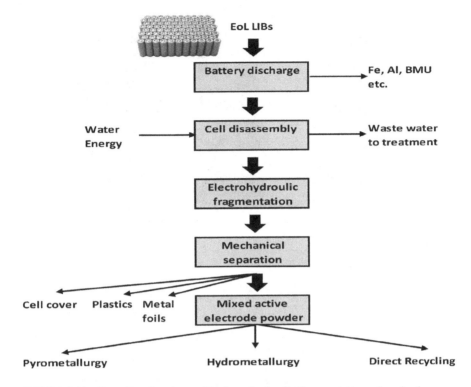

FIGURE 4.11   Recycling flowsheet with electrohydraulic fragmentation using shock waves.

4.5 CLASSIFICATION

Sieving and zigzag air classifiers can be used for classification after comminution. Air separation can be used in a zigzag classifier after separating the deactivated cells from the LIB scraps using a two-stage crushing process based on LCO.

4.6 BLACK MASS, CHEMICAL ANALYSIS, MORPHOLOGICAL INVESTIGATIONS

Three same-used Apple phone batteries (assembled in China) with 3.83 V, 6.55 Whr, and 1,715 mAh were manually dismantled to determine the material balance (Figure 4.12). The total average weight of S-LIB was 24.9919 g. Al current collector with CAM was 50.78%, Cu current collector with graphite was 39.17%, the black case cover (inside Al) was 3.64%, electrolytes, and separators were 3.01%, and the remaining were connection circuits and adhesive (Figure 4.13). About 90% of these S-LIBs were black mass.

Table 4.5 shows the chemical analysis of homogenized LCO, LFP, and NMC black mass after the physical treatment. Elemental chemical analysis can be performed using ICP-OES. Samples are digested 10–100 mg solids in a mixture of HCl: $H_2O_2 = 7/3$ by volume at a temperature of 80°C for 15 minutes. Metal concentrations

FIGURE 4.12   Apple phone batteries.

Weight distribution comparison

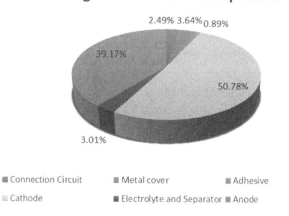

2.49% 3.64% 0.89%

39.17%

50.78%

3.01%

- Connection Circuit   ■ Metal cover   ■ Adhesive
- Cathode   ■ Electrolyte and Separator ■ Anode

FIGURE 4.13   Weight distribution of Apple phone batteries.

TABLE 4.5
Chemical Analysis of Some LIBs

| Element | LCO | Concentration (%) | |
		LFP	NMC
Co	31.17	<0.01	9.15
Li	3.27	2.57	4.01
Mn	0.09	0.02	6.58
Ni	0.13	<0.01	8.25
Al	1.02	0.21	1.24
Cu	0.98	1.94	0.54

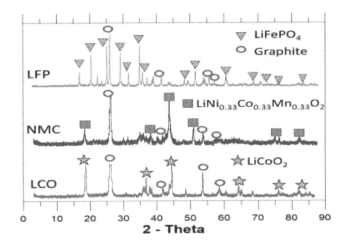

FIGURE 4.14   The XRD spectra of the LFP, NMC, and LCO black masses.

in the range of 0.5–25.0 mg/L are filtered before ICP analysis (Zhan et al., 2018). Widijatmoko et al. (2020) used −212 μm particles after calcination at 500°C to remove gases for 3 hours. In the digestion, $HNO_3 + H_2O_2$ and HCl with a microwave digester were used. The samples were analyzed by an ICP-MS.

Figure 4.14 illustrates the pretreated XRD spectra of the LCO, NMC, and LFP black masses. Only the components of the active cathode material and graphite were discovered, which is evidence of the effectiveness of the preliminary physical separation (Illes and Kekesi, 2022).

Morphological SEM-EDX investigations show the anodic and cathodic active materials bound to Cu and Al foils by binder PVDF and SBR-CMC. Liberation, particle shape, and sizes can be seen in SEM photos. Coarse particles are not liberated; fine particles are liberated active materials. The presence of fluorine atoms and the absence of Na atoms suggest the PVDF binder usage.

4.7   NaOH DISSOLUTION METHOD FOR Al REMOVAL

The recycling of the LIBs containing LFP and NMC was done using an aqueous solution that comprised NaOH salts without the use of strong acids or alkalis (He et al., 2019). By removing the Al-foil, the cathode materials might be peeled away from it. The Al-foil and electrode materials' resultant recovery efficiencies were nearly 100%. The cathode is leached using an alkaline NaOH solution in several proposed recycling methods to remove the cathode components from the aluminum foil. This method is effective because of the amphoteric nature of Al. Nan et al. (2006) used a 10% by-weight NaOH solution to separate the LCO-based CAMs from Al-foil as part of their procedure to retrieve the cathode active material. Approximately 98% of the Al-foil was dissolved using a NaOH solution with a 100 g/L S/L ratio, 5 hours of reaction time, and room temperature. The protective layer covering the collector's surface and the Al both dissolve when the Al-foil of the cathode is dissolved in a NaOH solution.

$$Al_2O_3 + 2NaOH + 3H_2O \rightarrow 2Na[Al(OH)_4] \qquad (4.2)$$

$$2Al + 2NaOH + 6H_2O \rightarrow 2Na[Al(OH)_4] + 3H_2 \qquad (4.3)$$

The simplicity of operation and high separation effectiveness of this technology are benefits. However, because of Al's ionic structure, recovery of Al is challenging. Alkali NaOH wastewater is also quite bad for the environment. Al can also be dissolved using a mixture of molten salts, $AlCl_3$ and NaCl.

4.8   BINDER REMOVAL

Figure 4.15 shows organic components (binder, electrolyte) attached to the graphite and $LiCoO_2$ particles (Zhang et al., 2014). Removal of these organic compounds is important for graphite and metal recovery from black mass. Graphite particles' attachment to the anodic Cu current collector electrode by binder is weaker than the $LiCoO_2$ attachment to the cathodic Al current collector electrode by a binder (Dai et al., 2019). Therefore, $LiCoO_2$ materials can be generally contaminated by graphite. It is difficult to remove/liberate binders from CAMs mechanically. Graphite material purity can not be improved, but pure cathode materials can be produced by flotation.

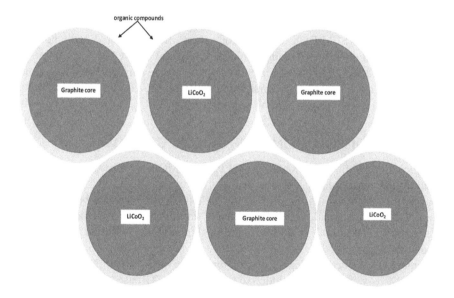

FIGURE 4.15   Organic compounds around graphite and $LiCoO_2$ particles before extraction.

4.8.1 SOLVENT DISSOLUTION PRETREATMENT FOR BINDER ELIMINATION

Surfaces of the Cu and Al current collectors exhibit a significant adhesion of electrode materials. Activated carbon, PVDF, and CAMs are used to prepare the cathode section before the slurry is applied to the Al surface. The separation of CAMs from an Al-foil surface via a quick and effective dissolution technique is the subject of current research.

After the classification stage and the separation stage, some released active elements are still linked to the current collectors, while other separated active materials are kept together by binders. As a result, the appropriate solvents are frequently used in the dissolving of the binders or the Al-foils. To separate the Al- and Cu-foils/films from the electrode active materials (LCO and graphite) in the S-LIB, solvent pretreatment is used. This procedure separates the active components by removing the additional binder material that strengthens the foil contact with it. On Al- and Cu-foil, the cathode and anode were joined with a binder. The active substance and metal foil can be separated by removing the binder. The electrode plate can be submerged at 100°C in N-methyl-2-pyrrolidine (NMP) to dissolve the PVDF binder and separate the Al- and Cu-foils, which is a well-known technique. He et al. (2019) used sharp nose pliers to manually separate the battery after discharging the LIB by soaking it in a NaCl solution. The cathode was divided into small pieces after the electrode plate, separator, metal case, and plastic had been removed. The cathode pieces were placed in NMP for 2 hours at 80°C, which caused the binder to dissolve and the Al-foil and cathode material to separate. The active components are often attached to batteries using a PVDF binder, while some batteries utilize PTFE binders, which necessitate the use of an appropriate organic solvent. Additionally, this process calls for the use of organic solvents heated to about 100°C, which raises the cost and results in hazardous waste products. Table 4.6 lists the solvents that have been used in the past to dissolve binders.

Contestabile et al. (2001) tested a laboratory-scale S-LIBs recycling process without the separation of anode and cathode electrodes. The battery rolls were treated with NMP at 100°C for 1 hour, and $LiCoO_2$ was effectively separated from their support substrate and recovered. The recovery of both Cu and Al in their metallic form was also achieved. Although this process was very convenient, the recovery effects of $LiCoO_2$ were demonstrated to be influenced by the adhesive agent and rolling method of electrodes. Li et al. (2014) reported that the CAMs were easily separated from an Al support in an NMP solution in 1 hour at 100°C. Many researchers have used other organic solvents such as N,N-dimethylacetamide (DMAC), (Zheng et al., 2016) N,N-dimethylformamide (DMF), (Xu et al., 2014) dimethyl sulfoxide (DMSO), (Bankole et al., 2013) and ionic liquid (Zeng et al., 2014) for the separation of the CAMs. For the separation of the CAMs, many researchers have employed additional organic solvents such as N,N-dimethylacetamide (DMAC), N,N-dimethylformamide (DMF), Xu et al. (2014), dimethyl sulfoxide (DMSO), Bankole et al. (2013), and ionic liquid (Zeng and Li, 2014). This method significantly streamlines the separation processes for Co and Al because it makes $LiCoO_2$ easily separate from their support substrate and recoverable. It is impossible to completely remove CAMs by a

TABLE 4.6

In S-LIB Recycling, Solvents Used for Binder Dissolutions

Solvent	Binder	Temp./ Time	Material Removed	Material Remained	References
NMP	PVDF (Solubility: 200 g/kg solvent; boiling point: 200°C) PVDF	<100°C 1 hour	$LiCoO_2$ and graphite	Al and Cu metals by filtration	Constestaible et al. (2001); Li et al. (2014)
DMAC	(Boiling point: 165°C)	10%	$LiCoO_2$	Al-foil	Liu et al. (2006)
DMF	Suitable for PVDF Not suitable for PVDF	60°C–70°C	NMC/LCO	Al-foil	Zhou et al. (2010); Xu et al. (2014)
TFA	PTFE acetic acid 15 v%, L/S: 8 mL/g with agitation (Boiling point: 71.8°C)	40°C, 3 hour	NMC	Al-foil	Zhang et al. (2014)
$AlCl_3$–NaCl Molten salt	PTFE, 1/10 g/mL	160°C 20 minutes	Cathode material	Al-foil	Wang et al. (2019b)

dissolving process; instead, additional processing is required, such as thermal treatment (calcination process) to burn off the C and PVDF residues. The organic solvent is also unsuitable for scaled-up recycling operations due to its high price and scarcity as per the polar and nonpolar nature.

Due to its cost advantage over NMP, N,N-dimethylacetamide (DMAC), which dissolves PVDF that is about 10%, was employed to separate the LCO cathode material from the aluminum foil (Liu et al., 2006). The cathode material was removed from the aluminum foil while the S/L ratios were kept between 1:4 and 1:5. LCO and the graphite conductive agents were then filtered from the solvent. Since DMAC has a 165°C boiling point, it can be evaporated by heating it at 120°C for 12 hours. The NMC and LCO cathode materials were separated from the Al-foil using N,N-dimethylformamide (DMF) (Song et al., 2013, 2014; Xu et al., 2014). Materials from the cathode have been submerged in DMF or a mixture of DMF and ethanol at 70°C.

A low-temperature $AlCl_3$-NaCl molten salt was used to separate the CAMs and the Al-foil (Wang et al., 2019b). The commonly used hydroxide (NaOH) or nitrate systems performed worse at peeling off than the $AlCl_3$-NaCl molten salt system. At 160°C for 20 minutes, a S/L ratio of 1:10 g/mL produced the best results. The PVDF organic binder would melt as a result of the heat storage of the phase transformation of the molten $AlCl_3$-NaCl due to a separation mechanism of the CAMs from the aluminum foil. In particular, the $AlCl_3$-NaCl mixture absorbs a significant amount of heat to change from a solid to a liquid compound when the heating temperature exceeds 153°C, which is the phase transformation temperature of the molten salt of $AlCl_3$ and NaCl. When the temperature reaches 160°C, PVDF melts, effectively separating the cathode active components from the aluminum foil.

The best solvent for PVDF, NMP, is frequently used for LCO separation, PVDF binders' dissolution, and Al-foil dissolution. Other suitable solvents include N,N-dimethylacetamide (DMAC), DMF, NeN-dimethyl sulfoxide (DMSO), ethanol, and molten $AlCl_3$-NaCl salt. With a S/L ratio of 1:10 g/mL at 60°C for 30 minutes, the peel-off efficiency in any of the solvents did not go above 10%. However, with the aid of the ultrasonic, the peel-off efficiency rose by at least sixfold in the following order: ethanol, DMSO, DMF, DMAC, and NMP. Peeling off at 100% was accomplished. Al-foil's cathode material was removed using solvents and ultrasonic treatment.

Trifluoroacetic acid (TFA), a potent carboxylic acid with a relatively low boiling point (71.8°C), was used to dissolve the polytetrafluoroethylene (PTFE) binders in the NMC-based LIBs (Zhang et al., 2014b). The optimum separation efficiency of the cathode material and the Al-foil was obtained with a 15 vol% TFA solution with an L/S ratio of 8 m/g, and a reaction temperature of 40°C for 180 minutes with agitation, and these factors were all tuned.

4.8.2 Ultrasonic-Assisted Separation for Binder Elimination

CAM is challenging to separate from Al-foil during the recycling of S-LIBs due to the PVDF binder's powerful adhesive force. Because of its powerful cavitation effect, ultrasonic treatment is thought to be one of the most efficient ways to remove cathode material from aluminum foil (Zhang et al. (2013); He et al., 2015; Li et al., 2015). Li et al. (2015) found that when mechanical agitation alone was utilized, the majority of the CAMs continued to attach to the surface of the collectors. This was found when examining the effects of agitation and ultrasonic treatment on the separation of CAMs. Only a portion of the CAMs were separated when only ultrasonic cleaning was employed. However, almost all of the CAMs could be removed from the collectors when both techniques were applied concurrently. This may be due to the cavitation effect of ultrasonic cleaning, which can produce more pressure to liquefy and scatter insoluble contaminants. The mechanical agitation's rinsing effect further encourages the separation of CAMs from the collection. He et al. (2015) explained that the dissolving of the binder and the cavitation effect caused by the ultrasound were the mechanisms by which CAMs were removed from Al-foil by ultrasonic cleaning. Based on this mechanism, the CAM's stripping effectiveness was 99% when NMP was used as the cleaning agent, 240 W of ultrasonic power, and 90 minutes of ultrasonic processing. By using ultrasonic cleaning to separate the CAM from the aluminum foil, there was less agglomeration, which aided the subsequent leaching procedures.

Yao et al. (2015) reported on an ultrasonic-assisted dissolving procedure. The cathode was first treated in an ultrasonic bath for 20 minutes after being submerged in an NMP solution for 1 hour at 80°C. The yield and separation effectiveness were said to have greatly increased. At room temperature, Yang et al. (2015) employed ultrasonic treatment to remove CAMs from an Al surface.

4.8.3 THERMAL (HEAT TREATMENT) PRETREATMENT (ROASTING/ PYROLYSIS/CALCINATION) FOR BINDER ELIMINATION

Thermal treatment can be used to remove the binder components (PVDF polymer) that hold the conductive C black and the CAMs together and fix them to the current collectors. The thermal treatment uses high-temperature binder breakdown to lower the bonding force between chopped CAM particles so they may be easily separated by screening (Bahgat et al., 2007; Zeng et al., 2015). PVDF has a low decomposition temperature. Roasting in air or oxygen, pyrolysis, calcination, and microwave treatment are examples of heat treatments. The benefits of microwave technology include quick heating rates, non-contact heating, simple power management, good material selectivity, uniform heating, and improved degradation kinetics.

In conventional LIBs heat treatment, between 100°C and 250°C, water is lost by evaporation resulting in dehydration. Between 320°C and 360°C, the long chain of molecules of organics is broken down into short-chain molecules releasing CO_2, CO, H_2O, and CH_4, phenols, and free C. More gases, similar to the one listed above, are also observed between 400°C and 600°C (Diaz et al., 2018; Feng et al., 2018). Thermostability analyses (TG and DTG) can be determined on spent graphite and CAMs. In this process, extremely toxic gases and smoke are released due to the combustion of C and organic polymers. Table 4.7 presents heat treatment temperature and its effect. Inorganic LCO structure is stable up to 900°C temperature. Heat treatment is performed between 300°C and 500°C to prevent graphite roasting and burning. This thermal treatment restores the crystal lattice of graphite by releasing internal stress. The structure of the treated graphite improves and becomes comparable to commercial graphite (Traore and Kelebek, 2023).

TABLE 4.7
Heat Treatment Effect for Binder Elimination

Heat Treatment Temperature °C

0–900	Inorganic metal oxides structure is stable
120–250	Dehydration, loss of water, electrolyte volatilization
320–360	Long chain organics is broken release CO_2, H_2O, CO, CH_4, phenols, and free C
335–555	PVDF binder decomposition and volatilization
300–500	**Roasting to prevent graphite loss**
>600°C	Melting Al foils
700–900	Carbon burning

Source: Zhang et al. (2018a, b), Wang et al. (2018), and Zhang et al. (2020).

The S-LIB is subjected to calcination pretreatment at temperatures between 150°C and 600°C to get rid of conductive carbon, acetylene black, and organic material. Additionally, the PVDF binder (which joins the active materials and metal foil) can be eliminated by calcination at 250°C–350°C, which lessens the adherence of the active materials on the Al- and Cu-foils. The cathode parts were cut into pieces and heated to between 550°C and 650°C in a tube furnace by Yang et al. (2016). Then, utilizing gravity separation, the cathode material and the current collector (Al-foil) were quickly separated. However, calcination treatment can produce harmful gases, necessitates pricey furnaces, and is energy-intensive (Bae and Kim, 2021).

Vibrating screening was used to separate the electrode materials from the Al current collectors after a two-step heat treatment that was carried out in a furnace at 300°C–500°C for 1 hour concentrated the electrode materials that contained Li and Co. The carbon and binders were then burned off at a temperature ranging from 500°C to 900°C for 0.5–2 hour to generate the LCO CAMs. After LCO was initially separated from the Al-foil, PVDF and the carbon powder in the cathode materials were removed with high-temperature calcining.

After LCO was originally separated from the Al-foil, PVDF and the carbon powder in the cathode materials were removed with high-temperature calcining (Liu et al., 2006). The dry powder was heated twice: once for 2 hours at 450°C and again for 5 hours at 600°C. Thermal processing of the S-LIBs was used to separate the LCO cathode components in two phases. The binders and the organic additives were burned during the first heat treatment stage, which was carried out at 150°C–500°C for 1 hour. For the second thermal treatment, carbon and the remaining unburned organics were removed by heating the material to 700°C–900°C for 1 hour. Thermal pretreatment was carried out to remove carbon and PVDF from the CAMs after drying at 60°C for 24 hours (Li et al., 2010; Meng et al., 2017; Natarajan et al., 2018; Zhang et al., 2014b). The cathode active components from the used batteries were calcined in a muffle furnace for 5 hours at 700°C before being cooled to room temperature.

Using a recycling process that included mechanical, thermal, hydrometallurgical, and sol-gel processes, Lee and Rhee (2002) were able to recover Co and Li from S-LIBs and create $LiCoO_2$ from leach liquid as CAMs. Li and Co-containing electrode materials can be concentrated using a two-step heating process. First, LIB samples were thermally processed for 1 hour at 100°C–150°C in a muffle furnace. A fast shredder was used to disintegrate the samples. The electrode materials were then separated from the current collectors by vibrating screening before a two-step thermal treatment was carried out in a furnace. The next step was to burn off C and the binder at a temperature between 500°C and 900°C for 0.5–2 hours to produce the CAM, $LiCoO_2$. Third, the gel was put inside a stainless steel crucible and calcined into powder in the air for 2 hours at a temperature between 500°C and 1,000°C after being leached of $LiCoO_2$ in an HNO_3 acid solution in a reactor.

Thermal treatment has the benefit of being straightforward and practical, but it also has the drawback of not being able to recover organic compounds, necessitating the installation of equipment for cleaning the smoke and gas produced by the combustion of C and organic compounds (Xu et al., 2008).

4.8.3.1 Roasting

Kim et al. (2004) roasted S-LIBs from cell phones in a furnace at 500°C and subjected them to froth flotation using 0.2 kg/t kerosene, 0.14 kg/t MIBC, and 10% pulp density, 2,500 rpm impeller speed, 92% of CAMs were recovered at a grade of 93%. After thermal treatment, it was also reported that a higher concentration of kerosene and pulp density did not affect the flotation efficiency and the grade of $LiCoO_2$.

Swoffer (2013) used roasting at 500°C in air/oxygen for 1 hour and froth flotation (using 1% kerosene and 0.5% MIBC) treatment at 10% solid at a pH of 5 (adjusted by acetic acid) for various types of LIBs. The process started with a wet hammer mill crushing under an inert atmosphere followed by sieving to separate black mass from coarse battery casings, Al, and Cu foils. Wang et al. (2018) decreased the C content of LCO from 42.7% to 16.5% after roasting at 450°C for 15 minutes. Roasting at 400°C causes contact angles of 53° and 19.2° for graphite and LCO samples, respectively. Using 300 g/t n-dodecane and 150 g/t MIBC, 97.7% Co, 93.7% Mn, 90.1% Cu, and 86.3% Al were recovered with flotation.

Roasting at 500°C for 2 hours followed by flotation achieved 99.8% metal oxide recovery at a grade of 98.5% and 97% graphite recovery (Zhan et al., 2020b).

Yao et al. (2016) peeled the CAMs from Al-foil using a vacuum furnace at 600°C for 2 hours. Sun and Qui (2011) retrieved CAMs by vacuum pyrolysis, which was conducted at a temperature of 600°C and a pressure of less than 1.0 kPa. A condenser and a gas collector, respectively, were used in this operation to collect volatile gases and gases that could not condense. LCO was easily peeled from Al-foil when the pyrolysis process was finished.

4.8.3.2 Pyrolysis

The process of pyrolysis, which involves the high-temperature decomposition of organic materials without the presence of oxygen, typically yields three products: pyrolytic oil, non-condensable gas, and solid carbonaceous residue. After thermal treatment of S-LIBs at 550°C. Around 99% of the binder was removed. Then, mixed electrolytes were pyrolysis at 500°C for 30 minutes before flotation (He et al., 2017c). About 26% of LCO was reported to the graphite concentrate due to the entrainment of fine particles during flotation. In addition to this, a residual pyrolytic C remained on the LCO surface, which is hydrophobic and floatable.

A subsequent work (Zhang et al., 2018a) investigated the removal of the pyrolytic carbon/products utilizing an ultrasonic treatment prior to the flotation stage. While there is no discernible change in the grade, ultrasonic treatment improves $LiCoO_2$ recovery by 12% and decreases the loss of metals to the graphite concentrate. To improve flotation efficiency, the ultrasonic treatment helped remove roughly 42.5% and 80% of the pyrolytic carbon and fluorobenzene, respectively. According to subsequent research, the cathode recovery rose to 96.9%, which corresponded to a decrease in cathode loss to the graphite concentrate of 2.09% (Zhang et al., 2019). After conventional flotation, the grade and recovery of $LiCoO_2$ were only 67.3% and 74.6%, respectively; however, pyrolysis pretreatment enhanced these numbers to 94.00% and 84.61%, respectively. The grade and recovery of $LiCoO_2$ were improved by additional ultrasonic treatment after pyrolysis to 93.9% and 96.9%, respectively (Zhang et al., 2018b, 2019).

According to Zhang et al. (2018b), ultrasonic treatment of pyrolysis can raise the grade and recovery of LCO to 93.9% and 96.9%, respectively. After heat treatment at 550°C, a method utilizing wet ball milling was also studied for the removal of pyrolysis byproducts and pyrolytic carbon. After ball milling, no observable change in the cathode's grade could be made. However, due to the removal of some hydrophobic pyrolysis residues and subsequent improvement of the cathode's hydrophilicity, the recovery increased from 83.8% to 91.4%. According to Zhang et al. (2020), the grade of cathode material in the graphite concentrate was as high as 15.77%, which points to the need for additional processing or multistage flotation.

The pyrolysis time, temperature, and heating rate all significantly influenced the flotation efficiency. The obtained grade of $LiCoO_2$ maintained above 86% at temperatures of 400°C and higher, regardless of the heating rate and roasting time. The associated recoveries are less than 90%. At a temperature of 550°C, which was suggested as the best temperature for pyrolysis, the maximum grade and recovery of $LiCoO_2$ concentrate were produced. According to the variation in pyrolysis time at this temperature, the most effective instance, which corresponds to the highest $LiCoO_2$ recovery of roughly 84% at a grade of 94.7%, takes place in 15 minutes. The variation in heating rate shows a decrease in recovery with an increase in heating rate, which has been linked to a low flotation stage separation efficiency due to insufficient breakdown of organic binders.

Sun and Qiu (2011) put out a cutting-edge technique to separate CAM utilizing vacuum pyrolysis in order to prevent the production of harmful gases during thermal treatment. The adhesion of the cathode material and collector was decreased throughout the pyrolysis process because the electrolyte and high molecular weight polymer binder evaporated or broke down into low molecular weight products. The cathode materials did not separate from the collectors when the pyrolysis temperature was less than 450°C. The separation efficiency increased with temperature when the temperature was between 500°C and 600°C. However, the Al-foil became brittle beyond 600°C, making it challenging to remove the collector from the cathode material. The separation of the CAMs from the Al collectors was achieved using a decreased thermal treatment technique, according to Yang et al. (2016). It has been demonstrated that adjusting the temperature of the reduction reaction makes it possible to isolate the current collectors from the CAMs. Additionally, the active cathode materials' molecular structures are altered during this process, which makes it easier to leach metals during the leaching phases. Thermal treatment has several benefits, including ease of use and high effectiveness. The binder and additives are thermally treated; however, this has the drawback of producing poisonous gas.

At 550°C, the impact of pyrolysis on $LiFePO_4$ batteries (LFP type) was also investigated. Prior to the flotation step, the goal was to evaluate the separation of residual electrolytes, the separator, and the PVDF. The recovery of the cathode material (in a fraction smaller than 0.25 mm) was found to be 80.3% after 2 hours of pyrolysis. An experiment using collectorless flotation (without kerosene) at pH 10 with MIBC and amylum at doses of 30 mg/L and 100 g/ton, respectively, produced a recovery of $LiFePO_4$ of 49.7% at a grade of 88.7%, indicating that this method does not provide any additional selectivity benefits despite the lack of a collector (Zhong et al., 2019).

4.9 EVALUATION OF VARIOUS LIB RECYCLING TECHNOLOGIES

All currently accessible industrial processes must undergo this hydrometallurgical and upstream pyrometallurgical primary treatment, hence the technical and financial evaluation of processes is mostly restricted to the pretreatment stage. Table 4.8 summarizes a comparison of several S-LIB processing pathways with six routes.

Table 4.8 does not display the preparatory steps (such as disassembly and discharge). However, they also significantly affect the entire recycling process. Only if a large electric mobile battery does not absolutely need to be manually disassembled before the first treatment step may a chosen process produce a long-term benefit. Automobile manufacturers use a wide range of designs and tactics from tiny, non-repairable throwaway batteries to modular, organized construction. As a result, there will be batteries in the future that robotics can deconstruct as well as batteries that cannot be disassembled, i.e. batteries that are securely bonded/glued. Therefore, the technological pretreatment plants must be configured for the anticipated range of future sizes and configurations. The discharge step—depending on how discharging is used in the process—could be a bottleneck in a recycling process. The manufacturer would hypothetically need to decode the access to the battery management system (BMS) in order to diagnose and unlock the high-voltage port, making this upstream sub-step technically challenging. The only way to open an EoL battery with high-voltage educated individuals and evade the safety measures, nevertheless, is to use the primary voltage terminals for ohmic discharge because EoL batteries are primarily electro-technically flawed. This discharge procedure requires a lot of time, space, and staff in addition to being logistically challenging, and the revenue

TABLE 4.8
Technical and Financial Comparison of the Current Pilot and Industrial LIB Recycling Routes

				Industrial Appicability ←		
Possible Process Flowsheet Combinations	1	2	3	4	5	6
Preprocessing 1	LIB Pyrolysis	LIB Pyrolysis	Mechanical	Mechanical	Mechanical	LIB Pyrolysis
Preprocessing 2		Mechanical	CAM Pyrolysis	CAM Pyrolysis		Mechanical
Main process 1	Pyrometallurgy	Pyrometallurgy		Pyrometallurgy		
Main process 2	Hydrometallurgy	Hydrometallurgy	Hydrometallurgy	Hydrometallurgy	Hydrometallurgy	Hydrometallurgy
Technical level	Industrial scale	Industrial scale	Pilot scale	Pilot scale	Lab scale	Lab scale
Disassembly	Not required	Not required	If necessary	If necessary	If necessary	Not required
Sensitivity to input contaminants	Low	Low	High	Low	High	High
Ni, Co, Cu losses in the intermediate products	Low	< 5%	40-60%	40-60%	40-60%	Low
Recovered metals	Co, Ni, Cu	Co, Ni, Cu	Co, Ni, Cu, Fe	Co, Ni, Cu, Fe	Co, Ni, Cu, Fe	Co, Ni, Cu, Fe
Recovered total product value	Good	Optimizded	Low	Low	Low	Optimized
Bottleneck for scalability for preprocessing	Only as large plant with minimum capacity > 50000 t/y	Good modular scalability	Explosion/fire Limited use of graphite-rich CAM in hydro	Explosion/fire Low Economic efficiency (high cost, low yield)	Explosion/fire HF emission Organic/ HF rich CAM can hardly be used in hydro	Explosion/fire Limited use of graphite-rich CAM in hydro
(Scale: 1 most costly; 6 least costly)						
Process additives (reagents, activated C, N_2, etc)	5	5	2	4	1	3
Energy consumption comparison	1	2	3	1	5	4
Other operating costs (maintanance, personnel, wear, etc.)	4	5	2	3	1	5
Capex preprocessing (t/y) and technical usage time (y)	4	3	1	1	2	1

from feeding residual charge into the power grid is secondary. The viability of the upstream discharge phase appears unrealistic when using conservative projections (e.g., roughly 250,000 EoL battery units in the EU in 2030), which has constrictive implications for the spread of the various recycling pathways outlined below. It is remarkable that only those process combinations (Routes 1 and 2) that integrate the stages of pyrolysis of the battery cell with subsequent pyrometallurgical pre-cleaning have attained practical significance after 15 years of international R&D work. This is mostly because impurities and undesired battery components, such as organic solvents, are eliminated in these two processes to prepare the material for hydrometallurgy's final refining. The extraction of the target metals Co, Ni, and Cu would be difficult if these contaminants weren't removed (see Table 4.9; Sojka et al., 2020).

TABLE 4.9
Potential Impurities and Their Consequences for Hydrometallurgical Refinement

Leach Contaminants	Generated problems in leaching	Troubleshooting
Electrolyte (EMC, DMC, etc.)	Organochlorine compounds and toxic gas formation; increased acid consumption, pre-wash costs	Evaporation/ decomposition by thermal pretreatment
Adhesive/binder (PVDF) (100% in AM)	Insoluble in acids and remains in the filter cake, increases in disposal costs	Decomposition by thermal pretreatment
Conducting electrolyte salt (LiPF$_6$) (100% in AM)	The HF$_{(g)}$ formation, fast corrosion to the equipment; LiF formation, thus, Li depletion in aqueous and metal-containing phase	Decomposition by thermal pretreatment
Mn	Increased operating costs because of necessary precipitation and disposal, as well as cross-contamination in Fe and Al fraction, more difficult recycling opportunity	Slagging pyrometallurgy
Al and Fe	Increased operating costs because of necessary precipitation and disposal of waste. Al reduces filtering efficiency.	Slagging pyrometallurgy
Plastic residues (separator, sleeve, cable covers, etc.)	Additional filter expenditure; disposal costs for incineration	Decomposition by thermal pretreatment
Graphite (C) (100% in AM)	Foam formation with an impact on the plant construction and cost of process additives	Reducing agent pyrometallurgy
LiFePO$_4$ (100% in AM)	Formation of stable Ni and Co PO$_4$'s, hazardous wastewater with PO$_4$, corrosion because of gas formation, HF favoritism, basic safety concerns, filtration difficulties	Slagging pyrometallurgy
Silicon (LiB Gen. Si Anode)	Gel formation increased filtration efforts and high waste disposal cost	Slagging pyrometallurgy

Because contaminants such as Cd, Pb, K, Zn, and La from the missorting of other battery systems or electronic components will essentially be contained in the mass production process, the pre-cleaning effect of pyrometallurgy further boosts the process resilience. Fe and Al will also be slagged here if they haven't been fully or completely removed from housings by mechanical processes. This reduces recycling efficiency and makes it more difficult to meet the 50% goal set by the Battery Directive (2006/66/EC). However, the plentiful graphite must also be utilized in order for LiB to reach this minimum level of recycling efficiency. For instance, it can be employed via material recovery, which has not yet been validated on an industrial scale, or as a substitute for reducing agents in pyrometallurgy, both of which are approved by EU/493/2012.

4.9.1 Pyrolysis Pretreatment Step Has Several Functions

On the one hand, it aims to entirely eliminate organic materials from hydrometallurgy, which is undesirable. In cutting-edge afterburning facilities, the organics can be controlled oxidized and even employed as a supplementary energy source. On the other hand, since the PVDF has irreversibly attached the target metals to the cathode, it should liberalize the active mass with those metals. The active mass will be challenging to detach from the electrode foil if this adhesive bond is not thermally decomposed. This link cannot be broken by pure mechanical stress, which causes significant loss of the target metal due to cross-contamination. After mechanical pretreatment alone, the cathode foils with CAM are present in numerous output fractions, increasing the dissipative CAM losses by up to 40%–60%. Pyrolysis also offers the benefit of carefully removing explosive and dangerous gas combinations from electrolytes and breakdown products, which increases process safety. Additionally, it ensures that the produced HF acid will be specifically neutralized. The equipment parts (feed, shredder, discharge, intermediate buffering, storage, etc.) must be explosion-proof and cold if LIBs are mechanically broken down in the first phase; however, they typically aren't. The HF gas condenses in the system segments and quickly corrodes and destroys them by HF acid when it comes into touch with any moisture that is present. Additionally, extremely volatile electrolytes can self-ignite when combined with ambient oxygen since cells can never be totally depleted of their residual energy. Additionally, a few smaller facilities operating in Europe that used this mechanical pretreatment were unable to demonstrate with certainty what happened to the evaporated electrolytes.

The overall economic efficiency of LIB recycling processes is determined by their cost composition, such as capital, general operating, and personnel costs, versus revenue generated from the quantity and quality of the recovered metals. The necessary investments and thus capital costs of the two pretreatments and two main treatment steps rise exponentially in the following order of ranking:

Mechanical treatment → Pyrolysis → Pyrometallurgy → Hydrometallurgy.

Following the following order, the complexity of the technologies is significantly rising, necessitating a larger scale for the plants for them to be economically viable.

The intermediate product active mass from the pretreatment of EoL batteries is currently primarily fed into these large industrial metallurgy plants, which process primarily primary and secondary raw materials with a suitable mixing ratio of secondary feedstock on a scale of 50,000–300,000 t/y, due to the economic benefit of scaling.

The maintenance costs per ton of treated LIB are inversely correlated to the capital costs, particularly in the case of mechanical pretreatment (Routes 3, 4, and 5), where the inevitable production of HF acid raises costs for maintenance and workplace safety as well as causes rapid total equipment failure and shortens depreciation periods to a few years. A major drawback of the pyrometallurgical Routes 1, 2, and 4 integrating the pyrometallurgical stage is the higher primary energy consumption. However, the lower operating costs in the hydrometallurgical phase offset these higher expenditures. The hydrometallurgical effort would need to be greatly enhanced if the active mass generated by shredding and classifying was not pre-cleaned by pyrolysis and/or pyrometallurgy, as in Routes 3, 4, and 5. It must be noted that although this hasn't been demonstrated on a wide scale, experts have indicated additional chemical consumption processes, gas production safeguards, additional precipitation steps, and disposal costs. Additionally, the mechanical pretreatment procedure, which is utilized to absorb the evaporated electrolytes from off-gas to meet the requirements of emission standards, consumes activated carbon at a high cost. Because of this, Routes 3, 4, and 5 have lower scaling and less expandability of the plants, higher personnel costs, higher off-gas cleaning costs, and additional maintenance and replacement requirements, all of which have a negative impact on overall profitability.

However, as the revenue originates from the recovery of Co, Ni, and Cu, additional recovered commodities in Routes 1/2/3/5/6 such as steel or aluminum, regardless of their differing quality brought about by various processing procedures, do not constitute an economically significant factor. This economic contribution is solely based on the recovery rates of the three target metals; therefore, the considerable losses brought on by cross-contamination during mechanical pretreatment result in serious drawbacks. Due to the dissipative losses of the commonly targeted metals throughout the process, the process selection of routes 3/4/5 in the household LiB business sector is uneconomical due to the leverage of the recoverable value with an average much greater Co content. Additionally, the gains from lithium and graphite extraction have only, at most, been demonstrated on a laboratory scale and have not yet contributed to profits in any of the current pilot scale or industrial recycling systems (Sojka et al., 2020).

In conclusion, the present review demonstrates that different chemical-physical reactions take place during the pretreatment of S-LIBs, both in the uncharged and partially charged states. These reactions always result in environmentally hazardous, human-toxic, extremely corrosive, process-relevant substances, and explosive gas mixtures, which must be treated carefully in the S-LIB pretreatment process. Generally, current recycling processes for S-LIBs can be divided into four different process phases, namely preparation, pretreatment, main treatment with pyrometallurgy, and hydrometallurgy, which can be combined into six different possible sequences. Currently, only two of these (Routes 1 and 2 in Table 4.7) have demonstrated their technical viability and industrial maturity and compete with each other after implementation in commercial enterprises. Both of them involve the sub-steps of pyrolysis and pyrometallurgy, which only appear to be disadvantageous considering their higher energy requirement. However, a hydrometallurgy process cannot refine an active mass obtained from simply mechanical pretreatment directly into the final product. Its impurity level and high load of organics and fluorine would result in unaffordably high additional operating expenses in hydrometallurgy without the pre-cleaning in the two thermal phases. Additionally, the hydrometallurgy input material needs to undergo thermal selection and homogenization because of the significantly disparate composition of the battery active masses (LCO, NMC, LMO, LFP, NCA, etc.). Due to the relatively low losses of the targeted metals (Co, Ni, and Cu), the operating costs of the entire process route are reduced. Additionally, the ecological balance is improved because no additional operational materials must be produced or used, unlike in other process routes. In the future, technical efforts will be made to further enhance or modify the potential process stages, particularly in order to produce additional recyclable materials (Li and graphite), save energy, or reduce losses. It is highlighted that the installation of a complete process chain, including hydrometallurgy specifically developed for the S-LIB sector, only becomes economically viable when the amount of active mass generated has expanded (due to market development) to a certain extent. There have been presented several cutting-edge lab-scale LIB recycling technologies. However, it still needs to be demonstrated that those lab-scale (sub-)processes are commercializable. The "Direct Recycling Process" and the use of segregated CAM for battery material precursor appear to be unfeasible in this context given the lack of product quality proof.

4.10   CHALLENGES AND FUTURE OUTLOOK

4.10.1   SORTING AND SEPARATION TECHNOLOGIES

The S-LIBs most likely come in a variety of sizes, shapes, and chemical compositions. Technologies for sorting and separating materials could improve recycling effectiveness (Chen et al., 2019):

> **Separation of LIBs Based on Different Chemistries:** S-LIBs are typically delivered to recycling plants without the internal chemical composition being known. The separation of LIBs based on chemistry would be made easier by battery manufacturers properly labeling LIBs. Then, it is more effective and efficient to sort and recycle single-chemistry LIBs. For instance, the Society of Automotive Engineers (SAE) has created labeling guidelines for LIBs (J2936). For energy storage devices, such as cell, battery, and pack-level goods, this standard offers labeling suggestions for the whole life span.
>
> **Material Separation:** The pretreatment of EoL LIBs is difficult due to the various chemistries and form factors of S-LIBs. It is necessary to develop a safe and efficient separation of battery components. For instance, auto-dismantling and separation are more practical if cell size and form are standardized to a few designs.

4.10.2 THE PYROMETALLURGICAL PROCESS

The most developed technology is the pyrometallurgical method, which has mostly been used in Europe and North America.

Slag Recycle: During the smelting process, most of the materials (graphite, separator, organic electrolyte, and plastics) are burned and not recovered. Slag including Li is produced. In most conventional pyrometallurgical processes, the Li and Al in slag are not recovered. However, the gradually increasing Li price renders this unsustainable; the Li must be recovered or not be sent to slag. Developing practical technologies to recover Li in the slag could be one of the important research directions for the pyrometallurgical process.

Adaptation to the Rapidly Developing LIB Industry (High Ni and Low Co): The LIB market is changing quickly. With the ultimate goal of "no-cobalt" cathodes, trends are shifting toward cathode materials with progressively higher Ni and decreasing Co content. For the pyrometallurgical process to be economically viable, Co concentrations must be fairly high. However, the business model will be challenged when Co concentrations are decreased. To enable pyrometallurgical processes or modify their business models for emerging generations of LIBs, particularly low- or no-Co cathode materials, innovation is required. One approach would be to improve roasting conditions so that different compounds can be purified and separated more easily. Additionally, using hydrometallurgical techniques to provide additional purification is another direction (some businesses are starting to do this).

Secondary Waste Treatment: Gases and solids are both part of the pyrometallurgical process' waste. The burning of batteries produces gas (mostly CO_2, VOCs, etc.), and non-recycled elements turn into slag. Before being released, gas is cleaned in the industry. However, CO_2 is released into the atmosphere immediately. The pyrometallurgical process will continue to yield economic rewards by recovering solid waste, such as graphite (in addition to Co, Ni, and Cu), or by turning that waste into valuable materials.

4.10.3 THE HYDROMETALLURGICAL PROCESS

The hydrometallurgical process is also a commercialized and marketable technology and has been primarily used in China.

Electrolyte and Anode Recovery: Due to its great value, recovering cathode material is currently the main focus of hydrometallurgical processing. Due to their poor value, other materials are not recovered or recycled. The economic viability of the recycling process will be further improved by the development of technologies that allow the recovery of the electrolyte and the graphite anode as high-value materials.

Recycling of LiFePO$_4$ Batteries: LiFePO$_4$ cathodes are commonly used in LIBs in e-buses. Additionally, several EVs use LIBs with LiFePO$_4$ cathodes, particularly in China. The inherent components of LiFePO$_4$ are inexpensive despite the material's cost. In actuality, LiFePO$_4$ synthesis is highly expensive. LiFePO$_4$ cathode-containing LIBs cannot be economically recycled by a hydrometallurgical process. If demonstrated to be physically possible, direct recycling of LiFePO$_4$ might be an option.

Secondary Waste Treatment: The hydrometallurgical process generates waste, which drives up the cost of recycling, in the form of water and chemicals from the leaching stage, co-precipitation, and washing. Further investigation is needed into ways to purify wastewater, reuse water, or minimize the amount of water utilized in the process in order to reduce or eliminate wastewater and related costs.

4.10.4 THE DIRECT RECYCLING PROCESS

The direct recycling process is still in the laboratory stage, and much work needs to be done in order to commercialize it.

Steps in Pre-processing to Obtain Pure Materials: The idea behind direct recycling is to regenerate and utilize cathode materials right away. However, conventional S-LIBs have a wide range of components (such as anode, cathode, Cu, Al, and polymers). One crucial area of research for direct recycling could be developing methods to automatically separate cathode materials from other materials and from one another. Another crucial research topic is the quality of the cathode materials that have been recovered.

Recover Other Materials in Addition to Cathode Materials: At the moment, the direct recycling process is mostly focused on cathode powder, which accounts for 30%–40% of material cost. The direct recycling process must recycle other materials as much as feasible, much like the pyrometallurgical and hydrometallurgical processes do.

Demonstrate Recovered Materials at Scale: For the industry to adopt the direct recycling method, recycling needs to be done at a particular level of scale in order to have a significant impact. Additionally, the industry must independently test the items covered.

Recover Mixed Cathode Materials: The S-LIBs may include different cathode materials. Furthermore, a given battery may use a mixture of cathode chemistries. This is especially challenging for the direct recycling process. Research into how to separate different cathode materials is critical, particularly in view of the many ratios of NMC coming led in the waste stream. Another option is trying to find a scenario in which the mixture can be utilized directly.

Combine Various Recycling Methods: Given that each recycling method has benefits and drawbacks, it can be required to combine various recycling methods to ensure the most efficient recycling. As an illustration, the hydrometallurgical process can be used to recover the materials used in cathodes, and direct recycling can be used to recover other materials.

In addition to the research needs outlined above, Chen et al. (2019) suggested the following areas for addressing:

- **Establish a Viable Business Model:** Due to poor collection rates, developing technology, and relatively low volumes, among other factors, LIB recycling has not yet developed a sustainable business model, unlike LABs. An evaluation tool for recycling strategies has been created by the Argonne National Laboratory (ANL), which simulates the financial and environmental effects of the recycling strategies indicated above. The model takes into account every step of the battery extraction process from the car through its final recycling using the user-selected process. A detailed comparison of various recycling processes is also provided by the model, which aids in determining the effects of a given recycling process on the economy and environment.
- **Local Recycling or Pretreating:** Because LIBs are hazardous, a significant portion of the expense of recycling is attributed to transportation. Only a few nations (the USA, Canada, France, Switzerland, Germany, Belgium, China, Japan, South Korea, and Singapore) around the world have recycling facilities. For instance, the east and west coasts of the USA are home to the majority of LIB recycling plants. S-LIBs in the middle of the nation have to be hauled a long way to be recycled. The establishment of dispersed recycling facilities across the nation will probably be justified for the best possible logistical and economic outcomes if S-LIBs reach significant quantities.
- **Design for Remanufacturing/Repurposing/Recycling:** EV battery designs are currently optimized for performance, safety, and cost. Remanufacturing/repurposing/recycling should play a more significant role in EV pack design.
- **Solid-State LIB Recycling:** For solid-state LIB, a lot of research and development is being done in order to significantly boost energy density. How to handle Li metal responsibly and securely will be the main difficulty for solid-state LIB recycling. The current recycling process includes many operations that are inappropriate for Li metal, such as discharging and shredding. Using a salt solution to discharge LIBs is a widely used technique for releasing the leftover energy. Li metal, however, can react with water in a very hostile manner. Li metal's softness also makes it easy to adhere to the shredder. Therefore, solid-state LIB recycling advances will also be required.

To create profitable business practices for recycling S-LIBs, collection, storage, logistics, and transportation are just as important as the technical aspects. Advancements in these enabling fields are the focus of the recently launched Recycling Prize (DOE).

Globally applicable laws and regulations must also be put into place for LIB recycling. The European Union (EU) has strict regulations governing the recycling of LIBs, and by 2030, recycling efficiencies must reach 50%. After August 2018, all EVs in China will have a unique ID that will enable tracking of the batteries from their initial production to their subsequent use and recycling. National guidelines for the collecting and recycling of large-format LIBs do not yet exist in the USA, nevertheless. Despite national laws prohibiting the recycling of LIBs. California is a national leader in recycling LIBs and has been a pioneer in pushing the electrification of vehicles. A new objective was set in 2016 by an EV action plan in California to create new business prospects for battery recycling.

In fact, the policy may be implemented for the production of LIBs, the collection and delivery of S-LIBs, recycling procedures, and the reuse of recycled materials. Requirements for module design (energy, size, and voltage), the joining mechanism (reversible to permit disassembly of packs), and adhesives could all be included in the manufacturing standardized recommendations and successfully promote recycling acceptance. Fewer varieties mean less labor-intensive and suitable disassembly and separation procedures for recycling.

Techniques for automation could be created. Regulations for the entire recycling life cycle, including collection, storage, logistics, and transportation as well as the actual recycling process in the facility, might also be tightened. Responsibility for S-LIBs treatment (manufacturers, distributors), collection rates, and recycling efficiencies could be among the policy's guiding principles or objectives. Manufacturers must collect and handle S-LIBs as a result of the EU's adoption of the extended producer responsibility (EPR) for LIBs across the entire union. Another benefit of a deposit-refund scheme is that it might persuade consumers to properly recycle or discard their EoL batteries. Recycled material performance is the main area of worry, thus it should be suggested to do extensive, dependable testing to provide assurances. The usage of recycled materials around the world could be facilitated by policies and rules for standardized evaluation of S-LIBs (Chen et al., 2019).

5 Hydrometallurgical Recycling of LIBs

ABBREVIATIONS

AB	Acetylene black
CAM	Cathode active material
EoL	End-of-life
EU	European Union
EW	Electrowinning
LAB	Lead acid battery
LFP	Lithium iron phosphate
LMO	Lithium manganese oxide
NMC	Nickel manganese cobalt
NMP	N-Metil-2-Pirrolidon
PFD	Process flow diagram
PLS	Pregnant leach solution
R&D	Research&Development
S/L	Solid/Liquid
S-LIB	Spend lithium-ion battery
SEM	Scanning electron microscope
SX	Solvent extraction

5.1 HYDROMETALLURGIC LEACH PROCESS

Leaching in an acid, alkaline, or ammonia-based medium and purifying techniques are chemical procedures related to dissolving the metallic component and recovering metal solutions that may be employed in the chemical or battery industries. Hydrometallurgy focuses on the recovery of metals and is the most commonly used method for Li extraction from spent lithium-ion batteries (S-LIBs). It dissolves the Li in the pretreated active materials with acids, bases, and ammonia to obtain Li^+ solutions from which Li can be extracted. Inorganic acids, such as H_2SO_4, HCl, and HNO_3, were generally used. Previous studies have shown that the structure of LIB the cathode active materials (CAMs) allows for simple Li dissolution via a deintercalation mechanism, whereas the subsequently oxidized transition metals generally require a reductant for efficient extraction. Heat or redox reactions using H_2O_2, H_2SO_3, NH_2OH, $Na_2S_2O_5$, Na_2SO_3, and $NaHSO_3$ (Meshram et al., 2015a; Sun and Qui, 2011; Vieceli et al., 2018; Zheng et al., 2016) were applied to increase the leaching efficiency. Among the redox agents, H_2O_2 is the most commonly used reductant due to its low cost and non-toxicity. H_2O_2 can increase the leaching reaction rate because of its strong reducibility. However, utilizing an acid with a low pH might cause the release of hazardous gases like Cl_2 from HCl and NO_x from HNO_3, which harms the environment.

DOI: 10.1201/9781003384557-5

Thus, tests are being conducted on weak organic acids like citric acid, oxalic acid, etc. Acids and bases can be used to leach Li compounds, which can then be produced through precipitation, solvent extraction (SX), electrowinning (EW), or selective adsorption.

The high recovery rates of Co, Ni, and Li (>95%), low energy consumption compared to pyrometallurgy, better selectivity, low impurities, and relatively low cost at low production capacities have all contributed to the hydrometallurgical process' increased interest in research and development (R&D) at research institutes, universities, and industry. But it also results in a sizable amount of wastewater creation. The recovered material might not perform as well as its virgin equivalents and several processes are necessary to satisfy specific efficiencies (Arshad et al., 2020). Many researchers used the low-cost and straightforward SX method with acid leaching to extract high-purity Co from S-LIBs. The hydrometallurgical process has four distinct steps: leaching with acids, bases, or ammonia; chemical precipitation/purification; SX; and electrochemical separation (EW). Initially, CAMs such as $LiCoO_2$, $LiNiO_2$, $LiMn_2O_4$, and $LiNi_{0.33}Mn_{0.33}Co_{0.33}O_2$ are dissolved by acid/alkaline/ammonia solutions, and then, the process proceeds to extract Co, Li, Ni, and Mn metals from the pregnant leach solution (PLS) by a suitable method. A typical hydrometallurgical process flowsheet of S-LIBs for the recovery of Co and Li is shown in Figure 5.1. The hydrometallurgical recovery flowsheet of S-LIBs is given in Figure 5.2.

Hydrometallurgical methods of S-LIBs recycling include firstly the pretreatment process (such as discharging, dismantling, separation, dissolution, ultrasonic washing, flotation, and heat treatment). After pretreatment, cathode, and anode active materials are separated. CAMs can be either acid-leached or bioleached. In acid leaching, inorganic or organic acids can be employed. Obtained PLS can be purified by chemical precipitation. SX and EW can be used for pure metal production.

The EU regulations and European Battery Directive force more battery recycling as well as recovery of specific individual elements, such as Li, Co, Ni, and Cu, which can be reused in new battery cells. This will likely emphasize the importance of hydrometallurgical LIB recycling methods due to their ability to recover a large variety of battery elements with high purity—either as separate operation or in combination with pyrometallurgical ones—when the unit operations are optimized accordingly (Zhang et al., 2018a).

The thriving lead-acid battery (LAB) recycling market can teach the future LIB recycling sector a number of valuable lessons. Because LABs are a reasonably standardized technology that is easy to disassemble and recycle, recycling costs are kept to a minimum. Unfortunately, these benefits are not expected to be applicable anytime soon for an emerging technology like electric-vehicle LIBs (Harper et al., 2019).

In comparison to the pyrometallurgical process, the hydrometallurgical method has four primary advantages: (i) the ability to manufacture high-purity materials, (ii) the ability to recover the majority of LIB components, (iii) the ability to operate at low temperatures, and (iv) fewer CO_2 emissions. The main drawbacks of the hydrometallurgical processes are

- the requirement for sorting, which increases the need for storage space and raises the cost and complexity of the process;
- the difficulty of separating some elements (Co, Ni, Mn, Fe, Cu, and Al) in the solution because of their similar properties, which can result in higher costs; and
- the cost of wastewater treatment and associated costs.

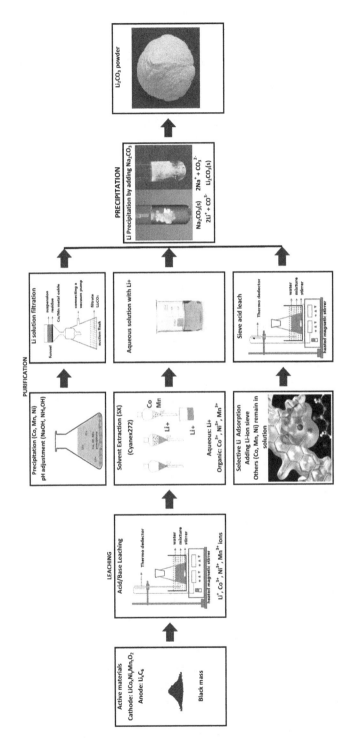

FIGURE 5.1 Hydrometallurgical S-LIB recycling steps.

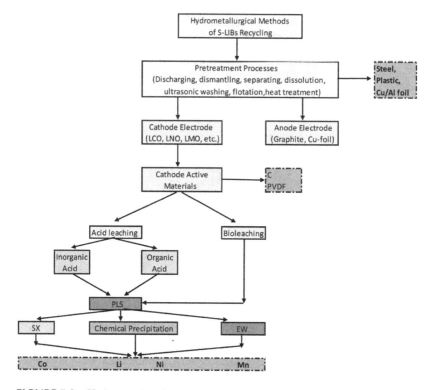

FIGURE 5.2   Hydrometallurgical recover flowsheet of S-LIBs.

A novel understanding of S-LIB leaching in the presence of mineral acids is presented in this chapter, which may serve as the foundation for future process improvements for developing hydrometallurgical battery recycling operations. Here, we discuss some recent technical advancements in hydrometallurgy related to leaching, solvent extraction, and chemical precipitation.

Hydrometallurgical recycling techniques are economically feasible for the majority of CAMs, especially for cathodes with high Co and Ni content. Due to the poor intrinsic value of the cathode material components (Fe and Mn), LFP and LMO present a hurdle for conventional business purposes.

5.2   INORGANIC ACID LEACH OF LIBs

Acid leaching is still commonly used due to its high recovery efficiency. Strong inorganic acid leaching may result in secondary pollution (excess acid solution and harmful gas emission) when used in leaching, however, organic acid leaching offers comparable leaching efficiency and has biodegradable properties.

Pyrometallurgy, hydrometallurgy, and direct recycling are the three major subcategories of typical S-LIB recycling processes. Hydrometallurgy can also be divided into reagent leaching and bioleaching (biometallurgy). Pyrometallurgy has not received as much attention as hydrometallurgy, which uses leaching and extraction as its primary processes, much like the treatment of other types of waste. S-LIB is typically processed through disassembly and/or mechanical shredding prior to metal recovery. Due to the presence of high-quality Al-foil and cathode materials, Co-based cathode scrap and its manufacturing scrap are regarded as the most valuable fraction to be recycled among the other S-LIB fractions after pretreatment.

Leaching, which dissolves the metallic fraction and recycled metal solutions for subsequent separation and Li and Co product recovery, is a step in the hydrometallurgical method of recycling metals from S-LIBs. One of the most popular and efficient techniques for removing valuable metals from S-LIBs is acid leaching. Inorganic acids, organic acids, alkaline solutions, and ammonia-ammonium salt systems are common leaching agents employed in the leaching process.

The primary goal of the leaching process is to achieve the highest percentage of Li, Co, and Ni dissolution in PLS under ideal conditions (lowest energy and reagent consumption, highest efficiency, and high solid content) while minimizing the content of other impurities. The process of leaching can be described as the diffusion of the desired solid content into the liquid, with heterogeneous reactions taking place at the interface between the solid particles and the liquid medium.

Recently, several researchers have worked on the acid-leaching process to extract metals from S-LIBs using different inorganic/mineral acids like H_2SO_4, (Sun and Qui, 2011; Chen et al., 2019; Granata et al., 2012; Gratz et al., 2014; Lee and Rhee, 2003; Meshram et al., 2015, 2015a; Pagnanelli et al., 2014; Sa et al., 2015; Sun and Qui, 2011; Zou et al., 2013), hydrochloric acid (HCl), (Guo et al., 2016; Guzolu et al., 2017; Shuva and Kurny, 2013b), nitric acid (HNO_3) (Lee and Rhee, 2002), phosphoric acid (H_3PO_4), (Chen et al., 2017; Pinna et al., 2017), and hydrofluoric acid (HF) (Suarez et al., 2017) were used as leaching agents have been used and studied extensively in the literature. These acids are strong mineral acids that speed up the leaching process and provide high rates of metal recovery from the S-LIBs. Inorganic acids are less expensive, but there are also some worries regarding their potential effects on the environment. For instance, the oxidation of HCl during the leaching process results in the production of dangerous $Cl_{2(g)}$ when HCl combines with $LiCoO_2$. Many researchers have employed reducing agents such as ethanol, hydrogen peroxide (H_2O_2), sodium hydrogen sulfite ($NaHSO_4$), and glucose ($C_6H_{12}O_6$) with leaching agents such as H_2SO_4, HCl, HNO_3, and H_3PO_4 to increase the leaching efficiency (Sun and Qui, 2011, Lee and Rhee, 2003, Meshram et al., 2015, 2015a; Pagnanelli et al., 2014; Sun and Qui, 2011; Yao et al., 2015; Zhao et al., 2019; Zhao and Zhang, 2020; Yao et al., 2015; Roy et al., 2021). The primary goal of the reducing agent is to lower the metal's level of oxidation and prepare it to dissolve in an acidic solution. The oxidation states of metals like Co and Mn, Co^{3+} and Mn^{4+}, are difficult to dissolve in acid solution; yet, the divalent forms of Co^{2+} and Mn^{2+} dissolve in acid solution rapidly and easily. High-valent Co or Mn in the solid phase is reduced to readily soluble Co^{2+} or Mn^{2+} with the aid of reducing agents. Leaching temperature, reaction time, leaching agent concentration, solid/liquid (S/L) ratio, and reducing agent concentration are the primary variables influencing the metal leaching process.

A typical leach flowsheet for S-LIB recycling with NMP addition for Al foil removal and calcination for the elimination of PVDF and C is given in Figure 5.3. SEM images of unleached and leached material residues are shown in Figure 5.4.

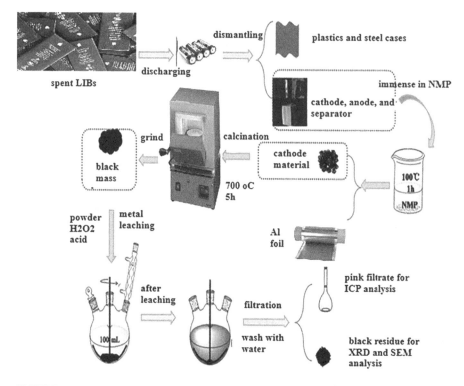

FIGURE 5.3 Process flow diagram for calcining S-LIB metals with NMP immersion and leaching acid.

FIGURE 5.4 SEM pictures of (a) S-LIB CAMs after calcination and grinding and (b) leach residues.

In contrast to the leaching residue, which is shown to have a smaller particle size distribution with diameters of around 2–5 mm in Figure 5.4b, the spent cathodic material is shown to be irregular and agglomerated in Figure 5.4a with particles sizes of approximately 5–12 mm (Li et al., 2015).

Figure 5.5 presents the XRD patterns of S-LIB CAMs after calcination and grinding and the residues after the acid-leaching process. Crystalline LiCoO$_2$ was easily recognized from the XRD data. The (003), (101), and (104) LiCoO$_2$ peaks, in particular, were visible in the XRD patterns of the powders after leaching, which is the difference between Figure 5.5a and b. The reaction between the used LiCoO$_2$ samples and the evaporating acid is responsible for this discrepancy.

5.2.1 H$_2$SO$_4$ Leaching of S-LIBs

As a frequently used mineral acid, H$_2$SO$_4$ can be ionized in an aqueous solution to form H$_3$O$^+$, HSO$_4^-$, and SO$_4^{2-}$. The ionization process can be described as follows:

$$H_2SO_4 + H_2O = H_3O^+ + HSO_4- \qquad (pKa_1 = -2.00) \qquad (5.1)$$

$$HSO_4- + H_2O = H_3O^+ + SO_4^{2-} \qquad (pKa_2 = 1.99) \qquad (5.2)$$

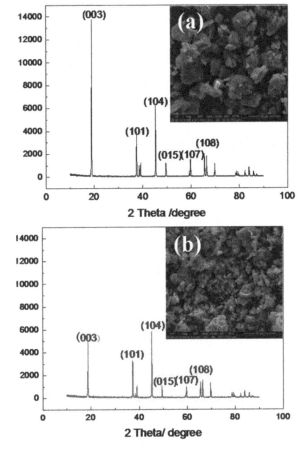

FIGURE 5.5 XRD patterns of (a) S-LIB CAMs after calcination and grinding and (b) leach residues.

The following equations describe the possible reactions of Co, Li, Mn, and Ni in an H_2SO_4 solution, with values of $\Delta G^0{}_{80}$ (Ferreira et al., 2009):

$$4\,LiCoO_{2(s)} + 3H_2SO_4 \rightarrow Co_3O_{4(s)} + 2Li_2SO_{4(aq)} + CoSO_{4(aq)} + 3H_2O + 1/2O_{2(g)} \quad (5.3)$$

$$Co_3O_{4(s)} + 3H_2SO_4 \rightarrow 3CoSO_{4(aq)} + 3H_2O + 1/2O_{2(g)} \qquad (5.4)$$

which results in the following global reaction:

$$4LiCoO_2 + 6H_2SO_4 \leftrightarrow 2Li_2SO_4 + 4CoSO_4 + 6H_2O + O_2 \;\; \Delta G^0{}_{80}: -608.96kJ \quad (5.5)$$

$$Li_2O + H_2SO_4 \leftrightarrow Li_2SO_4 + H_2O \qquad\qquad\qquad \Delta G^0{}_{80}: -306.69kJ \quad (5.6)$$

$$2Co_3O_4 + 6H_2SO_4 \leftrightarrow 6CoSO_4 + 6H_2O + O_2 \quad \Delta G^0{}_{80}: -188.42kJ \quad (5.7)$$

$$Li_2CO_3 + H_2SO_4 \leftrightarrow Li_2SO_4 + H_2O + CO_2 \quad \Delta G^0{}_{80}: -126.61kJ \quad (5.8)$$

$$CoO + H_2SO_4 \leftrightarrow CoSO_4 + H_2O \qquad\qquad \Delta G^0{}_{80}: -114.12kJ \quad (5.9)$$

$$2Co_2O_3 + 4H_2SO_4 \leftrightarrow 4CoSO_4 + 4H_2O + O_2 \;\; \Delta G^0{}_{80}: -94.17kJ \quad (5.10)$$

$$2LiMn_2O_4 + 5H_2SO_4 + H_2O_2 \rightarrow Li_2SO_4 + 4MnSO_4 + 2O_2 + 6H_2O \quad (5.11)$$

$$2LiNiO_2 + 3H_2SO_4 + H_2O_2 \rightarrow Li_2SO_4 + 2NiSO_4 + O_2 + 4H_2O \qquad (5.12)$$

$$12LiNi_{1/3}Co_{1/3}Mn_{1/3}O_{2(s)} + 18H_2SO_{4(aq)} \leftrightarrow 6Li_2SO_{4(aq)} + 4NiSO_{4(aq)}$$
$$+ 4CoSO_{4(aq)} + 4MnSO_{4(aq)} + 18H_2O_{(l)} + 3O_{2(g)} \qquad (5.13)$$

At all temperature ranges (20°C–80°C), the estimated values of ΔG^0 for H_2SO_4 in Eqs. (5.3–5.8) are negative. As a result, there is a strong possibility that the reactions will go in the direction of product creation.

After drying in an oven, $LiPF_6$ salt is kept in the cathode. Next, the cathodes are treated with an aqueous solution containing 1.0 M H_2SO_4 and 5 vol.% of H_2O_2, which is agitated ultrasonically for 1 hour at 60°C. According to Jo and Myung (2019), the $LiPF_6$ salt interacts with water, H_2SO_4 acid, and NaOH as follows:

$$LiPF_6 + H_2O \rightarrow LiF + 2HF + POF_3 \qquad (5.14)$$

$$2LiF + H_2SO_4 \rightarrow Li_2SO_4 + 2HF \qquad (5.15)$$

$$Li_2SO_4 + NaOH \rightarrow LiOH + Na_2SO_4 \qquad (5.16)$$

Na_2SO_4 is a residual impurity that may be eliminated easily because of its high solubility in water during the Li_2CO_3 final purification process. The cathode also contains PVDF as a binder and carbons as a conducting agent. The CAMs from the Al current collectors can be detached using the acid treatment, which contains 1.0 M H_2SO_4 with 5 vol.% of H_2O_2, and the active materials can be totally dissolved in the acidic solution. Black carbon powders that are floating on top of the solution and are being employed as conducting agents can be collected and dried during the process. The gathered powders may be heated in the air for 2 hours at 400°C.

Next, to separate the Li and transition metal present in the waste solution, 2.0 M NaOH solution can be added to the dark-red colored solution until the filtered waste solution becomes completely transparent using the following reaction:

$$CoSO_{4\ (aq)} + Li_2SO_{4\ (aq)} + 4NaOH \rightarrow Co(OH)_{2\ (s)} + 2LiOH_{\ (aq)} + 2Na_2SO_{4\ (aq)} \quad (5.17)$$

$Co(OH)_2$ and LiOH are the products of this reaction, which are easily distinguishable due to their distinct solubilities (12.8 g/100 mL for LiOH and 0.0032 g/100 mL for $Co(OH)_2$ at 25°C). The precipitates can then be filtered, and the end product can then be dried and rinsed with deionized water. Figure 5.6 shows the process details developed by Jo and Myung (2019) using H_2SO_4 leaching and NaOH precipitation of S-LIBs.

The carbonation process for Li_2CO_3 production can be summarized as follows:

$$CO_2 + H_2O \rightleftarrows H_2CO_3 \rightleftarrows HCO_3^- + H^+ \quad (5.18)$$

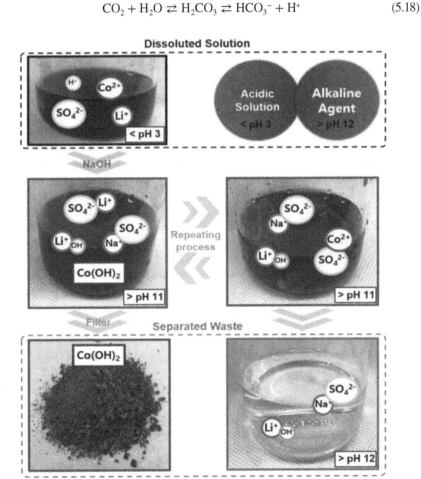

FIGURE 5.6 H_2SO_4 leaching and NaOH precipitation of S-LIBs (Jo and Myung, 2019).

$$Li^+ + HCO_3^- \rightleftarrows LiHCO_3 \tag{5.19}$$

$$2LiHCO_3 \rightarrow Li_2CO_3 + H_2O + CO_2 \tag{5.20}$$

Water, Eqs. (5.5) through (5.10), is the only waste product produced during the leaching stage. Sodium sulfate is produced during the precipitation of the metal hydroxides:

$$Li_2SO_4 + 2NaOH \rightarrow 2LiOH + Na_2SO_4 \tag{5.21}$$

$$CoSO_4 + 2NaOH + Co(OH)_2 + Na_2SO_4 \tag{5.22}$$

It is considered that the technique is an environmentally friendly way to recycle LIBs because it produces trash that is largely benign and relies heavily on acid-base chemistry. Table 5.1 displays some of the prior LIB leach results using only H_2SO_4 as the inorganic acid. Theoretically, for H_2SO_4 solutions with a smaller concentration range, only $LiCoO_2$ is a soluble component in battery scraps. As a result, H_2SO_4 with an appropriate concentration was selected as a more acceptable leaching reagent in comparison to HCl and HNO_3 in order to limit the volatilization pollution created by reactants and lower the solubility of Cu-foil in the acid-dissolving process.

Shin et al. (2005) examined the performance of metal leaching and created a technique for the recovery of Li and Co from the S-LIBs using H_2SO_4 acid and H_2O_2. The proposed method included a chemical leaching process and mechanical separation of metal-containing particles. In-depth research was done on the impacts of the leaching agent, particle size, and incineration as a leaching pretreatment. The separation of the metal-bearing particles from the trash in two phases of crushing and sieving was successful. Within 10 minutes of the leaching process, a complete recovery of the metals was achieved using H_2SO_4 acid and 15 vol.% H_2O_2. In a batch reactor, the leaching was done for a pulp concentration of 50 g/L at a temperature of 75°C and agitation of 300 rpm. Additionally, it was shown that burning $LiCoO_2$ particles to eliminate the carbon and organic binder prior to chemical leaching substantially reduced the leaching efficiency.

Nan et al. (2005) leached S-LIBs with alkali NaOH to remove $Al(OH)_3$. Then H_2SO_4 leach was performed. Co was precipitated as cobalt oxalate by adding ammonium oxalate. Optimum leaching was achieved using 3.0 M H_2SO_4 at 70°C, S/L: 1/5 and 6 hours leaching time. Li content increased from 0.16 to 2.06 g/L and Co content from $2.65*10^{-4}$ to 20.045 g/L at 98% Co and Li dissolution. Cu and Co were extracted by SX using Acorga 5640 and Cyanex 272 (di-(2,4,4 trimethyl pentyl) phosphinic acid) and obtained as $CuSO_4.5H_2O$ and $CoSO_4.6H_2O$. Li_2CO_3 was precipitated with Na_2CO_3.

Nan et al. (2006) looked at a method for recovering metal values from a blend of S-LIBs and Ni-metal hydride (NiMH) batteries. First, a device made especially for extracting S-LIB Fe-shells was employed. After the separation of the Al substrate by alkaline leach and electrolyte and the following heating treatment, the Fe-shells and the metal-mesh substrate in the broken-down materials were successfully separated using sieving. $LiCoO_2$, copper oxides, metal nickel, hydrogen storage alloy, and their oxides were among the black powder leftovers that were dissolved using 3.0 M

TABLE 5.1

Summary of Previous Research Results for Leaching of S-LIBs in Inorganic H_2SO_4 Acid

LIB Materials	Leach Reagents	Reducing Agent	T (°C)	t (min)	S/L Ratio (g/L)	Recovery (%)	Separating Method	Additives	Reference
$LiCoO_2$ batteries Co (22.7%–28.3%); Li (2.3%–2.7%); Ni (80.78%–0.89%); Fe (5.5–6.55)	H_2SO_4	15% H_2O_2	75	10	50	95% Co; 100% Li			Shin et al. (2005)
$LiCoO_2$ (Alkaline leach for Al) (Li 0.16 g/L; Co: 2.65 * 10^{-4} g/L)	3.0 M H_2SO_4		70	240	200	>90% Co; >90% Li; <10% Cu	Chemical deposition and SX	Acorga M5640, Cyanex 272 98% Co; 98% Li	Nan et al. (2005)
$LiCoO_2$, NiMH batteries (alkaline leach for Al)	3.0 M H_2SO_4	3% H_2O_2	70	300	67	> 94% for all metals	SX	Acorga M5640, Cyanex 272	Nan et al. (2006b)
$LiCoO_2$ (Two-step leach: Al with NaOH; acid leach) (Co: 43.3%; Li: 5%; Al: 10.2%; Cu: 0.7%)	4.0 % (v/v) H_2SO_4	1% (v) H_2O_2	40	60	33	100% Li; 97% Co	$CoSO_4,H_2O$ crystallization by evaporation		Ferreira et al. (2009)
$LiCoO_2$	2.0 M H_2SO_4	10% (v) H_2O_2	70	60	33	98.5% Co; 99.8% Li			Chen and Ho (2018)
$LiCoO_2$ cathode material (600°C vacuum evaporation, 30 minutes, 1 kPa)	2.0 M H_2SO_4	5% (v) H_2O_2	80	60	50	>99% Co; >99% Li			Sun and Qui (2011)
$LiCoO_2$ (Ultrasonic washing + leaching + SX)	2.0 M H_2SO_4	H_2O_2	25	240	20	98.3% Co (pH: 3)		1 M Cyanex 272 in kerosene	Takahashi et al. (2020)
$LiCoO_2$			50	240	20	98.4% Co (pH: 3)		96% Co ext. O/A: 1:1	Takahashi et al. (2020)

(Continued)

TABLE 5.1 (Continued)
Summary of Previous Research Results for Leaching of S-LIBs in Inorganic H₂SO₄ Acid

LIB Materials	Leach Reagents	Reducing Agent	T (°C)	t (min)	S/L Ratio (g/L)	Recovery (%)	Separating Method	Additives	Reference
LiCoO₂ (Nokia BLB-2 spent battery:) (27.5 LiCoO₂; 24.5% steel/Ni; 16% C; 14.5% Cu/Al) cathode-anode separate	4.0M H₂SO₄	H₂O₂	80	240		Co: 90%; CoSO₄: 92%	Precipitation with ethanol= 3:1; Rec: CoSO₄ 92%; Co(OH) 8%; 96% CuSO₄; 99% Al (Al(OH)₃)	C₂H₅OH; 90% Li₂SO₄	Aktaş et al. (2006)
LiCoO₂ (assisted by ultrasound; 40kHz, 50 W)	2.0M H₂SO₄	2.0% (v) H₂O₂	60	120	33	96.3% Co & 87.5% Li dissolution 94.7% Co & 71% Li recovery	CoC₂O₄.2H₂O & Li₂CO₃ precipitation with (NH₄)₂C₂O₄ & Na₂CO₃	Li₂CO₃; 71% Li (NH₄)₂C₂O₄; 94.7% Co	Zhu et al. (2012)
LiCoO₂ (anode + cathode, 4 mm) (Co: 23.67%; Li: 2.87%; Cu: 22.13%; Al: 4.3%; Ni: 0.26%)	2.0M H₂SO₄	5.0% (v) H₂O₂	75	60	100	70% Co; 99.1% Li	E_a Li: 32.4 and E_a Co: 59.81 kJ/mol	Cyanex 272, Exxsol D-80	Jha et al. (2013)
LiCoO₂ (supercritical extraction with CO₂; 5 minutes 75 bar)	2.0M H₂SO₄	4.0% (v) H₂O₂	75	5	50	95.5% Co		EW, 99.5% Co	Bertuol et al. (2016)
LiCoO₂ normal leach atmospheric pressure	2.0M H₂SO₄	8.0% (v) H₂O₂	75	60	50	98% Co			

(Continued)

TABLE 5.1 (Continued)
Summary of Previous Research Results for Leaching of S-LIBs in Inorganic H_2SO_4 Acid

LIB Materials	Leach Reagents	Reducing Agent	T (°C)	t (min)	S/L Ratio (g/L)	Recovery (%)	Separating Method	Additives	Reference
$LiCoO_2$ CAM (Co: 35.8%; Li: 6.5%; Mn: 11.6%; Ni: 10.06%)	1.0M H_2SO_4		95	240	50	50.2% Mn,; 96.2% Ni; 66.2% Co; 93.4% Li	E_a Li: 16.4; E_a Co: 7.4 and E_a Mn: 18.5 kJ/mol		Meshram et al (2015)
$LiCoO_2$	1.0M H_2SO_4	0.075M $NaHSO_3$	95	240	20	87.9% Mn,; 96.4% Ni; 91.6% Co; 96.7% Li	$CoC_2O_4.2H_2O$; pH:1.5; 50°C, 120 minutes. pH: 7.5 $MnCO_3$; pH: 9 $NiCO_3$	$Li_2CO_3\downarrow$ with Na_2CO_3; pH: 14	Meshram et al (2015a)
$LiCoO_2$ (Discharge, dehydrate, dry, and grind and leach of 16 mesh) (Co: 23.3%; Li: 2.7%; Ni: 1.4%; Cu: 12.2%; Al: 13.1%)	2.0M H_2SO_4	6.0% (v) H_2O_2	60	60	100	99% Co; 99.9% Cu	$Cu\downarrow$. Na_2S	Oxalic acid $CoC_2O_4.2H_2O\downarrow$ Calcination for Co_3O_4	Kang et al. (2010)
$LiCoO_2$ batteries (manual dismantling.anode + cathode leach)	6% (v/v) H_2SO_4	5% (v) H_2O_2	65	60	33	55% Al, 95% Li; 80% Co	$Al\downarrow$, NH_4OH precipi. (50°C at pH: 5)	Cyanex 272, Exxsol D-80	Dorella and Mansur (2007)
(mixed: 29.49% Co; 3.14% Li; 16.48% Cu, 8.02% Al)								0.72M Cyanex, T: 50°C, pH: 5.5	
$LiCoO_2$ (LIB production waste (Co: 60.2%; Li: 7.09%)	2.0M H_2SO_4	5% (v) H_2O_2	75	30	100	94% Li; 93% Co	Na_2CO_3 for Li, 99.99% Li	1.5M Cyanex 272 O/A:1.6, pH:5; 5.0M H_2SO_4; O/A: 5 stripping	Swain et al. (2007)

(Continued)

TABLE 5.1 (Continued)

Summary of Previous Research Results for Leaching of S-LIBs in Inorganic H_2SO_4 Acid

LIB Materials	Leach Reagents	Reducing Agent	T (°C)	t (min)	S/L Ratio (g/L)	Recovery (%)	Separating Method	Additives	Reference
$LiCoO_2$ (anode + cathode)(HT + 5% NaOH + Calcination + reductive leach) (Co: 26.77%; Li: 3.34%; Al: 5.95%; Cu: 1.24%; Fe: 3.76%; Mn: 1.1%)	4.0M H_2SO_4	10% (v) H_2O_2	85	120	100	96% Li; 95% Co	Sponified P507; 95% Co; 96% Li	$CoC_2O_4\downarrow$ (R: 93%, P: 99.9%)	Chen et al. (2011)
Mixed cathode materials (LCO, LMO, NMC, and LFP)	4.0M H_2SO_4	30% H_2O_2	70–80	120–180		100% Ni; Mn; Co; 80% Li_2Co_3	$Fe(OH)_3\downarrow$ pH>3; $Ni(OH)_2$, $Mn(OH)_2$, $Co(OH)_2\downarrow$ pH:11 $Ca(OH)_2$		Zou et al. (2013)
NMC(111)	2.0M H_2SO_4	4% H_2O_2	50	120	50	>98% metals	Na_2CO_3 for $Li_2CO_3\downarrow$	0.64 M Cyanex 272 for $CoSO_4$	Sattar et al. (2019)
(7.6% Li; 20.48% Co; 19.35% Ni; 19.47% Mn)	3.0M H_2SO_4	Ni;	90	30	50	92% Li and Ni; 68% Co; 34.8% Mn	$KMnO_4$ for Mn↓; $C_4H_8N_2O_2$ for Ni↓	pH: 5; O/A: 1/1; Na_2CO_3 for Li↓	
Cathode (ultrasonic agitation)	1.0M H_2SO_4	5% H_2O_2	60	60		38%			Jo and Myung (2019)
LIB scrap	H_2SO_4	H_2O_2				99% Li/Cu; 98% Co/Ni; 96% Mn; 94% Al	NaOH for Mn and Ni↓; Na_2CO_3 for $Li_2CO_3\downarrow$	D2EHPA for Mn, Al, Cu; Cyanex 272 for Co↓ $Na_2C_2O_4$	Atia et al. (2019)

(Continued)

TABLE 5.1 (Continued)
Summary of Previous Research Results for Leaching of S-LIBs in Inorganic H_2SO_4 Acid

LIB Materials	Leach Reagents	Reducing Agent	T (°C)	t (min)	S/L Ratio (g/L)	Recovery (%)	Separating Method	Additives	Reference
LIB scrap (LCO)	2g H_2SO_4/g powder	50% exc. glucose	90		100	Purity:98% Co; 97% Li		D2EHPA, Cyanex 272	Granata et al. (2012)
	1.5–2.0 gHCl/gr powder; 4.0–5.5 M HCl		90		100	99% Co, Mn, Ni, Al; 99% Li Recovery: 80% Li_2CO_3; 98% $CoCO_3$		O/A: 4; pH: 6; stripping with	
$LiCoO_2$	3.0 M H_2SO_4	04 g/g glucose	95	120	25	98% Co; 96% Li			Chen et al. (2018b)
	3.0 M H_2SO_4	sucrose				96% Co; 100% Li			
$LiFePO_4$ cathode material (Dilute alkali leach for Al removal)	0.3 M H_2SO_4	H_2O_2/Li: 2.07	60	120		96.85% Li; 0.027% Fe; 1.95% P	Precipitation with Na_3PO_4; 95.56% Li	$FePO_4$/C at 600°C $FePO_4$	Li et al. (2017)
(30.8% Fe; 3.85% Li; 17.05% P)		H_2SO_4/Li: 0.57							
$LiCoO_2$	2.0 M H_2SO_4	0.75 Cu/Co mol/mol	60		200	Full Co extraction			Chernyaev et al. (2021)
(38.1 Li; 25 Ni; 27.6 Mn; 260 Co; 34.1 Al; 132.8 Cu mg/g)		0.7 Al/Co mol/mol							
Spent LIBs	3.0 M H_2SO_4	1.6 mL/g H_2O_2	70	150	143		CoC_2O_4 purity 99.5%	D2EHPA for Cu and Mn and PC-88A	Wang et al. (2016)

(Continued)

TABLE 5.1 (Continued)
Summary of Previous Research Results for Leaching of S-LIBs in Inorganic H_2SO_4 Acid

LIB Materials	Leach Reagents	Reducing Agent	T (°C)	t (min)	S/L Ratio (g/L)	Recovery (%)	Separating Method	Additives	Reference
NMC (ultrasonic cleaning; cathode material) (7.42% Li; 10.84% Ni; 19.74% Co; 19.53% Mn)	1.0M H_2SO_4	1% (v) H_2O_2	40	60	40	>99% Li, Co, Mn, Ni	E_a: 64.98; 65.16; 66.12; 66.04 kJ/mol Li, Ni, Co, and Mn	Co and Ni; O/A: 1:1; pH: 2.6	He et al. (2017)
LiCoO$_2$ cathode material	2.0M H_2SO_4		25	1440	50		100% Li 79% Co; 73% Ni		Aaltonen et al. (2017)
	2.0M H_2SO_4	1% (v) H_2O_2	25	1440	50		100% Li 84% Co; 79% Ni		
(Co: 23%; Li: 3%;)	2.0M H_2SO_4	1% (v) H_2O_2	25	300	100		100% Li; 94% Co; 90% Ni		
S-LIB laptop batteries (18650 model, mixed brand) NMC	1.0M H_2SO_4	Fe-scrap (Fe^{2+}) (25 g(L))	70	150	10	99.1% Co; 94.9% Ni, 90.5% Li	Precipitation pH: 4 Fe↓ Zn for Co & Ni↓ 90% Co & Ni deposition (alloy) at 80°C	Li↓by Na$_3$PO$_4$/ Na$_2$CO$_3$	Ghassa et al. (2021)
NMC laptop batteries (18650 model, mixed brand)	0.57M H_2SO_4	Fe-scrap (Fe^{2+})	75	120	10	98.91% Co; 92.67% Li	E_a: 58.71; 37.45 kJ/ mol for Co and Li		Ghassa et al. (2020) (Continued)

TABLE 5.1 (*Continued*)
Summary of Previous Research Results for Leaching of S-LIBs in Inorganic H_2SO_4 Acid

LIB Materials	Leach Reagents	Reducing Agent	T (°C)	t (min)	S/L Ratio (g/L)	Recovery (%)	Separating Method	Additives	Reference
10.47% Co; 4.61% Li; 28.18% Ni; 16.30% Mn; 7.74% Al; 0.13% Fe	pH: 0.5	(6g(L))							
NMC111 cathode material (d_{50}: 7.5 µ)	1.0M H_2SO_4		80	120	17	35% Co; 37% Ni; 37% Mn; 90% Li			Partinen et al. (2023)
	1.0M HCl		80	120	17	35% Co; 37% Ni; 35% Mn; 90% Li			
	1.0M H_2SO_4+1.0M NaCl		80	120	17	55% Co; 58% Ni; 36% Mn; 95% Li			
S-LIB (mobile phone)	2.0M H_2SO_4	40g/L H_2O_2	60	120	20	96% Co; 94% Ni; 95% Mn; 98% Li	Co, Ni, Mn↓ by NaOH precipitation		Cuhadar et al. (2023)
(23.46% Co; 1.39% Li; 18.95% Mn; 42.7% C)									

H_2SO_4 + 3 wt.% H_2O_2 at 70°C, S/L = 1:15 for 300 minutes Co and Ni were extracted as their sulfates with 1.0 M Cyanex 272 at pH = 5.1–5.3 and 6.3–6.5, respectively, while Cu was extracted as $CuSO_4$ with 10 wt.% Acorga M5640 at pH = 1.5–1.7. According to the experimental findings, recovery rates for all metal levels exceeded 94%.

5.2.2 $H_2SO_4 + H_2O_2$ Leaching of S-LIBs

A reduction agent that has a lower standard electrode potential than $LiCoO_2$ is necessary to achieve the complete dissolution of $LiCoO_2$. 2.13 V vs. SHE (Eq. 5.23), which is a relatively high standard electrode potential for $LiCoO_2$ reduction, indicating that a number of elements, including Fe, Cu, and Al, which are already present in the raw material, may operate as reductants for $LiCoO_2$. O_2 and H_2O are the only decomposition products, thus theoretically no additional solution contaminants occur from the treatment (Chernyaev et al., 2021), although H_2O_2 (Eq. 5.25) has been commonly utilized as a reducing agent in hydrometallurgical LIB leaching (Meng et al., 2017):

$$LiCoO_2(s) + 4H^+ + 2e^- \rightarrow Li^+ + Co^{2+} + 2H_2O \ E_{eq} = 2.131V \text{ vs. SHE} \quad (5.23)$$

$$O_{2(g)} + 2H^+ + 2e^- \rightarrow H_2O_{2(aq)} \qquad\qquad E_{eq} = 0.584V \text{ vs. SHE} \quad (5.24)$$

$$H_2O_{2(aq)} + 2H^+ + 2e^- \rightarrow 2H_2O \qquad\qquad E_{eq} = 1.817V \text{ vs. SHE} \quad (5.25)$$

Dissolution of $LiCoO_2$, $LiMn_2O_4$, $LiNiO_2$, and NMC_{111} in the process of the reduction of Co^{3+} in the solid species to Co^{2+} in the aqueous phase belongs to a surface chemical reaction. The leaching process of Li compounds in H_2SO_4 solution could be represented as follows:

$$4LiCoO_{2(s)} + 3H_2SO_4 + H_2O_2 \rightarrow Co_3O_{4(s)} + 2Li_2SO_{4(aq)} + CoSO_{4(aq)} + 4H_2O + O_2 \quad (5.26)$$

$$Co_3O_4 + 3H_2SO_4 + H_2O_2 \rightarrow 3CoSO_{4(aq)} + 4H_2O + O_2 \qquad\qquad (5.27)$$

thus resulting in the following global reaction:

$$4LiCoO_{2(s)} + 6H_2SO_4 + 2H_2O_2 \rightarrow 4CoSO_{4(aq)} + 2Li_2SO_{4(aq)} + 8H_2O + 2O_2 \text{ or} \quad (5.28)$$

$$2LiCoO_{2(s)} + 3H_2SO_{4(aq)} + H_2O_{2(aq)} \rightarrow 2CoSO_{4(aq)} + Li_2SO_{4(aq)} + 4H_2O_{(g)} + O_{2(g)} \quad (5.29)$$

$$2LiMn_2O_4 + 5H_2SO_4 + H_2O_2 \rightarrow Li_2SO_4 + 4MnSO_4 + 2O_2 + 6H_2O \quad (5.30)$$

$$2LiNiO_2 + 3H_2SO_4 + H_2O_2 \rightarrow Li_2SO_4 + 2NiSO_4 + O_2 + 4H_2O \qquad (5.31)$$

$$6LiNi_{0.33}Mn_{0.33}Co_{0.33}O_2 + 9H_2SO_4 + H_2O_2 \rightarrow 2MnSO_4 + 2NiSO_4 + 2CoSO_4$$
$$+ 3Li_2SO_4 + 2O_2 + 10H_2O \qquad\qquad (5.32)$$

$$6LiNi_{1/3}Co_{1/3}Mn_{1/3}O_{2(s)} + 9H_2SO_{4(aq)} + 3H_2O_{2(aq)} \leftrightarrow 3Li_2SO_{4(aq)}$$
$$+ 2NiSO_{4(aq)} + 2CoSO_{4(aq)} + 2MnSO_{4(aq)} + 12H_2O_{(l)} + 3O_{2(g)} \qquad (5.33)$$

$$2LiFePO_4 + H_2SO_4 + H_2O_2 \rightarrow Li_2SO_4 + 2FePO_4\downarrow + 2H_2O \qquad (5.34)$$

In an acidic solution, H_2O_2 can typically act as an oxidant and a reductant, speeding up the leaching of Ni, Co, and Mn. H_2O_2 increases the dissolution of Co and Mn from the cathode scraps by reducing Co^{3+} and Mn^{4+} to Co^{2+} and Mn^{2+}, respectively (Zhang et al., 2015). The efficiency did not considerably rise when using more than 2.0% H_2O_2. Because H_2O_2 is unstable, the phenomenon is most likely a result of this. H_2O_2 can break down in solution when heated (Eq. 5.35), and an increase in concentration could hasten this process.

$$H_2O_2 \text{ Reduction: } H_2O_2 + 2H^+ + 2e^- \rightleftharpoons 2H_2O \quad E^0 = 1.76\,V \ \Delta G^0 = 339.63 \text{ kJ} \quad (5.35)$$

$$H_2O_{2(aq)} \rightarrow H_2O_{(aq)} + 1/2O_{2(g)} \quad (5.36)$$

$$2H^+ + O_2 + 2e^- \rightleftharpoons H_2O_2 \quad E^0 = 0.68\,V \quad (5.37)$$

Since the effective H_2O_2 grew slowly with the increase in its concentration during the leaching process at the condition of high H_2O_2 concentration (over 2.0%), there was no appreciable improvement in the leaching efficiency. Numerous other reductants have been researched because H_2O_2 synthesis requires a lot of energy and produces harmful by-products (Pertinen et al., 2023).

Ferreira et al. (2009) investigated Al, Co, Cu, and Li leaching from S-LIBs. Leaching of cathodes with NaOH was found very selective for the separation of Al, thus leaving all Co and the majority of Li in the solid phase. The recovery of Al (and Li) increased with the concentration of NaOH (1%–15% (w/w)), and it was found practically unaffected by the temperature range investigated (30°C–70°C). Eh-pH diagrams were used to explain such behavior. To prevent the formation of precipitates, it appears appropriate to do the procedure in stages while using a maximum of 10% (w/w) of NaOH. The use of a reductive agent like H_2O_2 was discovered to be crucial in the leaching of $LiCoO_2$ with H_2SO_4 in order to speed up the dissolution process and decrease the consumption of acid. In the presence of H_2O_2, 97% of the Co was leached in a single stage, and the effects of the operational variables temperature and H_2SO_4 concentration were determined to be insignificant. It was discovered that crystallization was sufficient to create a recycled mono-hydrated $CoSO_4$ product. The amount of water that is evaporated from the acid liquid will determine the purity grade of the deep pink crystal salt that is produced by crystallization. The best outcome in the current study was found for 85% of water evaporation.

Chen et al. (2018) investigated the recovery of valuable metals using the leaching agent H_2SO_4 and reducing agent H_2O_2. They obtained high reaction yields of 99.8% for Ni, 98.5% for Co, and 98.6% for Mn. Utilizing 2.0 M H_2SO_4, 10 vol.% H_2O_2, an L/S mass ratio of 30 mL/g, a leaching temperature of 70°C, and a leaching period of 120 minutes, the leaching reaction solution was created.

Sun and Qui (2011) combined the recovery of Co and Li from S-LIBs using a vacuum pyrolysis and hydrometallurgical approach. The cathode powder made of $LiCoO_2$ and CoO entirely peeled off of Al foils under the following experimental conditions: temperature of 600°C, vacuum evaporation period of 30 minutes, and residual gas pressure of 1.0 kPa, according to the results of vacuum pyrolysis of cathode material. Peeled $LiCoO_2$ could be leached using a 2.0 M H_2SO_4 acid solution at 80°C for 60 minutes at an S/L ratio of 50 g/L to recover over 99% of Co and Li. It is possible to scale up this technique to help reduce the environmental pollution of S-LIBs and it provides an effective approach to recycle valuable materials from them.

Takahashi et al. (2020) reported all three steps such as grinding, leaching, and SX with different leaching agents such as H_2SO_4, HCl, and HNO_3 with/without reducing agent H_2O_2 to enhance Co recovery from S-LIBs. The scrap was ground using a variety of methods, including hammer, knife, and ceramic balls, which were assessed while disassembling with Co-release in mind. To reduce the particle size distribution of ground material, ultrasonic washing was examined. The experimental conditions were kept constant for 240 minutes at a pH range of 3–5, a temperature range between 25°C and 50°C, and a S/L ratio of 1/5. Using 1.0 M Cyanex 272 in kerosene, selective Co separation was carried out. The only device to ground the batteries, according to the results, was the knife mill. $H_2SO_{4+}H_2O_2$ was used with the best efficiency for Co leaching at pH 3.0, S/L ratio of 1/5, and 50°C. In the oxidative medium, the efficiency of Co leaching was enhanced to 98.8%. Acid leaching, precipitation, and ultrasonic washing all helped to reduce the amount of energy and pollution used to recover Li and Co from S-LIBs. 91% of the Co was recovered after all hydrometallurgical processes, including leaching and solvent extraction.

Aktaş et al. (2006) investigated the viability of retrieving metallic values from S-LIBs using an ethanol-based precipitation method. The goal of this effort was to reduce the environmental impact of EoL batteries by maximizing the recovery of recyclable materials. Cu was recovered with a 96% recovery efficiency following digestion in H_2SO_4 acid as $CuSO_4.3H_2O$ with the addition of ethanol at a volume ratio of 3:1. Two actions were taken to regain Co. By using ethanol at a 3:1 volume ratio during the first step, 92% of the Co was recovered as $CoSO_4$. Ethyl alcohol promotes the precipitation of Co as $CoSO_4.H_2O$ by removing the water ligands from the Co^{2+} cation. The second stage was adding LiOH to raise the pH to 10 in order to precipitate the remaining Co as $Co(OH)_2$. By adding ethanol at a 3:1 volume ratio, Li that was still present in the solution was subsequently recovered as Li_2SO_4 with up to 90% recovery efficiency. With a 99% recovery efficiency, Al was recovered as $Al(OH)_3$. According to their concentrations in the solution, it was demonstrated that metals may be precipitated individually using the ethanol/sulfate precipitation method. The proposed method was shown to be conceptually simple.

Jha et al. (2013) investigated the recovery of Li and Co from mobile phone S-LIBs. 99.1% Li and 70.0% Co were recovered in 60 minutes after optimal leaching with 2.0 M H_2SO_4 acid and the addition of 5% H_2O_2 (v/v) at a pulp density of 100 g/L and 75°C. The H_2O_2 in the H_2SO_4 solution functions as an efficient reducing agent, increasing the percentage of metals that are leached. Calculations of the activation energy for both metals produced results of 32.4 and 59.81 kJ/mol for Li and Co, respectively.

To produce high-purity Co and Li products from the cathode material of LIBs, it is important to understand the chemical behavior of the metal in various oxidation states. In a 2012 study, Zhu et al. combined chemical precipitation with ultrasound-assisted leaching of $H_2SO_4+H_2O_2$ to recover Co and Li from S-LIBs. Based on the chemical behaviors of metal in various oxidation states, the CAM of S-LIBs was improved to get high-value-added Co and Li products. The active compounds that were removed from the cathode of S-LIBs were dissolved in a solution of 2.0 M H_2SO_4 and 2 vol.% H_2O_2, and by adding di-ammonium oxalate $(NH_4)_2C_2O_4$, they precipitated as $CoC_2O_4 2H_2O$ microparticles. The Li_2CO_3 precipitates were produced by adding Na_2CO_3 to the remaining filtrate after the $CoC_2O_4 2H_2O$ product had been collected by filtration. The experimental analysis demonstrates that in a solution of 2.0M H_2SO_4 and 2.0 vol.% H_2O_2, 96.3% of Co (by mass) and 87.5% of Li can be dissolved, and 94.7% of Co and 71.0% of Li can be recovered, respectively, in the form of $CoC_2O_4 2H_2O$ and Li_2CO_3.

Within the parameters of the experiment, a 2.0 M H_2SO_4 solution with 2.0% H_2O_2, a 33 g/L S:L ratio, a 2 hours leaching period, and a temperature of 60°C may leach 96.3% of Co and 87.5% of Li. Ammonium oxalate is used to precipitate the pearl-colored $CoC_2O_4 2H_2O$ powder at a temperature of 50°C for 1 hour, with an initial pH of 2, and a 1.2:1.0 molar ratio of ammonium oxalate to Co^{2+}. More than 94.7% of the Co element in the leach liquor was recovered during the chemical precipitation procedure. When the equilibrium pH is 10, the temperature is 50°C, the Li-ion concentration is 20 g/L, the reaction time is 1 hour, and the agitation speed is 300 rpm, after collecting the $CoC_2O_4 2H_2O$ product by filtration, the Li_2CO_3 precipitates are obtained by adding Na_2CO_3 in the left filtrate. This method allows for the recovery of 71.0% of Li as Li_2CO_3.

Observable changes in the $LiCoO_2$'s leaching efficiency brought on by ultrasonic irradiation. The leaching efficiencies of Co and Li utilizing ultrasound-assisted leaching are higher than those in H_2SO_4 using H_2O_2 as a reductant when the H_2SO_4 concentration is low (0.5 M). The leaching efficiencies of Co and Li using ultrasound-assisted leaching are less than those in H_2SO_4 or H_2O_2 when the H_2SO_4 concentration is high (over 1.0 M). It is most likely because a solution with little H_2SO_4 concentration can quickly produce H_2O_2 when exposed to ultrasonic waves. Cavitation happens when ultrasonic waves are introduced to an aqueous solution. According to Bankole et al. (2013) and Zhang et al. (2014), cavitation creates a "hot spot" with an exceptionally high temperature (over 5,000 K) and pressure (over 5.07 10^7 Pa). Free-radical reactions, such as those involving OH produced by the breakdown of H_2O, are conceivable in the heated cavities. Radical species like OH and H_2O_2 can be produced in the hot area. It is not favorable for the creation of H_2O_2 by ultrasonic irradiation since there is a correlation between a drop in water content and an increase in the concentration of H_2SO_4. Due to this, the equilibrium concentrations of H_2O_2 did not considerably rise, and ultrasound-assisted leaching of Co and Li had a lower leaching efficiency than H_2SO_4 or H_2O_2 (Zhu et al., 2012). The impact of ultrasonification with H_2SO_4 and $H_2SO_{4+}H_2O_2$ is depicted in Figure 5.7. According to Jo and Myung (2019), ultrasonication in this instance dramatically accelerated cathode disintegration.

According to the Jha et al. (2013) results, acid leaching with 2.0 M H_2SO_4 with the addition of 5 vol.% H_2O_2 reducing agent at an S/L ratio of 100 g/L at 75°C led to the recovery of 99.1% Li and 70.0% Co in 1 hour.

Bertuol et al. (2016) investigated the recovery of Co from S-LIBs using a supercritical CO_2 extraction with an $H_2SO_{4+}H_2O_2$ medium. Water is frequently present in a supercritical fluids extraction system in many environmental applications, either accidentally or as part of the original sample. The production and dissociation of carbonic acid causes water in contact with CO_2 to become acidic:

$$H_2O + CO_2 \leftrightarrow H_2CO_3 \leftrightarrow H^+ + HCO_3^- \qquad (5.38)$$

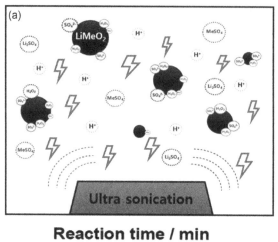

Reaction time / min

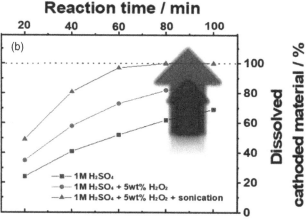

FIGURE 5.7 The effect of ultrasonication mechanism (a) with H_2SO_4 and $H_2SO_4 + H_2O_2$ (b).

This study examined CO recovery utilizing supercritical CO_2 with co-solvents (H_2SO_4 and H_2O_2) and acid leaching at atmospheric pressure, followed by an EW stage. Therefore, it is worthwhile to research the direct recovery of metals from solid wastes like S-LIBs using supercritical CO_2 modified with co-solvents. More effective reaction kinetics for the leaching process may arise from the conversion of metals from their solid forms to soluble ionic forms. The results demonstrated that increasing the amount of H_2O_2 (from 0% to 8% v/v) at atmospheric pressure favors Co leaching. With a reaction time reduction from 60 minutes to 5 minutes and a reduction in the amount of H_2O_2 required from 8% to 4% (v/v), supercritical conditions enable the extraction of more than 95% of the Co. Leach solution electrowinning yields a deposit with a Co content of 99.5% and a current efficiency of 96%.

Meshram et al. (2015, 2015a) investigated the recovery of important metals from CAMs of S-LIBs via leaching and kinetic factors. Results revealed that when the material was leached in 1.0 M H_2SO_4 at 95°C and 50 g/L pulp density for 240 minutes, the optimal leach recovery of 93.4% Li, 66.2% Co, 96.3% Ni, and 50.2% Mn was achieved. In addition, Meshram et al. (2015a) used sodium bisulfite ($NaHSO_3$) as a reducing agent to recover Li, Co, Ni, and Mn from the CAMs of S-LIBs. While adjusting factors including acid concentration, leaching period, temperature, and pulp density, the conditions for the dissolving of precious metals were maximized. It was discovered that 96.7% Li, 91.6% Co, 96.4% Ni, and 87.9% Mn were recovered in 240 minutes using 1.0 M H_2SO_4 and 0.075 M $NaHSO_3$ as reducing agents. 20 g/L of pulp density at 95°C. The kinetic model guided by the empirical logarithmic rate law best suited the kinetic data for the dissolution of metals like Li, Co, and Ni in the temperature range of 35°C–95°C. It was confirmed by XRD phase analyses and SEM-EDS of the untreated sample and the leach residues that the metals were leached by the diffusion of lixiviant on the surface of the substrate particles. By precipitating with oxalic acid, >98% of the Co was recovered as $CoC_2O_4 2H_2O$ from the leaching fluid. The Co-depleted solution precipitated $MnCO_3$, $NiCO_3$, and Li_2CO_3. By using this method, Mn and Ni could also be recovered as their carbonates together with high recovery of Li and Co in the solution and afterward in the form of carbonate and oxalate, respectively.

$CoSO_4$ recovery from S-LIBs by reductive leaching and SX with Cyanex 272 was studied by Kang et al. (2010). By mixing 2.0 M H_2SO_4 and 6 vol.% H_2O_2 at 60°C and 300 rpm with a S/L ratio of 100 g/L, Co was reductively leached from the 16 mesh fraction in 60 minutes, resulting in a Co concentration of 28 g/L and a leaching efficiency of >99%. By bringing the pH level to 6.5, metal ion contaminants like Cu, Fe, and Al precipitated as hydroxides from the solution. After equilibrating with 50% saponified 0.4 M Cyanex 272 at an equilibrium pH of 6, Co was then selectively extracted from the purified aqueous phase. The McCabe-Thiele plot predicted 99.9% Co extraction in a solution. The McCabe-Thiele plot projected that a two-stage counter-current operation with a 1/2 A/O ratio would extract 99.9% of the Co. Co/Li and Co/Ni extraction had separation factors that were very nearly 750 at pH 6. A solution of 96 g/L Co was created after the loaded organic phase was stripped with 2.0 M H_2SO_4, which allowed for the recovery of pure pigment grade $CoSO_4$ through evaporation and recrystallization. Overall, the S-LIBs could be used to recover 92% of the Co.

Through manual disassembly, acid leaching, and precipitation with NH_4OH and SX utilizing Cyanex 272 [bis(2,4,4-trimethylpentyl) phosphinic acid] as the extractant agent, the separation of the primary Al, Co, Pb, and Li metals contained in S-LIBs has been examined. Pb was only discovered in the anode of the battery, making it possible to physically separate it from the other metal species that were discovered to be more prevalent in the cathode. In the acid leaching process, the operational variables leach temperature, S/L ratio, H_2SO_4 concentration, and H_2O_2 concentration, which was utilized as the oxidizing agent, were examined. Leaching solutions with H_2O_2 were used, and the results showed that around 55% of Al, 80% of Co, and 95% of Li were leached from the CAM. Al was partially separated from Co and Li during the precipitation process at pH 5, which was achieved by adding NH_4OH to the leaching fluid to elevate the pH. Following filtration, the aqueous solution was put through liquid-liquid extraction using Cyanex 272, and according to Dorella and Mansur (2007), 85% of the Co was separated.

Swain et al. (2007) also looked into the recovery of a clean and commercially viable form of $CoSO_4$ solution by $H_2SO_4 + H_2O_2$ leach from waste CAMs produced during the production of LIBs. The 93% of Co and 94% of Li were leached at the optimal pulp density conditions of $100 g/L$, $2.0 M$ H_2SO_4, 5 vol.% of H_2O_2, 30 minutes of leaching time, and 75°C. In the SX investigation that followed, 85.42% of the Co was recovered from the leach liquor at pH 5.00 using $1.5 M$ Cyanex 272 as an extractant at an O/A ratio of 1.6. Using $0.5 M$ of Cyanex 272, an O/A ratio of 1, and a feed pH of 5.35, the remaining Co was completely recovered from the raffinate. Following that, $0.1 M$ Na_2CO_3 was used to remove the co-extracted Li from the co-loaded organic. Finally, the Co-loaded organic was stripped using H_2SO_4 to provide a 99.99% pure $CoSO_4$ solution.

To recover CoC_2O_4 from S-LIBs, a hydrometallurgical procedure utilizing alkali leaching, reductive H_2SO_4 leaching, SX, and chemical cobalt oxalate deposition has been developed. The CAM powder material was first leached with a 5-weight percent NaOH solution to remove Al selectively, and the residues were then leached with a 4.0-millimeter percent $H_2SO_4 + 10 wt.\%$ H_2O_2 solution. Under ideal conditions of S/L ratio 1:10, leaching time of 120 minutes, and temperature of 85°C, Co leaching efficiency was 95% and Li leaching efficiency was 96%. By altering the pH level, the impurity ions Fe^{3+}, Cu^{2+}, and Mn^{2+} in the leaching liquid were precipitated. Using saponified P507 (2-ethylhexyl phosphonic acid mono-2-ethylhexyl ester), Co^{2+} was then selectively recovered from the purified aqueous phase, and chemically deposited as oxalate from the strip liquor with a yield of 93% and purity >99.9%. This method works well for recovering priceless metals from S-LIBs and is easy on the environment (Chen et al., 2011).

The following steps make up the proposed flowsheet: Battery disassembly to remove steel crusts, crushing and thermal treatment to separate the spent battery powder from plastics, Al-, and Cu-foils; selective leaching of Al with 5% NaOH, calcination to remove the carbon and organic binder; dissolution of the powder with $4.0 M$ $H_2SO_4 + 10$ vol.% H_2O_2; selective chemical precipitation of iron as sodium jarosite, manganese as MnO_2, and Cu as $Cu(OH)_2$ from the leach liquor under controlled pH conditions; selective SX of Co from Ni and Li with 25 wt.% P507 in kerosene; and stripping of Co with $3.0 M$ H_2SO_4 and precipitation of cobalt oxalate at pH 1.5 using ammonium oxalate (Chen et al., 2011).

To recover all used commercial LIBs, Zou et al. (2013) created a novel method that combines the synthesis of $LiNi_{0.33}Mn_{0.33}Co_{0.33}O_2$ with our cycle process. This process does not segregate the individual components (Mn, Ni, and Co). $LiCoO_2$, $LiMn_2O_4$, $LiNi_{0.33}Mn_{0.33}Co_{0.33}O_2$, and $LiFePO_4$ are a few examples of cathode materials from various battery chemistries that can be recovered with high efficiency and either converted into commercially viable materials or used directly to create new materials. Experiments reveal that over 80% of Li is recycled as Li_2CO_3, and analysis results suggest that almost 100% of Ni, Mn, and Co are recovered. The synthesized $LiNi_{0.33}Mn_{0.33}Co_{0.33}O_2$ exhibits outstanding electrochemical performance, according to the electrochemical performance tests. The ideal environment was $4.0 M$ $H_2SO_4 + 30$ vol.% H_2O_2.

Despite the hydrometallurgical recycling of $LiCoO_2$ CAMs from S-LIBs has been extensively researched, studies on $LiNi_xCo_yMn_zO_2$ type CAMs are rare. As a result, Sattar et al. (2019) concentrated on the leaching of $LiNi_xCo_yMn_zO_2$ CAM for resource recovery of all the critical and rare metals from S-LIBs using $H_2SO_{4+}H_2O_2$. The CAM powder contained 7.6% Li, 20.48% Co, 19.47% Mn, and 19.35% Ni. Leaching at 90°C in $3.0 M$ H_2SO_4 without H_2O_2 allowed for the maximum leaching of 92% Li and Ni, 68% Co, and 34.8% Mn. By adding 4 vol.% H_2O_2, leaching efficiencies of metals were shown to improve within 30 minutes and reach >98% even at a lower temperature, 50°C. Following that, Mn and Ni were selectively precipitated from the leaching liquid using the appropriate precipitants, $KMnO_4$ and $C_4H_8N_2O_2$, respectively. A two-stage SX process was then used to recover a very pure solution of $CoSO_4$ using 0.64 M Cyanex 272 (50% saponified) at an equilibrium pH of 5.0 and a 1:1 O:A ratio. Finally, Li could precipitate at a $Li^+:Na_2CO_3$ ratio of 1.2:1.0, and a process flowsheet for recycling used LIBs has been suggested.

In 1.0 M H_2SO_4 aqueous solution containing 5 vol.% H_2O_2, Jo and Myung (2019) treated the collected cathodes from the disassembled cells and ultrasonically stirred them for 60 minutes at 60°C. By extending the reaction time, adding H_2O_2, and applying extra sonication, CAMs were more easily dissolved. Using the decreasing H_2O_2 additive, complete dissolution happened in under 100 minutes, and the solution turned dark crimson. The cathode materials and H_2SO_4 are most likely to react as follows:

$$2LiCoO_{2\,(s)} + 3H_2SO_{4\,(aq)} + H_2O \rightarrow 2CoSO_{4\,(aq)} + Li_2SO_{4\,(aq)} + 3H_2O + 1/2O_2 \quad (5.39)$$

The addition of the H_2O_2 reducing agent increases the reaction rate because it reduces the Co^{3+} to Co^{2+} form in $LiCoO_2$, facilitating the formation of $CoSO_4$. Otherwise, additional reaction time is needed to finish the dissolution of $LiCoO_2$.

A more advanced hydrometallurgical procedure for recycling LIBs was presented by Atia et al. in 2019. To create a representative electrode active material, EoL batteries were first put through a physical pretreatment process. By using liquid-liquid extractions, an inventive Li-Na separation, and Fe-precipitation, the resulting $H_2SO_4 + H_2O_2$ leachate PLS was purified to yield valuable compounds. High-grade graphite, Co_3O_4 (purity 83%), CoC_2O_4 (purity 96%), NiO (purity 89%), and Li_2CO_3 (purity 99.8%) are a few of these items. The quantitative recovery rate for graphite ranged from 80% to 85% for Co, 90% for Ni, and 72% for Li, depending on the recovery technique. According to the zero-waste principle, secondary streams were also valorized to produce Na_2SO_4 (purity 96%) and $MnCoFe_2O_4$ magnetic nano-sorbents. Recycled Co_3O_4 and NiO were used as conversion-type anode materials for advanced LIBs displaying promising performances in order to close the loop. During leaching Co, Li, and Mn extraction was more than 99% and Al, Mn, and Fe extraction was more than 94%. Mn, Al, and Cu were extracted with D2EHPA + kerosene in two-stage SX, Co was extracted with Cyanex 272 + kerosene in one-stage SX. Ni was precipitated with NaOH as $Na(OH)_2$ and Li was precipitated with Na_2CO_3 as Li_2CO_3 at 95°C. NiO and Co_3O_4 can be produced by heat treatment.

To recover Li and Co from LIBs, Granata et al. (2012) examined two hydrometallurgical procedures (H_2SO_4 and HCl leach). Co and Li were taken out of an input material that came through a massive mechanical pretreatment process at a recycling facility in Northern Italy. Using 2 g/g of H_2SO_4 and more than 50% glucose, a reducing agent that may also be food factory waste, this powder was successfully leached. At pH 5.0, hydroxides of Fe, Al, and Cu were precipitated out and (partially) eliminated. High-purity $CoCO_3$ (47%, w/w of cobalt) was precipitated during the SX process. A product containing 36%–37% (w/w) of Co was produced by carrying out the same procedure without SX. Li was obtained by crystallization with a 98% purity yield (yield). Although SX makes it possible to produce products of a high degree of purity, it is also one of the most expensive hydrometallurgical operations. According to process simulations, the process with SX produces economically sound outputs (gross margin and payback time) that are superior to those of the process without SX for the same input flow rate. If low-purity $CoCO_3$ (made without SX) could be sold for 18 $/kg, processes would exhibit the same economic indices for a feed flow rate of at least 250 t/y.

For the mitigation or prevention of potential threats toward environmental contamination and public health, as well as for the conservation of priceless metals, it is important to recover metal values from S-LIBs. Chen et al. (2018) suggested employing organics as reductants in an H_2SO_4 medium to investigate the probability of different metals leaching from used cathodic materials ($LiCoO_2$) of S-LIBs. According to the leaching results, about 98% Co and 96% Li can be leached under the optimal experimental conditions of reaction temperature: 95°C, reaction time: 120 minutes, reductive agent dosage: 0.4 g/g, slurry density: 25 g/L, the concentration of sulfuric acid: 3 mol/L in H_2SO_4 + glucose leaching system. Under ideal leaching circumstances, an H_2SO_4 + sucrose leaching system can produce results that are comparable (96% Co and 100% Li). Only 54% of Co can be dissolved in the H_2SO_4 + cellulose leaching system under optimal leaching conditions, despite complete leaching of Li (100%). Last but not least, various characterization techniques. such as UV-Vis, FT-IR, SEM, and XRD, were used for the preliminary investigation of reductive leaching reactions employing organic as a reductant in an H_2SO_4 medium. All of the leaching and characterization results show that sucrose and glucose both work well as reductants during leaching, while cellulose needs to undergo additional degradation to produce organic compounds with low molecular weights in order to operate satisfactorily.

Li et al. (2017b) provided a selective leaching method to recover Li, Fe, and P from the CAMs of used lithium iron phosphate ($LiFePO_4$) batteries. It was discovered that, in contrast to the conventional method of using excess inorganic acid to leach all the elements into solution, Li could be selectively leached into solution while Fe and P could remain in leaching residue as $FePO_4$. This was accomplished by using stoichiometric H_2SO_4 at a low concentration as the leachate and H_2O_2 as an oxidant. Leaching rates of 96.85% for Li, 0.027% for Fe, and 1.95% for P were observed at the ideal circumstances (0.3 M H_2SO_4, H_2O_2/Li molar ratio 2.07, H_2SO_4/Li molar ratio 0.57, 60°C, and 120 minutes). Then, by adding Na_3PO_4 as a precipitant, the solution's Li was recovered. Under test circumstances, a total of 95.56% Li was precipitated and recovered as Li_3PO_4. Additionally, by burning the leaching residue at 600°C for 240 minutes to eliminate carbon slag, the $FePO_4$ in the residue was directly recovered. Additionally, the leaching solution was concentrated by evaporation and cleaned up by adding an alkali (NaOH) solution. Following that, Na_3PO_4 was added to the leaching solution to precipitate and recover Li as the product Li_3PO_4. Finally, the mother solution was reused to start the next cycle of cathode plate alkali leaching. This industrially applied study presents a straightforward, efficient, and economical method for recycling used $LiFePO_4$ batteries.

For the purpose of recycling Co in S-LIBs, Wang et al. (2016) used a hydrometallurgical technique. The optimal experimental parameters for Co leaching in the H_2SO_4 + H_2O_2 system were 3.0 M H_2SO_4 concentration, S/L ratio of 1:7, and H_2O_2

dosage of 1.6 mL/g for 2.5 hours at 70°C. The best experimental conditions involved D2EHPA and PC-88A saponification rates of 20% and 30%, respectively, sulfonated kerosene volume of 70%, oil-water (O/A) ratio of 1:1, and extraction time of 10 minutes. Two extractions were used, the first at a pH of 2.70 and the second at a pH of 2.60, both utilizing D2EHPA to get rid of the Cu and Mn ions. To efficiently separate the Co and Ni ions after the extraction operation, PC-88A was used to further extract the leaching solution and maintain the pH at 4.25. The Co ions were then separated by oxalic acid, and $CoC_2O_4.2H_2O$ was produced. Co is as pure as 99.50%.

He et al. (2017) investigated a safe leaching method for recovering Li, Ni, Co, and Mn from cathode scraps and spent LIBS made of $LiNi_{1/3}Co_{1/3}Mn_{1/3}O_2$. Eh–pH diagrams were used to determine ideal leaching conditions. Operating variables impacting the leaching efficiencies for Li, Ni, Co, and Mn from $LiNi_{1/3}Co_{1/3}Mn_{1/3}O_2$, such as the H_2SO_4 concentration, temperature, H_2O_2 concentration, stirring speed, and pulp density, were investigated to determine the most effective conditions for leaching. The leaching efficiencies for Li, Co, Ni, and Mn reached 99.7% under the optimized conditions of 1.0M H_2SO_4, 1 vol.% H_2O_2, 400 rpm stirring speed, 40 g/L pulp density, and 60 minutes leaching time at 40°C leach temperature. The leaching kinetics of $LiNi_{1/3}Co_{1/3}Mn_{1/3}O_2$ were found to be significantly faster than those of $LiCoO_2$. Based on the variation in the weight fraction of the metal in the residue, the "cubic rate law" was modified as follows: $\theta(1-f)^{1/3} = (1-kt/r_0\rho)$, which may best describe the leaching kinetics. The activation energies were found to be 64.98, 65.16, 66.12, and 66.04 kJ/mol for Li, Ni, Co, and Mn, respectively, demonstrating that the leaching process was controlled by the rate of surface chemical reactions. Finally, a straightforward method for recovering precious metals from cathode scraps and using LIBs based on $LiNi_{1/3}Co_{1/3}Mn_{1/3}O_2$ was suggested.

Aaltonen et al. (2017) investigated how valuable metals from S-LIBs could be recycled. They tested various reducing agents (hydrogen peroxide (H_2O_2), glucose ($C_6H_{12}O_6$), and ascorbic acid ($C_6H_8O_6$), as well as different types of acids (2.0M citric ($C_6H_8O_7$), 1.0M oxalic ($C_2H_2O_4$), 2.0M sulfuric (H_2SO_4), 4.0M hydrochloric (HCl), and 1.0M nitric (HNO_3) acid). The average amounts of Co, Li, and Ni, Cu, Mn, Al, and Fe in the crushed and sieved material were 23% (w/w), 3% (w/w), 1%–5% (w/w), respectively. Results showed that mineral acids (4.0M HCl and 2.0M H_2SO_4 with 1% (v/v) H_2O_2) often produced greater yields than organic acids, with a nearly full dissolving of Li, Co, and Ni at 25°C with these mineral acids. When utilizing $C_6H_8O_6$ as a reducing agent (10% g/g scraps) at 80°C, practically all of the Co and Li may be leached out in H_2SO_4 (2.0 M), according to additional leaching studies done with H_2SO_4 media and various reducing agents with a slurry density of 10% (w/v). With and without the addition of a reducing agent (H_2O_2), the mineral acids, particularly 2.0M H_2SO_4 and 4.0M HCl, were shown to be the most successful for Li leaching from the industrially-crushed LIB batch tested, with H_2O_2 having a favorable impact on metals extraction. As the later divalent metals are known to precipitate as oxalates, it was also demonstrated that $C_2H_2O_4$ was the most selective leaching medium between Li and Ni/Co.

The layered lattices that make up the NMC structure are made of transition Ni, Mn, and Co metals, and O_2 and Li ions can be found in the spaces in between these lattices. Thus, Li can easily be dissolved from this deintercalation mechanism. Transition metals within lattice structure in LCO and NMC_{111} can not be leached more than 30 to 40% without using a reductant usage (Joulie et al., 2014; Partinen et al., 2023). Mn has a different leaching behavior than Co, Ni, and Li. Up to 30 minutes Co, Li, and Ni leaching increased and then remained constant. Mn dissolution firstly increases and then remains constant at 37% for 30°C leaching temperature. Increasing temperature reduced Mn dissolution and increased MnO_2 precipitation with increasing leaching time.

Pertinent et al. (2023) investigated how chlorides affected the leaching mechanism for NMC_{111} ($LiNi_{1/3}Mn_{1/3}Co_{1/3}O_2$). They performed studies in chloride (HCl), mixed sulfate–chloride (H_2SO_4–NaCl), and pure sulfate solution (H_2SO_4, reference), in the absence of external reductants. The results indicate that there aren't any significant differences in the leaching performance between studied different lixiviants at 30°C (approximately 75% for Li, 35% for Co, Ni, Mn), and raising the temperature to 80°C mostly enhanced the dissolution of Li (>90% in all lixiviants). However, it was discovered that Mn behaved differently when it came to leaching; in sulfate solutions, Mn precipitated as MnO_2 after 120 minutes (at 80°C), leaving only 6% of the metal in the solution, whereas in chloride-containing solutions, dissolved Mn remained soluble (30%–40% extraction). Chlorides aided in the extraction of Co and Ni, which resulted in extraction rates of 55% for Co and 58% for Ni at 80°C in the mixed H_2SO_4-NaCl system. It is proposed that the empirically confirmed reaction between Cl⁻ ions, and NMC_{111}, which produces Cl_2 and simultaneous NMC dissolution, can be used to explain the rise in dissolutions. The results showed that when operating at higher temperatures, hydrometallurgical battery recycling processes may have trouble extracting Mn from the solution. The presence of chloride ions can help to overcome these challenges by preventing unintentional precipitation of manganese compounds, which results in the production of Cl_2 gas. The inclusion of additional reductants is required to ensure high metal leaching efficiency in such chloride-containing systems, though, as the final extractions were low even in the most efficient lixiviant (55% Co, 58% Ni, and 39% Mn at 80°C).

Cuhadar et al. (2023) concentrated on the utilization of mineral processing techniques for recycling LIBs for mobile phones made of lithium nickel manganese cobalt. Pretreatment measures are used to improve overall healing. To recover metals and plastics, a mechanical procedure combining gravity, magnetic separation, and flotation was created. The results of the enrichment tests in the coarse fraction demonstrated that reverse flotation may be used to produce polymers with natural hydrophobicity. H_2SO_4 leaching was treated to a black mass in the presence of H_2O_2 in order to recover Li, Co, Ni, and Mn. By altering the leaching parameters, the dissolution conditions were improved. With 2.0 M H_2SO_4 concentration, 60°C temperature, 120 minutes leaching time, and 40 g/L H_2O_2 dosage 96% Co, 94% Ni, 95% Mn, and 98% Li were all dissolved. Co and Ni were partially separated from Mn during the precipitation process at pH 7.7 after NaOH was applied to the leachate to raise the pH. The suggested flowsheet for the complimentary recycling of components in S-LIBs is straightforward and extremely efficient.

High metal recovery was reported with the successful application of the sulfuric acid/reductant system scaleable to up to 30 kg S-LIB per experiment (Chan et al., 2021; Chen et al., 2019; Sa et al., 2015; Zou et al., 2013). Heat treatment is a different reductant that can directly convert Co^{3+} to Co^{2+} and improve leaching conditions. The $LiCoO_2$ peaks vanish during pyrolysis at 600°C, indicating the reduction of Co^{3+} into Co^{2+}, which was supported by XRD (Zhang et al., 2020a).

5.2.3   H$_2$SO$_4$ + IRON SCRAP (OBTAINED FROM LIBS SHELL) REDUCTIVE LEACHING

Fe^{3+} potentially transfers its reductive power to the LiCoO$_2$ dissolution.

$$Fe^{3+} + e- \rightarrow Fe^{2+} \; E_{eq} = 0.667 \, V \; vs. \; SHE \qquad (5.40)$$

$$Fe^{2+} + LiCoO_2 + 4H^+ \rightarrow Li^+ + Co_{2+} + Fe_{3+} + 2H_2O$$
$$\Delta G = 141 \, kJ \qquad (5.41)$$

Fe-scrap (FeSO$_4$.7H$_2$O), a sustainable, affordable, and effective reducing agent for Co, Ni, and Li leaching from S-LIBs, was first introduced by Ghassa et al. (2020, 2021). In this approach, ferrous ions (Fe^{2+}) from the dissolving of Fe-scrap were released to create a reducing leaching environment. In the leaching process, the impact of Fe-scrap concentration, solid content, and reaction time were examined. The findings demonstrated that employing Fe-scrap considerably boosts both Co and Ni dissolving while having no effect on Li recovery. At a Fe-scrap content of 25 g/l, a S/L ratio of 1/10, and a leaching period of 150 minutes, the maximum recoveries of Co (99.1%), Ni (94.9%), and Li (90.5%) were attained.

By converting the Co^{3+} to Co^{2+}, the Fe^{2+} addition creates a decreasing leaching environment and improves the Co recovery. The highest Co and Li recoveries may be attained at 500 rpm, pH of 0.5 (0.57 mol/L H$_2$SO$_4$), 6 g/L Fe^{2+} concentration, and 75°C temperature. One of the main benefits of ferrous leaching of LIBs is the ability to run the leaching process at low H$_2$SO$_4$ molarity. The leaching rate constants are relatively high during the first 20 minutes, according to the kinetics modeling. Co and Li both have activation energies of 58.71 and 37.45 kJ/mol, respectively. At the test with the best conditions, Fe-scrap (the shells of 18,650 LIBs) took the place of the ferrous ions. The results show that adding 6 g/L of Fe-scrap to the leaching environment greatly boosts Co recovery. In the presence of Fe-scrap, the maximum Co and Li recoveries were 98.91% and 92.67%, respectively. It should be emphasized that while Li dissolution is not a reduction reaction, the addition of Fe-scrap had no impact on Li recovery. An environmentally beneficial method to boost the metals recovery from LIBs is to use Fe-scrap as a reductant. Therefore, it is advised against removing iron connectors and shells from used LIBs prior to leaching.

In systems for leaching S-LIB, ferrous ions provide a reducing environment. For instance, Co and Ni can dissolve in the presence of Fe^{2+} according to the following oxidation-reduction reactions (Eqs. 5.42 and 5.43):

$$Co^{3+}_{(s)} + Fe^{2+}_{(aq)} \rightarrow Fe^{3+}_{(aq)} + Co^{2+}_{(aq)} \qquad (5.42)$$

$$Ni^{3+}_{(s)} + Fe^{2+}_{(aq)} \rightarrow Fe^{3+}_{(aq)} + Ni^{2+}_{(aq)} \qquad (5.43)$$

1.0 mol Fe^{2+} should be introduced to the leaching reactor to leach 1.0 mol Co or Ni based on the aforementioned processes. This implies that a substantial amount of ferrous sulfate needs to be required for recycling LIBs, which may have an impact on the process's economics. Therefore, a cost-effective way to create ferrous ions must be found. According to Eq. (5.44), Fe-scrap can be dissolved in H$_2$SO$_4$ to form the Fe^{2+} ions.

$$Fe + H_2SO_4 \rightarrow Fe^{2+} + SO_4^{2-} + 2H^+ + 2e^- \qquad (5.44)$$

The Co and Ni are reduced by the Fe^{2+} ions created in the aforementioned process. The major phases of Co and Ni in LIBs, Li-Co oxide, and Li-Ni-oxide, respectively, leach faster as a result of Eqs. (42) and (43).

$$2LiCoO_2 + Fe^{2+} + 3SO_4^{2-} + 8H^+ \rightarrow Fe^{3+} + 2CoSO_4 + Li_2SO_4 + 4H_2O \quad (5.45)$$

$$2LiNiO_2 + Fe^{2+} + 2SO_4^{2-} + 8H^+ \rightarrow Fe^{3+} + Ni_2SO_4 + Li_2SO_4 + 4H_2O \quad (5.46)$$

The following equation shows lithium-nickel-manganese-cobalt oxide (NMC) dissolution in the presence of ferrous and H_2SO_4 (Ghassa, et al., 2020):

$$2FeSO_4 + 6Li(Co_{1/3}Ni_{1/3}Mn_{1/3})O_2 + 10H_2O_2 \rightarrow Fe_2(SO_4)_3 + 2CoSO_4 + 2NiSO_4$$
$$+ 2MnSO_4 + 3Li_2SO_4 + 12H_2O \quad (5.47)$$

According to the above reaction, the Fe^{2+} oxidizes to Fe^{3+} ions and Co^{3+} reduces to Co^{2+} which is a soluble form of Co.

The Fe-scrap obtained from LIBs shells dissolves in diluted H_2SO_4 according to Eq. (5.48):

$$Fe + H_2SO_4 \rightarrow Fe^{2+} + Sn^{3+} \quad (5.48)$$

Fe^{2+} concentration can be determined using potassium dichromate titration after the PLS has been filtered. By adding $SnCl_2$, all ferric ions were converted to ferrous ions in order to detect the Fe^{3+} (Eq. 5.49):

$$Fe^{3+} + Sn^{2+} \rightarrow Fe^{2+} + Sn^{3+} \quad (5.49)$$

$HgCl_2$ can be added to the solution to precipitate the excess Sn as $SnCl_4$ (Eq. 5.50). Dichromate is used to titrate the solution once more. The concentration of ferric ions in the solution is shown by the difference between the first and second titrations. The findings showed that all scrap dissolved as Fe^{2+}, and there was no Fe^{3+} concentration in the reagent that was formed.

$$Sn^{2+} + 2HgCl_2 \rightarrow SnCl_4 + Hg_2Cl_2 \quad (5.50)$$

The medium pH was then controlled to eliminate the added Fe from PLS. H_2O_2 is introduced to the PLS before Fe-precipitation to oxidize the Fe^{2+} to Fe^{3+} precipitates at the lower pH values and stop other metals from precipitating concurrently. The results showed that Fe was removed efficiently at a pH of 4 in the form of Fe_2O_3, without other metals' co-precipitation. Co, Ni, and Mn precipitate without crystallization as sulfide forms at a pH of 8.5. Almost all Co and Ni co-precipitated pHs between 4 and 8.5 while Li was stable. Mn co-precipitates with Co and Ni. This makes the product useless. Therefore, Mn precipitation should be prevented by using Zn precipitation. Following the elimination of Fe, Co, and Ni are recovered from PLS using a Zn-scrap cementation technique. The findings show that Co-Ni may be made by Zn-cementation without the presence of Mn, Li, or Fe impurities.

Eqs. (5.51–5.56) show the half-reactions for the reduction of Co, Ni, Zn, Mn, and Fe to their metallic forms. As these reactions show, the reduction potential for Co^{2+}

and Ni^{2+} are higher than Zn; therefore, they can be substituted by Zn. On the other hand, the Mn reduction potential is lower than all other metals, so it remains unreacted in solution.

$$Co^{2+} + 2e^- \rightarrow Co \qquad\qquad E^0_{Co^{2+}/Co} = 0.28 \text{ V} \qquad\qquad (5.51)$$

$$Ni^{2+} + 2e^- \rightarrow Ni \qquad\qquad E^0_{Ni^{2+}/Ni} = 0.257 \text{ V} \qquad\qquad (5.52)$$

$$Zn^{2+} + 2e^- \rightarrow Zn \qquad\qquad E^0_{Zn^{2+}/Zn} = 0.7618 \text{ V} \qquad\qquad (5.53)$$

$$Mn^{2+} + 2e^- \rightarrow Mn \qquad\qquad E^0_{Mn^{2+}/Mn} = 1.185 \text{ V} \qquad\qquad (5.54)$$

$$Fe^{2+} + 2e^- \rightarrow Fe \qquad\qquad E^0 = -0.4 \text{ V} \qquad\qquad (5.55)$$

$$Fe^{3+} + e^- \rightarrow Fe^{2+} \qquad\qquad E^0 = 0.68 \text{ V} \qquad\qquad (5.56)$$

The total reactions for Co and Ni cementation on Zn are:

$$Co^{2+} + Zn \rightarrow Co + Zn^{2+} \qquad\qquad (5.57)$$

$$Ni^{2+} + Zn \rightarrow Ni + Zn^{2+} \qquad\qquad (5.58)$$

The Gibbs free energies for Co and Ni in the presence of Zn were determined to be −92.97 and −97.41 kJ/mol by Ghassa et al. in 2021. These low Gibbs free energies show that spontaneous deposition of Co and Ni occurs. Lastly, a conceptual process flow diagram (PFD) for recycling discarded LIBs is suggested. This PFD is a comprehensive flowsheet to recycle all valuable metals from LIBs and covers grinding, physical separation, leaching, and metals recovery through PLS (Ghassa et al., 2021).

5.2.4 H_2SO_4 AND SCRAP AL AND CU CURRENT COLLECTOR MATERIAL AS REDUCING AGENT

Scrap Al and Cu can be used as reductants in LIBs leaching with H_2SO_4. Half-cell balancing reactions for Cu, Al, and H are given as follows:

$$Cu^{2+} + 2e^- \rightarrow Cu \qquad E_{eq} = 0.337 \text{ V vs. SHE} \qquad (5.59)$$

$$2H^+ + 2e^- \rightarrow H_2 \qquad E_{eq} = 0.000 \text{ V vs. SHE} \qquad (5.60)$$

$$Al^{3+} + 3e^- \rightarrow Al \qquad E_{eq} = -1.681 \text{ V vs. SHE} \qquad (5.61)$$

The reaction of Cu and Al scraps with LCO can be given as follows:

$$2LiCoO_2 + Cu + 8H^+ \rightarrow 2Li^+ + Cu^{2+} + 2Co^{2+} + 4H_2O \quad \Delta G = -350 \text{ kJ} \quad (5.62)$$

$$3LiCoO_2 + Al + 12H+ \rightarrow 3Li^+ + Al^{3+} + 3Co^{2+} + 6H_2O \quad \Delta G = -1100 \text{ kJ} \quad (5.63)$$

$$2Al + 3H_2SO_4 \rightarrow Al_2(SO_4)_3 + 3H_2 \qquad\qquad \Delta G = -1054 \text{ kJ} (5.64)$$

$$2LiCoO_2 + H_{2(g)} + 3H_2SO_4 \rightarrow Li_2SO_{4(aq)} + 2CoSO_{4(aq)} + 4H_2O \quad \Delta G = -531 \text{ kJ} (5.65)$$

The use of scrap Cu and Al current collector materials as reductants in LIB waste leaching with H_2SO_4 was studied by Chernyaev et al. in 2021. At 60°C, 2.0 M H_2SO_4 concentration, and 200 g/L S/L ratio, under industrially relevant process conditions, they leached pretreated $LiCoO_2$-rich battery waste concentrate. Both factors have separate, linear effects on the extraction of cobalt and the use of acid. According to the model, 100 g of sieved industrial battery waste concentrate must contain either 11 g of Cu (0.75 Cu/Co, mol/mol), 4.8 g of Al (0.7 Al/Co, mol/mol), or a combination of the two to fully extract Co. Al was shown to affect Co leaching, but due to associated side reactions including excess H_2 generation, it was less effective (47%) than Cu (66%) in terms of current efficiency. In contrast to Cu, which only operates through Fe^{3+} reduction, Al has a variety of chemical pathways that it can use to reduce $LiCoO_2$ in parallel or in series, including H_2 generation, Cu^{2+} cementation, and/or Fe^{3+} reduction. These findings suggest that the CO_2 footprint of the battery leaching step might be reduced by at least 500 kg of CO_2 per ton of recycled Co by utilizing Cu scrap instead of the more common H_2O_2. Al, on the other hand, despite being promising, is less appealing because of the difficulties associated with removing it during the subsequent solution purification.

5.2.5 H_2SO_4 + GLUCOSE

Atia et al. (2019) leached LIBs with $H_2SO_4 + C_6H_{12}O_6$. The reaction between LCO and H_2SO_4 with a glucose-reducing agent is given as follows:

$$24LiCoO_2 + 36H_2SO_4 + C_6H_{12}O_6 = 24CoSO_4 + 12Li_2SO_4 + 6CO_2 + 42H_2O \quad (5.66)$$

H_2SO_4 was used as 2.0–2.5 g/g powder and glucose as 50%–100% stoichiometric excess at a 10% S/L ratio and temperatures of 30°C and 90°C. Granata et al. (2012) also used glucose as a reductant for H_2SO_4 and HCl leaching.

5.3 H_2SO_3 LEACH

Zhang et al. (1998) leached $LiCoO_2$ using 1–6% sulfurous at 60°C and 1/100 S/L ratio for 30 minutes. At 4% H_2SO_3 dosage, 65% Co and 63% Li dissolutions were achieved. Both Co and Li dissolutions were low.

5.4   HCl LEACH

Although up to now, such claims have frequently been made without assessing the level of Cl_2, chloride-based solutions can operate as reductants by producing chlorine gas in the process (Takacova et al., 2016). Additionally, S-LIB often contains metallic Al, Fe, and Cu particles that can serve as reductants toward the cathode material, and the presence of chloride ions enhances the reductive properties of these metals (particularly Cu). Chloride ions prevent the formation of Cl_2 gas while stabilizing Cu^+ species, or cuprous chloride complexes, in the solution. This enables the reductive leaching of battery cathode materials. It is hypothesized that the presence of chlorides can facilitate a reduction process in which Cl^- ions are oxidized to produce chlorine gas, releasing electrons to allow active material leaching (E = 1.29 V vs. SHE at 80°C), (Eq. 5.67) (Pertinen et al., 2023).

$$Cl_{2(g)} + 2e^-_{(aq)} \leftrightarrow 2Cl^-_{(aq)} \tag{5.67}$$

The installation of specialized leach equipment is required for the HCl process to treat chlorine (Cl_2) produced by HCl, which will result in much higher recycling costs. Due to its toxic properties, chlorine gas is also frequently linked to workplace and environmental risks. However, with the right mitigation measures, the Cl_2 formed can be safely collected and used as an oxidant in other process steps, potentially reducing operating costs and the overall process' use of oxidative chemicals.

Equations given below describe the possible reactions of Li and Co in an HCl solution in the presence/absence of H_2O_2, with values of ΔG^0_{80}:

$$2LiCoO_2 + 8HCl \leftrightarrow 2LiCl + 2CoCl_2 + 4H_2O + Cl_{2(g)} \quad \Delta G^0_{80}: -358.60 kJ \tag{5.68}$$

$$2\ LiCoO_2 + 6H^+ + H_2O_2 \leftrightarrow 2Co^{2+} + O_2 + 2Li^+ + 4H_2O$$

$$Li_2O + 2HCl \leftrightarrow 2LiCl + H_2O \quad\quad\quad\quad\quad \Delta G^0_{80}: -195.87\ kJ \tag{5.69}$$

$$CoO + 2HCl \leftrightarrow CoCl_2 + H_2O \quad\quad\quad\quad \Delta G^0_{80}: -42.95\ kJ \tag{5.70}$$

$$Co_2O_3 + 6HCl \leftrightarrow 2CoCl_2 + 3H_2O + Cl_2 \quad \Delta G^0_{80}: -34.02\ kJ \tag{5.71}$$

$$Li_2CO_3 + 2HCl \leftrightarrow 2LiCl + H_2O + CO_2 \quad \Delta G^0_{80}: -28.12\ kJ \tag{5.72}$$

$$Co_3O_4 + 8HCl \leftrightarrow 3CoCl_2 + 4H_2O + Cl_2 \quad \Delta G^0_{80}: +24.05\ kJ \tag{5.73}$$

Except for the reaction (5.73), all estimated values of G_0 for HCl are negative over the entire temperature range (20°C–80°C). As a result, there is a strong possibility that the reactions will go in the direction of product creation. The converse direction is probably true for the reaction (5.73), which has a positive value for G_0. The likelihood of Li and Co extraction with both leaching agents increases with temperature from room temperature to 80°C if just values of G_0 are considered, omitting kinetics, system complexity, and side effects variables. The reaction between NMC and HCl is given as follows (Partinent et al., 2023):

$$3LiN_{1/3}M_{1/3}Co_{1/3}O_{2(s)} + 12HCl_{(aq)} \leftrightarrow 3LiCl_{(aq)} + NiCl_{2(aq)} + MnCl_{2(aq)}$$
$$+ CoCl_{2(aq)} + 6H_2O_{(l)} + 1.5\ Cl_{2(g)} \tag{5.74}$$

The resulting Eh-pH diagrams for the systems $Co-S-H_2O$ and $Co-Cl-H_2O$ at 80°C are displayed in Figure 5.8a,b, where it is possible to observe the ion stability in an aqueous solution of H_2SO_4 and HCl. Co is present in soluble forms as Co^{2+} in H_2SO_4

and is stable to a pH of 5.3 or lower. Co exists in HCl as chlorine compounds that are stable to a pH of 5.7. Co from both leaching agents hydrolytically precipitates when the pH value is raised. Practically speaking, this means that free acid may be partially consumed after Co and Li have been leached into the solution. Throughout the entire acidic pH range, Li is still soluble as Li^+ (Ferreira et al., 2009; Meshram et al., 2014). By selecting appropriate leaching conditions, Co-precipitation should be prevented. Thermodynamic analysis demonstrates that the extraction of Co and Li in both HCl and H_2SO_4 leaching agents is theoretically feasible.

The earlier LIB leaching experiments with HCl in the presence and absence of H_2O_2 are summarized in Table 5.2. As an alternative to S-LIB, certain investigations have also been done utilizing pure cathode material powders, most notably $LiCoO_2$ and various forms of NMC ($LiNi_xMn_yCo_zO_2$), where the stoichiometric ratios of Ni, Mn, and Co are denoted by a corresponding number (e.g., NMC_{111} or NMC_{811}). Chlorides have a considerable effect on the way NMC_{111} cathode material leaches. The dissolving results demonstrate that, at a low temperature of 30°C, the choice of lixiviant has no discernible influence on the leaching efficiencies of Li, Co, Ni, or Mn. Although Mn precipitates out of H_2SO_4 solutions at higher temperatures, especially at 80°C, both HCl and H_2SO_4-NaCl solutions produced significant amounts of Cl_2 gas, supporting the theory that chlorine gas evolves as a result of interactions between battery cathode materials and Cl^- ions. Due to the higher cathode material reactivity of chloride salt solutions compared to HCl, enhanced Co and Ni dissolutions arise from this chlorine gas development, particularly in H_2SO_4-NaCl solutions.

Zhang et al. (1998) leached $LiCoO_2$ active material was peeled off from the Al substrate using 1.0–6.0 M HCl acid at 80°C and 1/100 S/L ratio for 30 minutes. At 1.0 M HCl dosage, 65% Co and 92% Li dissolutions were achieved. At 4.0 M HCl concentration, 80°C of leach temperature, 60 minutes leach period and 1/10 S/L ratio, 99% Li and Co were extracted, and 17 g Co and 1.7 g/L Li were dissolved. Under ideal 5.0 M HCl solution conditions, including a temperature of 95°C, a reaction duration of 70 minutes, and a S/L ratio of 10 g/L, Guzolu et al. (2017) recovered 98% of Li and nearly 99% of Co. NaOH used in the precipitation procedure, which was carried out at pH 12.5 for 1 hour at 55°C. Co and Li were recovered at rates that were higher than 99%, it was discovered. To recycle the S-LIBs, Shuva and Kurny (2013b) used HCl and H_2O_2 as leaching and reducing agents. 3.0 M HCl, S/L ratio = 1:20 mg/mL, 80°C, 60 minutes, and 3.5 vol.% H_2O_2 were found to be the ideal conditions for leaching Li and Co. According to reports, 89% of Co and Li were leached under ideal circumstances. With rising temperature, HCl concentration, duration, and (S/L) ratio, $LiCoO_2$ dissolving efficiency increased. Using SX and PC-88A, Co and Li were precipitated as carbonate after $LiCoO_2$ was leached with HCl (Zhang et al., 1998).

Guzolu et al. (2017) used chemical techniques to explore Li and Co recovery from S-LIBs. This experiment employed a procedure that included ultrasonic cleaning, leaching, and precipitation. While HCl was used as a leaching agent, ultrasonic cleaning reduced energy expenditure and emissions. Under ideal conditions of 5.0 M HCl solution, 95°C, a reaction period of 70 minutes, and an S/L ratio of 10 g/L, 98% of Li and nearly 99% of Co were produced. In this procedure, the leaching solution was first precipitated to obtain Ni, Cu, Fe, Al, Co, and Mn using NaOH at a pH of

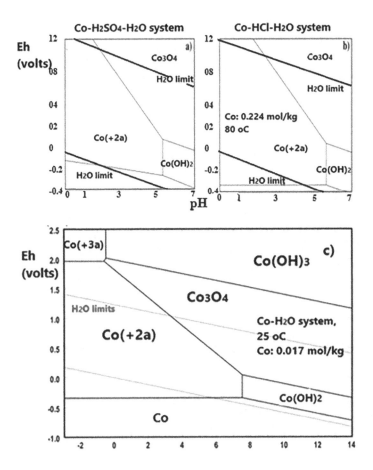

FIGURE 5.8 Eh–pH diagrams at 80°C (a) Co–S–H$_2$O (b) Co–Cl–H$_2$O system and at 25°C Co–H$_2$O system.

12.5 and a reaction duration of 60 minutes at a temperature of 55°C. All metal recoveries were greater than 99%. Li loss in the trials on precipitation was just 18.34%. By adding saturated Na$_2$CO$_3$ solution to the leftover filtrate from the initial precipitation step, white Li$_2$CO$_3$ was precipitated in the following step. Li powder that was retrieved had a 95% purity level.

Li et al. (2009) used ultrasonic washing for the first time as an alternative process to improve the recovery efficiency of Co and decrease energy use and pollution. They recovered Co from spent LIBs using a combination of crushing, ultrasonic washing, HCl acid leaching, and precipitation. To remove the electrode materials from their support substrate, S-LIBs were crushed with a 12 mm screen, and the undersized products were then placed into an ultrasonic washing container. A 2 mm screen was used to filter the washed materials in order to obtain the underflow products or recovered electrodes. And 92% of the Co was transferred to the recovered electrodes, where it made up 28% of the mass and the remaining 2% was made up of impurities such as Al, Fe, and Cu. Cu, Al, and Fe were among the precious materials left in the 2–12 mm products, and they were presented as thin sheets that were simple to separate.

TABLE 5.2

Previous LIB Leaching Studies with HCl in the Presence and Absence of H_2O_2

LIB Materials	Leach Reagents	Reducing Agent	T (°C)	t (minutes)	S/L ratio (g/L)	Recovery (%)	Separating Method	Additives	Reference
$LiCoO_2$ batteries	4.0M HCl		80	60	100	99% Co and Li	SX	PC-88A	Zhang et al. (1998)
	1.0–6.0M H_2SO_3		60	60	100	65% Li			
$LiCoO_2$ (NMP + ultrasonic treatment (30 minutes, 40 Hz, 100 W)	5.0M HCl		95	70	10	~99%; 98% Li	NaOH precipitation pH: 12.5, 1 hour T: 55°C	Li precipitation Purity: 95%	Guzolu et al. (2017)
$LiCoO_2$ (cylindrical 18,650 size) (NMP for PVDF dissolution, 100°C, 1 hour) (Li: 1.3%; Co: 7.23%)	4.0M HCl		80	60		$Co(OH)_2\downarrow$ Co_3O_4 450°C, 3 hours	Change pH values	Co_3O_4, Li_2CO_3, 700°C, 20 hours for LCO	Contestabile et al. (2001)
$LiCoO_2$ (Ultrasonic washing + screening + leaching + precipitation)	4.0M HCl		80	120	100	99% Co; 97% Li	NaOH precipitation		Li et al (2009)
S-LIBs (Anode and cathode from EVs and computers) (Heat treatment (400°C, 1 hour) + leaching + purification + precipitation)	1.5M HCl		60	60	100	100% Li/ Cu/Al	4.5 < pH < 6.0		Yang et al. (2019)

(Continued)

TABLE 5.2 (Continued)
Previous LIB Leaching Studies with HCl in the Presence and Absence of H_2O_2

LIB Materials	Leach Reagents	Reducing Agent	T (°C)	t (minutes)	S/L ratio (g/L)	Recovery (%)	Separating Method	Additives	Reference
S-LIB (wet shredding + sieving + roasting 500°C, 1 hour) (11.73% Co; 0.58% Li; 8.48% Mn; 0.26% Ni)	1.75 M HCl		50	90	20	>99% Co and Mn	Mn precipitated with NaOCl with 1.5 times stoich., pH: 1., 30 minutes		Barik et al. (2017)
$LiNi_{0.8}Co_{0.15}Al0_{.05}O_2$ (NCA cathode)	4.0 M HCl		90	1,080	50	100% Li, Ni, Co, Al	pH: 3 NaClO ClO/Co:3 $Co_2O_3.3H_2O$	NaOH pH:11 $Ni(OH)_2$	Joulie et al. (2014)
$LiCoO_2$, $LiMn_2O_4$, LiCoNiMnO	4.0 M HCl		80	60	20	Purity: 96.97% Li, 98.23% Mn, 97.43% Co		NaOH, Na_2CO_3	Wang et al. (2009)
$LiCoO_2$ (22.43% Co; 3.65% Li; 1.54% Ni; 1.33% Cu; 0.72% Al; 1.49% Mn; 1.27% Fe)	2.0 M HCl/ 2.0 M H_2SO_4	3.5% (v)	60	90	50	100% L, and Co	95% Co precipitated at pH: 11–12		Takacova et al. (2016)
$LiCoO_2$	3.0 M HCl	H_2O_2	80	60	5	>89% Co; >89% Li		93% Li in leach liquor	Shuva and Kurny (2013b)
$LiCoO_2$ cathode material	4.0 M HCl		25	1,440	50	91% Li; 85% Co			Aaltonen et al. (2017)

(Continued)

TABLE 5.2 (*Continued*)
Previous LIB Leaching Studies with HCl in the Presence and Absence of H_2O_2

LIB Materials	Leach Reagents	Reducing Agent	T (°C)	t (minutes)	S/L ratio (g/L)	Recovery (%)	Separating Method	Additives	Reference
	4.0 M HCl	1% (v) H_2O_2	25	1,440	50		98% Li; 92% Co		Aaltonen et al. (2017)
Scraped $LiCoO_2$	3.0 M HCl	5% H_2O_2	60			90% Co, 30% Li	Precipitation	NaOH, Na_2CO_3	Qadir and Gulshan (2018)
NMC_{811}	4.0 M HCl		25		120				Xuan et al. (2019)

(Li: 7.36%; Co: 6.28%; Ni: 46.01%; Mn: 5.4%; O: 34.95%)

The recovered electrodes were agitated while being leached with 4.0 M HCl for 2.0 hours at 80°C. In recovered electrodes, 99% of the Co and 97% of the Li could be dissolved. By chemical precipitation, the contaminants may be eliminated at pH 4.5–6.0 with negligible loss of Co. Scaling up this method is possible for recycling S-LIBs.

Yang et al.'s (2019) research on recycling Li and graphite from S-LIBs shows that it may significantly reduce the lack of Li resources, fully utilize used anode graphite, and safeguard the environment. In this investigation, the used graphite was initially gathered using a two-stage calcination process. Second, the collected graphite is subjected to simple acid leaching to convert about 100% of the Li, Cu, and Al in it into leach liquor under the ideal conditions of 1.5 M HCl, 60 minutes, and an S/L ratio of 100 g/L. Third, 99.9% Al and 99.9% Cu was removed from leach liquor by first increasing pH to 7 and then to 9, and the Li was recovered by adding Na_2CO_3 to generate high-purity (>99%) Li_2CO_3. The regenerated graphite has outstanding cycle performance at a high rate of 372 mA/g and is found to have a high initial specific capacity at rates of 37.2 mA/g (591 mAh/g), 74.4 mA/g (510 mAh/g), and 186 mA/g (335 mAh/g). It also has a high retention ratio of 97.9% after 100 cycles. This method allows for the comprehensive usage of anode material from S-LIBs and allows for the recovery of Cu and Li as well as the regeneration of graphite.

H_2SO_4, HNO_3, and HCl acids have all been studied for NCA cathode leaching. It has been noted that the nature of the species with the best leaching efficiency in HCl solution has a major impact on the leaching phase. The presence of chloride ions to facilitate the dissolution has been attributed to these variations in acidic conditions. This observation has been attributed to the chloride ions' capacity to undermine the development of a surface layer. It has been discovered that 4.0 M HCl, 90°C, 18 hours, and a 5% (w/v) S/L ratio are the ideal leaching conditions. All of the cathode material's precious metals are leached away under these experimental circumstances. The valuable metals of the cathode material can be recovered from the LIBs using the material recovery procedure. By oxidative precipitation, it has been recovered for the Co. It has been established that pH = 3 and a molar ratio of 3 for $Co_2O_3.3H_2O$ precipitation are the ideal working conditions. When it comes to the Ni recovery, the second stage involves adding NaOH until the pH is at 11. According to the experimental findings, Co and Ni recovery efficiency are 100% and 99.99%, respectively. $Co(OH)_2$ and $Ni(OH)_2$ have purity values of 90.25% and 96.36% by weight, respectively. The presence of Ni in Co_2O_3, $3H_2O$ precipitate, Co in $Ni(OH)_2$ precipitate, and Al is what causes these low purities. The low purity of the recovered materials, which need a post-treatment to be refined, is the fundamental disadvantage of recovery and element separation by hydroxide precipitation. The leach solution is primarily made up of Li at the end of the procedure, which can be recovered as a carbonate or phosphate salt (Joulie et al., 2014).

The S-LIBs' CAMs were processed by experiments to separate and retrieve metal values including Co, Mn, Ni, and Li (Wang et al., 2009). With a 4.0 M HCl solution, an 80°C leaching temperature, a 60 minutes leaching period, and a 0.02 mg/L S/L ratio, it was possible to achieve a leaching efficiency of more than 99% of Co, Mn, Ni, and Li. For the mixture's recovery process, the Mn in the leaching liquid was first selectively reacted with a $KMnO_4$ reagent and nearly finished. The Mn was then recovered as MnO_2 and manganese hydroxide. Second, using dimethylglyoxime, the

Ni in the leaching fluid was selectively extracted and almost completely removed. Thirdly, the selective precipitation of the $Co(OH)_2$ was made possible by adding the aqueous solution to the 1.0 M NaOH solution to achieve pH = 11. With the addition of a saturated Na_2CO_3 solution, the remaining Li in the aqueous solution was easily recovered as Li_2CO_3 precipitated. Li, Mn, Co, and Ni recovery powders were each 96.97%, 98.23%, 96.94%, and 97.43% by weight pure, respectively.

Using H_2SO_4 and HCl, Takacova et al. (2016) explored the recovery of Co and Li from CAM of used mobile phone and laptop LIBs. The sample's Co, Li, Ni, and Cu concentrations were 22.43%, 3.65%, 1.54%, and 1.33%, respectively, for the −0.71 mm size fraction. It was discovered that utilizing HCl as a leaching agent is preferable to employing H_2SO_4. The ideal leaching conditions for Co and Li recovery were 2.0 M HCl, 60°C–80°C, and a 90-minute leaching period. It was discovered that Co extraction in H_2SO_4 takes place twice. The process is governed by the pace of a chemical reaction in the initial period, which lasts for 15–20 minutes after leaching starts ($E_{a(Co)}$ = 43–48 kJ/mol). The value of $E_{a(Co)}$ = 3–3.5 kJ/mol indicates that the process switches to diffusion control during the second time period. In the case of HCl, the rate of the chemical reaction, $E_{a(Co)}$ = 40–44 kJ/mol, governs Co extraction in the first time period. The process changes to a mixed mechanism in the second time frame, with $E_{a(Co)}$ = 20–26 kJ/mol. In both time periods, Li extraction is governed by diffusion or takes place in mixed mode. $E_{a(Li)}$ in both leaching agents ranges from 2 to 20 kJ/mol. The analysis of the fine structure supported the theory that cobalt extraction is reliant on lithium extraction from the active mass, which is predominantly made of $LiCoO_2$, and that cobalt extraction modifies the internal structure of the active mass.

Aaltonen et al. (2017) investigated different types of acids (2.0 M citric ($C_6H_8O_7$), 1.0 M oxalic ($C_2H_2O_4$), 2.0 M sulfuric (H_2SO_4), 4.0 M hydrochloric (HCl), and 1.0 M nitric (HNO_3) acid) and reducing agents (hydrogen peroxide (H_2O_2), glucose ($C_6H_{12}O_6$), and ascorbic acid ($C_6H_8O_6$)) were selected for investigating the recovery of valuable metals from S-LIBs. The material that had been crushed and sieved typically comprised 23% Co, 3% Li, and 1%–5% Ni, Cu, Mn, Al, and Fe. At 25°C with a slurry density of 5% (w/v), the results showed that mineral acids (4.0 M HCl and 2.0 M H_2SO_4 with 1% (v/v) H_2O_2) produced typically better yields than organic acids. Li, Co, and Ni nearly completely dissolved. Co and Li yields were both 92% and 98%, respectively.

Qadir and Gulshan (2018) scraped active electrode components from the laptop LIB off of the polyethylene separators and the Cu current collector. The electrode material was discovered to be linked to the severely damaged Al current collectors. NaOH was used to cure it, and Al_2O_3 was later recovered. Graphite was entirely removed during the leaching of $LiCoO_2$ using 3.0 M HCl and 5% H_2O_2 at 60°C from the scraped active electrode materials ($LiCoO_2$ and graphite). NaOH precipitated $Co(OH)_2$, which was then changed into Co_3O_4 by the addition of $Co(OH)_2$. The residual solution was processed with saturated Na_2CO_3 to produce high-purity crystals of Li_2CO_3 as a precipitate. Co and Li recovered 99% and 30%, respectively. Li_2CO_3 with Co_3O_4 were mixed in stoichiometric proportions and calcined around 950°C with an air supply to achieve $LiCoO_2$ successfully.

In their investigation of leaching NMC_{811} in a 4.0 M HCl solution, Xuan et al. (2019) reported the generation of Cl_2 gas in line with the dissolving of cathode material. The reactions that occurred when $LiNi_{0.8}Mn_{0.1}Co_{0.1}O_2$ (NMC_{811}), one of the most promising positive electrodes for the next LIBs, were leached by HCl were examined in this work. This work demonstrates that Li leaching behavior differs significantly from that of the NMC_{811}-contained Ni, Co, and Mn due to Li dissolving occurring more quickly. Leaching kinetic data analysis showed that NMC_{811} dissolution happens in two stages. The first step involves the transformation of NMC into a new phase with reduced Li ($2.8 < n < 3.6$): First step:

$$LiMO_{2,(s)} + \frac{4n-4}{2n-1}HCl_{(l)} \rightleftharpoons \frac{2n-2}{2n-1}LiCl_{(l)} + \frac{n-1}{2n-1}MCl_{2,(l)}$$

$$+ \frac{n}{2n-1}Li_{\frac{1}{n}}MO_{2,(s)} + \frac{2n-2}{2n-1}H_2O_{(l)} \tag{5.75}$$

where M = Co, Ni, Mn. In the second step, the new phase is dissolved (limiting step): Second step:

$$Li_{\frac{1}{n}}MO_{2,(s)} + 4HCl_{(l)} \rightleftharpoons \frac{1}{n}LiCl_{(l)} + MCl_{2,(i)} + 2H_2O_{(l)} + \left(1 - \frac{1}{2n}\right)Cl_{2,(g)} \tag{5.76}$$

Finally, the overall reaction of NMC_{811} leaching by hydrochloric acid can be written as:

$$2LiMO_{2(s)} + 8HCl_{(l)} \rightleftharpoons 2LiCl_{(l)} + 2MCl_{2(l)} + 4H_2O_{(l)} + Cl_{2(g)} \tag{5.77}$$

Later, they expanded this study to include additional NMC-type cathode chemistries as well, and they came to the conclusion that the cathode composition significantly influences the leaching kinetics, with leaching rates falling in the order of $NMC_{811} > NMC_{622} \gg NMC_{532} > NMC_{111}$ (Xuan et al., 2021).

Pyrometallurgical and hydrometallurgical processes were integrated in a patentable manner. The S-LIBs were first calcined, sieved, and then treated with HCl for a dissolution etching process to produce metal and metal oxide-containing ash. After filtering, metals Cu and Co were sorted out using a membrane electrolysis technique, and the solution was then combined with carbonate ions to create Li_2CO_3.

5.5 HNO₃ LEACH

Some researchers studied HNO_3 leaching of LIBs. Table 5.3 presents the optimum results of this investigation. Castillo et al. (2002) leached LMN spent batteries using 2.0 M HNO_3 at 80°C for 120 minutes and achieved full Li and 95% Mn extraction. PLS was subjected to the NaOH precipitation until a pH of 5.2 to precipitate Fe. Then pH was increased to 10 to precipitate Mn. The final filtrate contained Li solution.

Aaltonen et al. (2017) investigated LCO leaching with 4.0 M HNO_3 with and without 1% vol. H_2O_2 at 25°C. Without H_2O_2 85% Li and 75% Co and with H_2O_2 90% Li and 83% Co extractions were achieved in 24 hours.

The recovery of Co and Li from S-LIBs and the synthesis of $LiCoO_2$ from leach liquid as CAMs have been accomplished through the application of a recycling process involving mechanical, thermal, hydrometallurgical, and sol-gel processes. In the Lee and Rhee (2002, 2003) method, LIB samples were thermally treated twice at temperatures between 100°C and 150°C, then disassembled with a high-speed shredder and categorized into sizes between 1 and 10 nm. Finally, the $LiCoO_2$ electrode material was separated from the current collectors by vibrating screening and burning off carbon and binder. The resulting $LiCoO_2$ was then utilized to create $LiCoO_2$ electrode material by leaching it with an HNO_3 solution. According to Lee and Rhee (2002), Li and Co were leached with over 95% recovery rate with the addition of 1.7% v H_2O_2, and leaching efficiency increased with increasing temperature and leaching agent HNO_3 concentration. It was discovered that 1.0M HNO_3, 10–20 g/L initial S/L ratio, 75°C temperature, and 1.7% vol. H_2O_2 addition were the ideal conditions for recovering Co and Li. The decrease of Co^{3+} to Co^{2+} was used to explain the dissolving process. Additionally, they determined the activation energies for Co and Li, which were 12.5 and 11.4 kcal/mol, respectively.

The molar ratio of Li to Co in the leach liquor is adjusted to 1.1 by adding a fresh $LiNO_3$ solution after leaching utilized $LiCoO_2$ with HNO_3. Then, to create a gelatinous precursor, 1.0M citric acid solution at 100% stoichiometry is added. It is possible to successfully produce pure crystalline $LiCoO_2$ by calcining the precursor for 24 hours at 950°C. The resulting crystalline powders had particle sizes of 20 μ and a specific surface area of 30 cm²/g, respectively. The charge-discharge capacity and cycle performance of the $LiCoO_2$ powder are found to be favorable for a CAM.

5.6 H_3PO_4 LEACH

Phosphoric acid is polyprotic, and therefore it presents several dissociation reactions in aqueous solution, each species acting as an acid or as a base. These three equilibriums of dissociation enable us to see that the first proton is readily detached even at an acidic pH, indicating that the H_3PO_4 is a moderately strong acid. The second dissociation pK (7.2) has buffer action and the third H^+ is dissociated in alkaline medium probable dissolution reactions and calculated $\Delta G_{2°C}$ values of $LiCoO_2$ with phosphoric acid in reductive media were proposed by Pinna et al. (2017). Table 5.4 summarizes previous H_3PO_4 leach results.

$$3H_3PO_4 + 2LiCoO_2 + H_2O_2 \rightarrow 2Li^+ + 2Co^{+2} + 2PO_4^{-3} + 4H_2O + O_2 \quad (5.78)$$

$$K_1 = 5.7 \times 10^{-3} \qquad \Delta G_{25} = -656.1\,kJ$$

$$3H_3PO_4 + 2LiCoO_2 + H_2O_2 \rightarrow 2Li^+ + 2Co^{+2} + 3HPO_4^{-2} + 4H_2O + O_2 \quad (5.79)$$

$$K_2 = 6.2 \times 10^{-8} \qquad \Delta G_{25} = -866.8\,kJ$$

$$3H_3PO_4 + LiCoO_2 + 1/2H_2O_2 \rightarrow Li^+ + Co^{+2} + 3H_2PO_4^- + 2H_2O + 1/2O_2 \quad (5.80)$$

$$K_3 = 2.2\ 10^{-13} \qquad \Delta G_{25} = -1398.7\,kJ$$

TABLE 5.3
Previous HNO₃ Leach Results of LIBs

LMN Spent Battery	2.0MHNO₃		80	120	100% Li;95% Mn	Change pH Values	NaOH	
(Cylinderical:1.5% Li;0.1% Co;5.4% Ni;9.6% Mn, 34% Fe)					pH:5.2 Fe↓, pH:10 Mn		Calcined:1,000°C, 2 hour	Castillo et al. (2002)
LiCoO₂ cathode material	4.0 MHNO₃		25	1,440	50		85% Li; 75% Co	Aaltonen et al. (2017)
	4.0 MHNO₃	1% (v) H₂O₂	25	1,440	50		90% Li;83% Co	Aaltonen et al. (2017)
LiCoO₂ (18,650 cylindrical size) (reductive leaching)	1.0MHNO₃	1.7% (v) H₂O₂	75	30	20	~99% Co	Amorphous citrate precursor	Lee and Rhee (2003)
Thermal treatment 1st:150°C, 1hour, 2nd:700–900°C, 1hour							E :11.4 and12.5 kcal/ mol Li & Co	
Li, Mn, Ni (cylindrical spent battery)	1.0MHNO₃	1.7% (v) H₂O₂	75	60	20	95% Co & Li Precipitation(citric acid)	Pure LiCoO₂ was obtained	Lee and Rhee (2002)
LIBs	1.0 MHNO₃	1% H₂O₂	80	60	50	100% Co & Li		Li et al. (2011)

Nayaka et al. (2016) used mild phosphoric acid and a sustainable recycling method to recover the important metals. The CAM was reportedly leached in $0.7 M$ H_3PO_4 and 4 vol.% H_2O_2 at $40°C$ for 60 minutes, recovering 99.7% Co and 99.9% Li. Calculated values for Co and Li's leaching kinetics and activation energy were 7.3 and 10.2 kJ/mol, respectively.

According to Pinna et al. (2017), a technique that used H_3PO_4 acid as the leaching agent and H_2O_2 as the reducing agent effectively dissolved $LiCoO_2$ by nearly 99%. Temperature of $90°C$, H_3PO_4 concentration of 2% v/v, H_2O_2 concentration of 2% v/v, reaction time of 60 minutes, stirring speed of 330 rpm, and S/L ratio of 8 g/L were the ideal conditions for the dissolution process. Li_3PO_4 recovery was 88%, and the solid's purity was 98.3%. Additionally, 99% of Co is CoC_2O_4. The outcomes demonstrated that the extraction of Li and Co using the reducing dissolution process with H_3PO_4 acid is a successful method, obtaining dissolution values near 100%. In addition, a recovery of 88% of Li^+ as Li_3PO_4 and 99% of Co^{2+} as CoC_2O_4, with purities of 98.3% and 97.8% was found, respectively.

Meng et al. (2017) used a mixture solution of glucose and phosphoric acid to establish a hydrometallurgical leaching process for recovering Co and Li from LCO cathode material collected from S-LIBs. In the beginning cathode material was separated from the anode material and NaOH was leached for Al removal. Then, PVDF, acetylene black (AB), and carbon material were calcined at $650°C$ for 2 hours. In the end, the remaining CAM was leached with H_3PO_4 in the presence of glucose as a reductant. With $1.5 M$ H_3PO_4 acid and 0.02 mol/L glucose at $80°C$, a leaching rate of around 98% Co and almost 100% Li is produced in about 2 hours. Glucose was converted into monocarboxylic acid during the leaching process, resulting in a decrease of Co^{3+} to Co^{2+}. After being leached, Co in solution was recovered as Co-oxalate. In this study, $LiCoO_2$ is dissolved using glucose as a reductant and phosphoric acid as a chelating agent.

In this study, the leaching of used $LiCoO_2$ materials uses H_3PO_4 acid as a chelating agent. Chelating agents drain Li and Co from the cathode material during the complexation process. However, Co is present as the unstable Co^{3+} ion in used $LiCoO_2$, which is difficult to dissolve. Glucose was utilized in this study as a reductant to convert Co^{3+} to Co^{2+}, which is more stable and dissolvable in the acid solution. To create monocarboxylic acids, such as gluconic acid ($C_6H_{12}O_7$), tartaric acid ($C_4H_6O_6$), oxalic acid ($C_2H_2O_4$), and formic acid (CH_2O_2), glucose can be sequentially oxidized and degraded throughout the leaching process. Figure 5.9 depicts a potential glucose oxidative leaching pathway during the H_3PO_4 leaching procedure. Co^{3+} was leached from the glucose during oxidation and then appeared as Co^{2+}-phosphate. Li was present during the leaching process as lithium phosphate (Li_3PO_4) (Meng et al., 2017).

The selective extraction of Li from mixed types of LIBs ($LiCoO_2$, $LiMn_2O_4$, $LiFePO_4$, and $LiCo_{1/3}Mn_{1/3}Ni_{1/3}O_2$) was established by Chen et al. (2018) utilizing mild phosphoric acid as an effective leaching agent. The LCO, LMO, LFP, and NCM leaching systems, respectively, may provide high yields of 100%, 92.86%, 97.57%, and 98.94% Li at each optimal leaching condition. In moderate phosphoric acidic media under the same leaching circumstances, only a very small quantity of other metals (i.e. Co, Mn, Fe, and Ni) will dissolve.

Hydrometallurgical Recycling of LIBs   163

TABLE 5.4
Previous H_3PO_4 and HF Leach Results of LIBs

LIB Materials	Leach Reagents	Reducing Agent	T (°C)	t (minutes)	S/L Ratio (g/L)	Recovery (%)	Separating Method	Additives	Reference
$LiCoO_2$	0.7M H_3PO_4	4% (v) H_2O_2	40	60	50	99.7% Co; 99.9% Li			Nayaka et al (2016)
$LiCoO_2$ (NaCl discharge 48 hours, calcined 300°C, (Li: 4.8%, Co: 41.5%, Mn: 2.1%)	2% (v/v) H_3PO_4	2% (v) H_2O_2	90	60	8	99% Co	%99Co, CoC_2O_4; 97.8% purity		Pinna et al. (2017)
	0.7M H_3PO_4	4% (v) H_2O_2	40	60	2		%88 Li, Li_3PO_4; 98.3% purity		
LCO (Co: 59.6%, Li: 7%); LMO (Li: 4%, Mn: 58.4%); LFP (Li: 4.3%); NMC (Co: 11.7%, Li: 6.2%)	1.0M H_3PO_4	4% (v) H_2O_2	40	10	33	100%, 92.86%, 97.57%, and 98.94% Li LCO, LMO, LFP, NMC, respectively	MnC_2O_4, NiC_2O_2	Li_3PO_4	Chen et al. (2018a)
$LiCoO_2$, BL–5CA, Nokia S-LIB (NaOH leach+Calcination 650°C, 2 hours) after calcination: 5.77% Co, 58.03% Li	1.5M H_3PO_4	0.02 M Glucose	80	120	0.2	98% Co; 100% Li	$CoC_2O_4\downarrow$		Meng et al. (2017)
$LiCoO_2$	15% (v/v) HF		95	120	20	98% Co			Suarez et al. (2017)

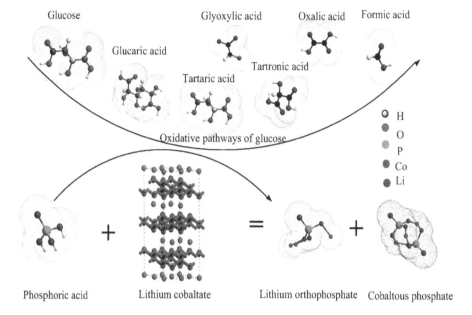

FIGURE 5.9 A possible leaching process of $LiCoO_2$ material in the phosphoric acid and glucose solution.

Fitting results from leaching kinetics indicate that the leaching of Li from waste CAMs is controlled by chemical reaction and internal diffusion, with an apparent activation energy of 37.74, 21.16, 27.47, and 21.86 kJ/mol for waste cathode material of $LiCoO_2$, $LiMn_2O_4$, $LiFePO_4$ and $LiCo_{1/3}Mn_{1/3}Ni_{1/3}O_2$, respectively. After purification and precipitation operations, relatively pure lithium phosphate (Li_3PO_4, with purity of 98.4%) can be produced. In comparison to earlier studies, it is claimed that this approach may be an effective choice for recovering Li from various types of used cathode materials, with little negative effects on the environment and little waste generated.

5.7   HF LEACH

Suarez et al. (2017) recovered Li and Co metals using hydrofluoric (HF) acid leaching. In this experiment, the effects of dissolution parameters including temperature, reaction duration, S/L ratio, stirring speed, and HF concentration were examined. According to the findings, increasing the HF concentration, temperature, and reaction time can facilitate $LiCoO_2$'s leaching process. Co and Li had recovery rates of 98% and 80%, respectively.

5.8 $NH_2.OH.HCl$ LEACH

Zhang et al. (1998) leached $LiCoO_2$ at 1.0–6.0 M hydroxylamine hydrochloride concentration at 80°C, 1/100 S/L ratio for 30 minutes. At 1.0 M $NH_2.OH.HCl$ dosage of 92% Li and 96% Co dissolutions were obtained.

5.9   ALKALINE LEACHING

Due to its selective leaching and ability to skip expensive separation or purification processes, alkali leaching has attracted interest. Because it can create stable ammonia complexes with metals like Ni, Co, and Cu, an ammonia-based system is employed. Ammonia and/or ammonium sulfate were used as a leaching solution and sulfites as the reducing agent by Zheng et al. (2017) and Chen et al. (2018). Table 5.5 summarizes the high overall leaching efficiency of Ni, Co, and Li from both tests; Mn had a distinct leaching pattern. According to Chen et al. (2018), the concentration of $(NH_4)_2SO_3$ had the greatest influence on the leaching efficiency of Mn, with 0.75 M being the ideal concentration.

Due to the creation of ammine complexes, ammonia has been widely utilized as an effective reagent to selectively leach Cu, Ni, and Co from various waste materials and low-grade ore. The following processes result in the formation of Co^{2+} and Ni^{2+} ammine complex ions at correct pH levels:

$$Ni^{2+} + nNH^3 \leftrightarrow Ni(NH_3)_n^{2+} \qquad (5.81)$$

$$Co^{2+} + nNH^3 \leftrightarrow Co(NH_3)_n^{2+} \qquad (5.82)$$

$Co(NH_3)_6^{3+}$, $Co(NH_3)_5^{2+}$, and $Co(NH_3)_4^{2+}$ are the main soluble species in the solution over the pH range of 9–11, while $Ni(NH_3)_5^{2+}$ is the predominant species in the range of pH from 8.5 to 10.5, according to the Eh-pH diagrams of the Co-NH_3-H_2O system and Ni-NH_3-H_2O system. The reaction kinetics are extremely sluggish even though the dissolution of high valence state Co oxides in ammonia solution is thermodynamically advantageous (Ku et al., 2016). Li, Ni, Co, and Mn are present in the oxidation states +1, +2, +3, and +4 in Li(Ni, Co, Mn)$_{1/3}O_2$. Because Co^{2+} is easily dissolved in solution, Co^{3+} in higher valence should always be reduced to Co^{2+}.

Zheng et al. (2017) used an ammonia-ammonium sulfate solution and sodium sulfite as a reductant to selectively leach off Ni, Co, and Li from cathode scrap powder. More than 98.6% of the first-step leaching solution's selectivity went to Li, Co, and Ni, whereas just 1.36% went to Mn. 89.8% Ni, 95.3% Li, 80.7% Co, and only 4.3% Mn could be leached away under ideal conditions with 4.0 M NH_3, 1.5 M $(NH_4)_2SO_4$, and 0.5 M Na_2SO_3 at 80°C, 10 g/L pulp density, and 300 minutes. The overall leaching of Ni, Co, and Li achieved 94.8%, 88.4%, and 96.7%, respectively, after the two-step leaching method, but the leaching rate for Mn was just 6.34%. With only 1.9% of dangerous impurity elements, the final solution's total selectivity of Ni, Co, and Li was over 98%. It was discovered that the kinetics of Ni, Co, and Li leaching after the chemical process was controlled, with their respective activation energies being 77.93, 87.92, and 83.30 kJ/mol.

When the CAM was calcined at 550°C, Chen et al. (2018) discovered that the organic compounds could be eliminated. The layer structure of $LiCoO_2$ collapsed, as evidenced by the phase characterization results showing the appearance of a new phase of Co_3O_4. In the spinel structure of $LiMn_2O_4$, Mn's valence also rose, forming $Li_4Mn_5O_{12}$. In the $(NH_4)_2SO_4$-$(NH_4)_2SO_3$ solution, it was possible to recover Ni, Co, Mn, and Li from CAM. 98% Ni, 98% Li, 81% Co, and 92% Mn may be leached out

TABLE 5.5

Previous Leach Study Results with Alkaline and Amonia Solutions

LIB Materials	Leach Reagents	Reducing Agent	T (°C)	t (min)	S/L ratio (g/L)	Recovery (%)	Separating Method	Additives	Reference
LMO + NMC mixed (cathode material) Ni: 15.3%; Co: 6%; Mn: 14.3%	NH_4OH + Buffer: $(NH_4)_2CO_3$ (3 M; 1.5 M; 3M)	$(NH_2)_2S$	80	60	1	94% Co; 100% Cu; 37% Ni; Al & Mn 0%	E_a: 57.4 and 60.4 kJ/mol for Ni and Co		Ku et al. (2016)
$LiCoO_2$ (mobile phone) (discharging: 10% NaCl, 36hours)	3.0M $(NH_4)_2SO_4$ + 0.75M $(NH_4)_2SO_3$		60		83	98% Ni; 98% Li; 81% Co; 92% Mn			Chen et al. (2018)
Thermal pretreatment 300°C, 1 hour + 550°C calcining 0.5hour (5.78% Li; 41.3% Co; 3.75% Ni; 11.4% Mn; 0.16% Al)									
LIB-NMC cathode scrap (550°C, 1 hour heating for Al removal)	4.0M $NH_3 + 1.5M$ $(NH_4)_2SO_4$	0.5M Na_2SO_3	80	300	10	89.8% Ni; 95.3% Li; 80.7% Co, 4.3% Mn	E_a: 77.93; 87.92; 83.30 kJ/mol for Ni, Co, Li		Zheng et al. (2017)

(Continued)

TABLE 5.5 (Continued)
Previous Leach Study Results with Alkaline and Amonia Solutions

LIB Materials	Leach Reagents	Reducing Agent	T (°C)	t (min)	S/L ratio (g/L)	Recovery (%)	Separating Method	Additives	Reference
(7.18% Li; 19.53% Co; 19.74% Ni; 18.63% Mn; 0.21% Al)	Two-step leaching					94.8% Ni; 96.7% Li; 88.4% Co; 6.34% Mn			
NMC(532) (EV LIBs)	367.5 g/L NH$_3$,H$_2$O	63.24 g/L H$_2$O$_2$				81.2% Li; 96.3% Co	NiSO$_4$, CoSO$_4$	NaOH, Na$_2$CO$_3$; Li$_2$CO$_3\downarrow$	Wang et al. (2017)
(Ni: 15.3%; Co: 6%; Mn: 14.3%; Cu: 0.3%)	140 g/L NH$_4$HCO					96.4% Ni			
	0.5 M HCl				10				

under ideal circumstances with 3.0 M $(NH_4)_2SO_4$, 0.75 M $(NH_4)_2SO_3$, and a S/L ratio of 83 g/L. The ammoniacal leaching behaviors of Mn, however, demonstrated remarkably distinct characteristics from those of other metals. It was discovered that the production of the double salts $(NH_4)_2Mn(SO_3)2H_2O$ and $(NH_4)_2Mn(SO_4)_26H_2O$) caused the leaching efficiency of Mn to be drastically reduced to 4% with an increase in $(NH_4)_2SO_3$ co

A hydrometallurgical technique was created by Wang et al. (2018) to extract Li, Ni, and Co from S-LIB powders. $NH_3H_2ONH_4HCO_3$ solutions were used to selectively leach 81.2% Li, 96.4% Ni, and 96.3% Co from the pretreatment S-LIB powders in the presence of H_2O_2. Using Mn-type Li-ion-sieves, 99.9% of the lithium was then successfully extracted from leaching solutions. The selectivity coefficients of Li^+ to Ni^{2+} and Li^+ to Co^{2+} reached 212.11 and 983.89, respectively, while the maximum amount of Li adsorbed was 31.62 mg/g. Li_2CO_3, $NiSO_4$, and $CoSO_4$ could also be produced as products, and ammonia could be recycled in a closed loop. This environmentally friendly method for extracting precious metals from used LIBs will promote resource conservation and recycling.

5.10   AMMONIA LEACH

It is possible to express the relationship between the base NH_3 and its conjugate acid NH_4^+ from a pH buffer as follows:

$$4NH_3 + H^+ \leftrightarrow NH^{4+} \tag{5.83}$$

Consequently, to determine the pH of an ammonia-ammonium system at ambient temperature,

$$pH = 9.26 + \log([NH_3]/[NH_4^+]) \tag{5.84}$$

The initial addition of ammonium sulfite significantly changes the pH of the binary (ammonia + ammonium sulfite) system because the pH of an ammoniacal solution depends on the ratio of NH_3 to NH_4^+. However, the pH of the ternary system (ammonia + ammonium sulfite + ammonium carbonate) does not considerably change once the third component containing NH_4^+ is present in an ammoniacal solution. In particular, ammonium carbonate can function as a pH buffer to ensure that the leaching solution's pH changes minimally so that stable Ni, Co, and Cu ammonia complexes can form.

The following reactions result in the formation of complex ions like $Ni(NH_3)_6^{2+}$, $Co(NH_3)_6^{2+}$, and $Cu(NH_3)_4^{2+}$ when an excessive amount of ammonia is added to a solution containing Ni^{2+}, Co^{2+}, and Cu^{2+}.

$$Ni^{2+} + 6NH_3 \leftrightarrow Ni(NH_3)_6^{2+} \qquad K_{sp}: 5.5 \times 10^8 \tag{5.85}$$

$$Co^{2+} + 6NH_3 \leftrightarrow Co(NH_3)_6^{2+} \qquad K_{sp}: 1.3 \times 10^5 \tag{5.86}$$

$$Cu^{2+} + 4NH_3 \leftrightarrow Cu(NH_3)_4^{2+} \qquad K_{sp}: 1.1 \times 1013 \text{ at } 25°C \tag{5.87}$$

Before disassembling, S-LIBs must be completely discharged, and Ni and Co should have an oxidation state greater than 2+. As a result, it is challenging to extrapolate the Ni and Co leaching efficiency from the formation constant values for $Ni(NH_3)_6^{2+}$ and $Co(NH_3)_6^{2+}$. A significant amount of Ni and Co in the LIB cathode can, however, be leached out in the presence of the reducing agent in ammoniacal solutions, as predicted by the high values of the formation constants. On the other hand, thermodynamics favors Mn in an ammonia solution to create the matching hydroxides or oxides. Manganese oxide is transformed into manganese carbonate through an intermediate manganese ammine complex in the presence of ammonium carbonate, even though Mn can form unstable ammonia complexes (Ku et al., 2016).

The cathode active materials are a composite of $LiMn_2O_4$, $LiCo_xMn_yNizO_2$, Al_2O_3, and C, whereas the leach residue is made up of $LiNi_xMn_yCo_zO_2$, $LiMn_2O_4$, Al_2O_3, $MnCO_3$, and Mn oxides, according to the SEM-EDS and XRD examinations of treated CAM powders. Co recovery through ammoniacal leaching is thought to outperform acid leaching by lowering the cost of adding NaOH to the leaching solution to raise its pH and by eliminating the need to separate Mn and Al (Ku et al., 2016).

5.11 ACETONITRILE + NO₂BF₄ LEACHING

Venkatraman et al. (2004) studied Li extraction from the layered $LiNi_{1-y-z}Co_yMn_zO_2$ using the oxidant NO_2BF_4 in an acetonitrile medium. While all the Li could be removed in 30 minutes from the Co-rich $LiNi_{1-y}Co_yO_2$ ($0.5 \leqslant y \leqslant 1$), it takes longer (6–48 hours) to extract all the Li from the Ni-rich $LiNi_{1-y}Co_yO_2$ ($0 \leqslant y < 0.5$) and $LiNi_{1-y}Mn_yO_2$ ($y = 0.25$ and 0.5). In the case of $LiNi_{0.5-0.5y}Mn_{0.5-0.5y}Co_yO_2$ samples, the time required to extract all the Li decreases with increasing Co content. The slow Li extraction rate in the Ni-rich $LiNi_{1-y}Co_yO_2$, $LiNi_{1-z}Mn_zO_2$, and Co-poor $LiNi_{0.5-0.5y}Mn_{0.5-0.5y}Co_yO_2$ systems is attributed to a considerable cation disorder as indicated by the Rietveld analysis of the X-ray diffraction data. In the $LiNi_{1-y}Mn_yO_2$ and Co-poor $LiNi_{0.5-0.5y}Mn_{0.5-0.5y}Co_yO_2$ systems, a larger inter-slab space for lithium in the structure is found to help the Li extraction. Table 5.6 shows the summary of the time required for the quantitative extraction of Li and the structure of the end members.

5.12 SUMMARY

It seems that hydrometallurgical processing is more beneficial than pyrometallurgical processing of active materials of S-LIB recycling. The use of inorganic acids in industrial LIB recycling is more acceptable from an economical and ecological point of view. The best metallurgical performance can be achieved by inorganic H_2SO_4 acid in the presence of a reductant. H_2O_2 is the most commonly used reductant for LIB recycling. H_2O_2 is not cheap and environment friendly, the use of Fe-scrap, Cu, and Al current collector materials obtained from LIB recycling was tested and found acceptable. The use of HCl generates $Cl_{2(g)}$ and HNO_3 $NO_{X(g)}$ problems. More than 95% Co, Li, and other possible metal dissolutions can be achieved with the $H_2SO_4 + H_2O_2$ combination. H_2SO_4 dosage can change between 1.0 and 4.0 M, H_2O_2 concentration 1–10 vol.%, leach temperature 25°C–95°C, leaching time 30–120 minutes, S/L 25–200 g/L. Leaching at room temperature generally requires long

TABLE 5.6

Time Required to Extract All the Li from $LiNi_{1-y-z}Co_yMn_zO_2$ and the Structure of the End Members, $Ni_{1-y-z}Co_yMn_zO_{2-\delta}$

Cathode Type	Time for Li Extraction	Structure of $Ni_{1-y-z}Co_yMn_zO_{2-\delta}$
$LiCoO_2$	Less than 0.25 hour	P3
$LiCo_{0.9}Ni_{0.1}O_2$	0.5 hour	P3
$LiCo_{0.8}Ni_{0.2}O_2$	0.5 hour	P3
$LiCo_{0.7}Ni_{0.3}O_2$	0.5 hour	P3
$LiCo_{0.5}Ni_{0.5}O_2$	0.5 hour	-
$LiCo_{0.3}Ni_{0.7}O_2$	6 hours	O3ⁱ
$LiCo_{0.15}Ni_{0.85}O_2$	12 hours	O3ⁱ
$LiNiO_2$	48 hours	O3ⁱ
$LiNi_{0.75}Co_{0.25}Mn_zO_2$	36 hours	O3ⁱ
$LiNi_{0.5}Co_{0.5}Mn_zO_2$	36 hours	O3
$LiNi_{0.33}Co_{0.33}Mn_{0.33}O_2$	1 hour	O1
$LiNi_{0.425}Co_{0.15}Mn_{0.425}O_2$	18 hours	O3 + O1

leaching times. Leaching performance depends on LIB active material chemistry. The use of ultrasonication-assisted leaching seems to be beneficial for metallurgical performance. After leaching purification can be carried out using precipitation with alkaline ($NaOH$, NH_4OH, etc.), alcohol (ethanol), oxalic acid, or solvent extraction with Cyanex 272, D2EHPA, Acoga, PC-88A, etc. After SX acid stripping is applied. $CoC_2O_4.2H_2O$, $Co(OH)_2$, and $CoSO_4$ cobalt compounds or Li_2CO_3 with Na_2CO_3, Li_2PO_4 with Na_2PO_4, or $LiOH$ with $NaOH$ lithium compounds can be produced for new battery material production. From cobalt oxalate Co_3O_4 can be produced by the calcination process. $LiCoO_2$ can be produced from Co_3O_4 and Li_2CO_3 calcination. For economic LIB recycling, the leaching temperature should be close to room temperature, the leaching time should be short (1–2 hours), the leaching reagent dosage should be low (1.0–2.0 M) without reductant or minimum dosage reductant, and high percent solid (>200 g/L). Leach areductant reagents should be environmentally friendly. Wastewater and waste solids should be as much neutral as possible. Recovering metals such as Co, Li, Ni, Mn, Cu, Al, and Fe and non-metal graphite is possible from S-LIBs with appropriate pretreatment and hydrometallurgical flowsheet. It seems that in the near future, we will see more hydrometallurgical LIB recycling plants in the world to recover secondary Co and Li. Globally, these critical metals will be more valuable for electrified transportation sectors.

6 Organic Acid Leaching and Bioleaching of LIBs

ABBREVIATIONS

AA	Ascorbic acid
AcA	Acetic acid
AAS	Atomic absorption spectrophotometry
CA	Citric acid
CAM	Cathode active material
DMG	Dimethylglyoxime
EDS	Energy-dispersive spectroscopy
FA	Formic acid
ICP	Inductively coupled plasma
IUPAC	International Union of Pure and Applied Chemistry
LCO	$LiCoO_2$
LFP	Lithium iron phosphate
MA	Malic acid
NMC	Nickel manganese cobalt
NMP	*N*-methyl pyrrolidone
OA	Oxalic acid
PTFE	Polytetrafluoroethylene
PVDF	Polyvinylidene fluoride
SA	Succinic acid
SEM	Scanning electron microscope
S-LIB	Spent lithium-ion battery
SX	Solvent extraction
TCA	Tricholoric acid
TFA	Trifluoroacetic acid
XRD	X-ray diffraction

6.1 ORGANIC ACIDS USED IN HYDROMETALLURGY

An organic substance with acidic characteristics is known as an organic acid. The most prevalent organic acids are carboxylic acids, which get their acidity from the carboxyl group in their structure, COOH. Sulfonic acids, which belong to the SO_2OH group, are comparatively more potent acids. Alcohols can function as acids when they have –OH, but they are typically quite weak. The acid's acidity is determined by the conjugate base's relative stability. The thiol group –SH, the enol group, and the phenol group are other groups that can impart acidity, but typically weakly.

DOI: 10.1201/9781003384557-6

Organic molecules with these groups are typically referred to be organic acids in biological systems.

In contrast to strong mineral acids, which totally dissociate in water, organic acids are typically weak acids. Higher molecular mass organic acids, like benzoic acid, are insoluble in molecular (neutral) form, but lower molecular mass organic acids, such as formic and lactic acids, are miscible in water.

Strong mineral acids like HCl or combinations of HCl and HF are far more reactive with metals than simple organic acids like formic or acetic acids. Because of this, organic acids are employed in situations requiring prolonged leaching times or high temperatures. Oxalic and citric acids are used to remove rust. Organic acids have the same ability to dissolve iron oxides as stronger mineral acids, but without causing base metal deterioration. They might be able to bind the metal ions in the dissociated state, accelerating elimination. Numerous more sophisticated organic acids with hydroxyl or carboxyl groups are produced by biological processes.

Figure 6.1 shows the general structure of two organic acids. Carboxylic acid and sulfonic acid left to right. Each molecule's acidic hydrogen is gray in hue. Table 6.1 lists the characteristics of organic acids, structural covalent bond formulations, manufacturing processes, and current pricing that have previously been employed in leaching experiments.

6.2 ORGANIC ACID LEACH OF S-LIBs

Many researchers have used some environment-friendly organic acids (e.g., citric acid, (Chen et al., 2015, 2016a; Nayaka et al., 2015; Yao et al., 2015) DL-malic acid, (Li et al., 2010) ascorbic acid, (Nayaka et al., 2015) oxalic acid, (Sun and Qui, 2012; Zeng et al., 2015) succinic acid, (Li et al., 2015), and trichloroacetic acid (Zhang et al., 2015)) as leachants in S-LIBs recycling. Reductants like H_2O_2, $NaHSO_3$, and glucose were frequently added to the solution during the leaching process to speed up the recovery rate of various metals (Gratz et al., 2014; Guo et al., 2016, Li et al., 2015; Sa et al., 2015; Sun et al., 2017; Zhang et al., 2015; Zou et al., 2013). Low leaching selectivity inorganic acids quickly leach practically all metals from cathode scrap while selectively leaching the desired metals, and the subsequent solvent extraction (SX) stages play a significant role in the metal recovery (Bankole et al., 2013). Unlike inorganic acids, organic acids can be used during cathode ray tube production as a leachate, reductant, precipitant, or chelating agent. Oxalic acid was employed by Zeng et al. (2015) and Sun and Qiu (2012) as a leachate and precipitant to remove Co and Li from $LiCoO_2$. To treat the cathode active material (CAM), which had already been isolated from Al foil using N-methyl pyrrolidone (NMP) dissolving at 80°C, Yao et al. (2015) used citric acid as both leaching and chelating agents.

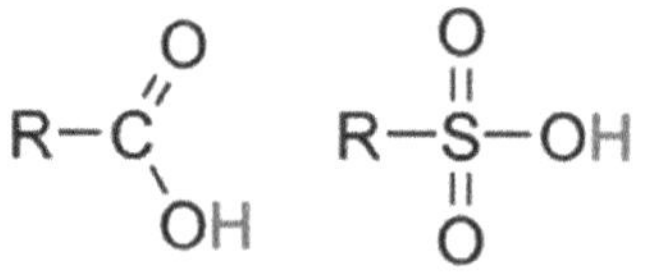

$$R-C\overset{\displaystyle O}{\underset{\displaystyle OH}{}} \qquad R-\overset{\displaystyle O}{\underset{\displaystyle O}{S}}-OH$$

FIGURE 6.1   Structure of carboxylic acid and sulfonic acid.

TABLE 6.1
Some Organic Acids Used in Leaching

Common Name	IUPAC Name	Formula	Chemical Covalent Bond Structure
Formic acid (FA) (Molar mass: 46.025 g/mol Density: 1.22 g/mL Boiling point: 100.8°C P_{Ka}: 3.745	**Methanoic acid** *(Produced from methanol or byproduct of acetic acid Miscible with water Colorless liquid ten times stronger than acetic acid)*	HCO_2H CH_2O_2	
Low toxic Concentrated acid is corrosive to the skin **Price: 300–850 $/t 85% Ind. grade**	*Producers: BASF, Eastman Chem. Corp., LC Industrial, Feicheng Acid Chemicals)*		Cyclic dimer of formic acid; dashed **green** lines represent hydrogen bonds
Acetic acid (AcA) (Molar mass: 60.052 g/mol Density: 1.049 g/mL Boiling point: 118°C P_{Ka}: 4.756 Concentrated acid is corrosive to the skin **Price: 400–520 $/t 99% glacial grade**	**Ethanoic acid** *(Produced from bacterial fermentation Miscible in water Colorless liquid Vinegar-like odor Midely corrosive to Fe, Mg, and Zn to form metal-acetate + $H_{2(g)}$) Producers: Celanese, BP, Millenium Chem., Sterling Chem., Samsung, Eastman)*	CH_3CO_2H CH_3COOH	
Oxalic acid (OA) (Molar mass: 90.034 g/mol Density: 1.90 g/mL Melting point: 190°C P_{Ka}: 1.27, 4.27 **Price: 300–800 $/t 99.6% grade**	***Ethanedionic acid*** *(Production: oxidizing sucrose using nitric acid in the presence of a small amount of vanadium pentoxide as a catalyst Stronger than acetic acid Chelating and reducing agent for metals)*	Anhydrous HO_2CCO_2H $C_2H_2O_4$ Dihydrate form $C_2H_2O_4.2H_2O$	

(Continued)

TABLE 6.1 (*Continued*)
Some Organic Acids Used in Leaching

Common Name	IUPAC Name	Formula	Chemical Covalent Bond Structure
Ascorbic acid (AA) (Molar mass: 176.124 g/mol Density: 1.65 g/mL Melting point: 190°C P_{Ka}: 4.10, 11.67 **Price: 3,000–6,000 $/t 99% grade)**	**Hexuronic acid** *(Production: From glucose via the Reichstein process White or light yellow solid Solubility in water: 330 g/L Midely acidic solution Mild reducing agent 80% is produced in China*	$C_6H_8O_6$	
Succinic acid (SA) (Molar mass: 118.088 g/mol Density: 1.565 g/mL Boiling point: 235°C P_{Ka}: 4.2, 5.6 **Price: 1,700–3,000 $/t 99% grade)**	*(Production: Hydrogenation of maleic acid Odorless solid Highly acidic teste Irritant to skin and eyes)*	$C_4H_6O_2$	
Citric acid (CA) (Molar mass: 192.123 g/mol Density: 1,666 g/mL Boiling point: 310°C P_{Ka}: 3.13, 4.76, 6.39 **Price: 450–800 $/t 99.5% grade)**	**Citrus fruit acid** (Weak organic acid White solid Weak acid Odorless Soluble in water 80% (80°C–90°C) Acidifier Chelating agent, binding metals by making them soluble More than 50% is produced in China)	Anhydrous $C_6H_8O_7$ Monohydrate $C_6H_8O_7 . H_2O$	

(Continued)

TABLE 6.1 *(Continued)*
Some Organic Acids Used in Leaching

Common Name	IUPAC Name	Formula	Chemical Covalent Bond Structure
Malic acid (MA) (Molar mass: 134.09 g/mol Density: 1,609 g/mL Melting point: 130°C P_{Ka}: 3.40, 5.20 **Price: 1,000–5,000 $/t 99% grade)**	**Apple acid** (Racemic malic acid is produced industrially by the double hydration of maleic anhydride Colorless L-malic acid is the naturally occurring form, whereas a mixture of L- and D-malic acid is produced synthetically)	$C_4H_6O_5$	
Lactic acid (LA) (Molar mass: 90.078 g/mol Boiling point: 122°C P_{Ka}: 3.86, 15.1 **Price: 1,000–3,000 $/t 80% grade)**	**Milk acid** (Produced from fermentation of carbohydrates White solid Miscible in water ten times more acidic than acetic acid)	$H_3H_6O_3$	
Trifluoroacetic acid (Molar mass: 114.023 g/mol Density: 1,489 g/mL Boiling point: 72.4°C P_{Ka}: 0.52 **Price: 13,000– 14,000 $/t)**	**TFA** (Produced from electrofluorination of acetyl chloride or acetic anhydride, followed by hydrolysis of the resulting trifluoroacetyl fluoride Colorless liquid, Highly corrosive Stronger acid than acetic acid	$C_2HF_3O_2$	

(Continued)

TABLE 6.1 (*Continued*)
Some Organic Acids Used in Leaching

Common Name	IUPAC Name	Formula	Chemical Covalent Bond Structure
Trichloroacetic acid (Molar mass: 163.39 g/mol Density: 1,63 g/mL Boiling point: 196°C P_{Ka}: 0.66 **Price: 1,500– 20,000 \$/t)**	**TCA** (Produced by Hell–Volhard– Zelinsky halogenation. White powder)	$C_2HCl_3O_2$	

Source: Compiled from www.wikipedia.org.

Thermal treatment at 600°C is necessary to achieve efficient separation of the Al foil and cathode material because NMP can only dissolve polar organic binders (such as polyvinylidene fluoride, PVDF), rather than the highly nonpolar binders like polytetra-fluoroethylene (PTFE) (Yao et al., 2015; Zhang et al., 2014). The resulting NMP-contained solution, however, is relatively challenging to recycle and must ultimately be disposed of as organic liquid waste. Additionally, the aforementioned hydrometallurgical methods typically call for manually disassembling the battery and then either dissolving and burning the binders to get cathode powder for additional recycling or peeling the cathode components off the aluminum foil. Li et al. (2015); Zhang et al. (2015); Sa et al. (2015); Gratz et al. (2014); Zou et al. (2013); Guo et al. (2016). There is a dearth of research on cathode scrap that has undergone efficient mechanical processing (such as separation and shredding) or such scrap from the fabrication of LIBs. Leaching selectivity and closed-materials-loop are therefore crucial factors to take into account when creating an efficient method for S-LIBs recycling that is close to the industry.

6.2.1 Reductive Citrate/Citric Acid Leaching

The citric acid ($C_6H_8O_7$) was used for LIBs recycling with different reductants (such as H_2O_2, tea waste, and D-glucose). Table 6.2 lists some of the prior leach findings for various metallurgical performances and optimal circumstances. Citric acid (CA) dosage changed from 0.5 to 1.5 M, H_2O_2 dosage from 1.0 to 1.5 vol.% at a high temperature (between 80°C and 90°C) and 20 g/L solid ratio for 20–120 minutes.

Li et al. (2018) looked at the kinetics of leaching as well as the procedure for recycling mixed-cathode materials from S-LIBs. The recycling of used mixed-cathode materials ($LiCoO_2$, $LiCo_{1/3}Ni_{1/3}Mn_{1/3}O_2$, and $LiMn_2O_4$) has been suggested as a "grave-to-cradle" process. The procedure involves leaching CAMs with citric acid and hydrogen peroxide (H_2O_2), and then resynthesizing a cathode material from the

TABLE 6.2

Previous Organic acid Leach Results

LIB Materials	Leach Reagents	Reducing Agent	T (°C)	t (min)	S/L Ratio (g/L)	Recovery (%)	Separating Method	Additives	Reference
LCO, NCM, LMO mixed	0.5 M $C_6H_8O_7$ (Citric acid)	1.5% (v) H_2O_2	90	60	20	99.1% Li; 99.8% Co; 98.7% Ni; 95.2% Mn	E_a: 66.9; 86.6; 49.5; and 45.2 for Li, Co, Ni, and Mn		Li et al. (2018)
$LiCoO_2$ (Li: 6.81%, Co: 58.79%, Ni: 0.58%, Al: 0.71%)	1.5 M $C_6H_8O_7$	1(v) H_2O_2	90	80	20		Precipitation $CaC_2H_4.2H_2O$ with $H_2C_2O_4$	Li_3PO_4 with H_3PO_4	Chen et al. (2015)
LCO battery	1.5 M $C_6H_8O_7$	0.4 g/g tea waste	80	120	3		98% Li; 96% Co	99% $CoC_2O_4.2H_2O$; 93% Li_3PO_4	Chen et al. (2015b)
LCO battery	1.5 M $C_6H_8O_7.2H_2O$	0.5 g/g D-glucose	80	120	20	99% Li; 91% Ni; 92% Co; 94% Mn			Chen et al. (2016)
Spent LIBs (Li: 4.4%; Co: 53.8%; Ni: 0.8%; Mn: 0.97%)	1.25 M $C_6H_8O_7.H_2O$	1% H_2O_2	90	30	20	>90% Co; 100% Li			Li et al. (2010a)

(Continued)

TABLE 6.2 (*Continued*)
Previous Organic acid Leach Results

LIB Materials	Leach Reagents	Reducing Agent	T (°C)	t (min)	S/L Ratio (g/L)	Recovery (%)	Separating Method	Additives	Reference
(NMP, 100°C for 1 hour; Calcine 700°C, 5 hour pretreatment)									
Spent LIBs	2.0 M $C_6H_8O_7$	2% H_2O_2	80	90	3.3	95% Co; 97% Ni; 94% Mn; 99% Li			Chen et al. (2014)
Spent LIBs									
$LiCoO_2$ (manual dismantling, NMP immersion, calcination)	1.25 M $C_6H_8O_7.H_2O$	1.0% (v) H_2O_2	90	35	20	98% Li; 90,2% Co	$H_2C_2O_4$: Co^{2+} = 1:1.05 for Co	Na_3PO_4 or Li	Fan et al. (2016)
(6.78% Li; 58.65% Co; 0.69% Al: 0.54% Ni)	circulatory leaching	0.9% (v) H_2O_2	90	35	16.7	90% Li; 80% Co	99.5% Co precipitation	90.2% Li recovery	
$LiCoO_2$ (cathode active material (laptop LIBs)	1.25 M $C_6H_8O_7.H_2O$	1.0 % (v) H_2O_2	90	30	20	99% Li & Co			Li et al. (2013)
(NMP, 100°C, 1 hour to separate Al foil; calcination 700°C, 5 hours to	$C_4H_5O_6$ (DL-malic)	2.0% (v) H_2O_2	90	40		99% Li & Co			
eliminate/burn C and PVDF; grinding 2 hours)	$C_4H_7NO_4$ (L-aspartic)	4.0% (v) H_2O_2	90	120	10	60% Li & Co			

(Continued)

TABLE 6.2 (Continued)
Previous Organic acid Leach Results

LIB Materials	Leach Reagents	Reducing Agent	T (°C)	t (min)	S/L Ratio (g/L)	Recovery (%)	Separating Method	Additives	Reference
Spent LIBs (LiCoO$_2$) (Li: 4.4%, Co: 53.8%) (Discharge + NMP + calcination 700°C)	1.5 M DL-Malic	2% H_2O_2	90	40	20	90% Co; 100% Li			Li et al. (2010)
LiCoO$_2$ cathode material (NMP+ultrasonication for Al foil detachment+ calcination 700°C, 2 hours for PVDF and C removal)	1.0 M IDA (iminodiacetic acid)	0.02 M ascorbic acid (AA)	80	360	0.2	99% Li; 91% Co			Nayaka et al. (2015)
	1.0 M Maleic acid (MA)	acid (AA)	80	360	0.2	100% Li; 97% Co			Sohn et al. (2006)
LiCoO$_2$	3.0 M Oxalic acid		80	60	50	99% Li; 96% Co			Zeng et al. (2015)
	2.0 M H_2SO_4	10% (v)	75	75	50				
LiCoO$_2$ cathode material (24.53% Co; 3.52% Li; 2.45% Cu; 0.83% Al; 0.25% Fe)	1.0 M Oxalic acid		95	150	15	97% Co; 98% Li	CoC_2O_4; $Li_2Co_2O_4$/ $LiHCO_4$		

(Continued)

TABLE 6.2 (Continued)
Previous Organic acid Leach Results

LIB Materials	Leach Reagents	Reducing Agent	T (°C)	t (min)	S/L Ratio (g/L)	Recovery (%)	Separating Method	Additives	Reference
LiCoO$_2$ cathode material	1.0 M Oxalic acid	15% (v) H$_2$O$_2$	80	120	50	98% LiCoO$_2$			Sun and Qui (2012)
(Vacuum pyrolysis, 1 kPa; 600°C, 30 minutes)		(no effect)							
LiCoO$_2$ cathode material	1.0 M Oxalic acid		25	1440	50	74% Li	Co in precipitate		Aaltonen et al. (2017)
LiCoO$_2$ cathode material	1.0 M Oxalic acid	1% H$_2$O$_2$	25	1440	50	79% Li	Co in precipitate		Aaltonen et al. (2017)
NMC cathode material NMP + Calcination)	0.6 M Oxalic acid		75	10	2	85% Li; 98% Ni, Co, Mn			Zhang et al. (2018)
NMC cathode material (bioleaching)	0.04 M Oxalic acid		30	648 h		100% Li; Ni; Co; 90% Mn			Horeh et al. (2016)
LiFePO$_4$ (Discharge + NMP, 1 hour 100°C + calcination 400°C, 1 h)	0.3 M Oxalic acid		80	60	6	98% Li; 92% Fe			Li et al., (2018a)

(Continued)

TABLE 6.2 (*Continued*)
Previous Organic acid Leach Results

LIB Materials	Leach Reagents	Reducing Agent	T (°C)	t (min)	S/L Ratio (g/L)	Recovery (%)	Separating Method	Additives	Reference
LiCoO$_2$ cathode material	2.0 M Citric acid		25	1440	0.5	62% Li; 41% Co			Aaltonen et al. (2017)
LiCoO$_2$ cathode material	2.0 M Citric acid	1% H$_2$O$_2$	25	1440	0.5	65% Li; 46% Co			Aaltonen et al. (2017)
LiCoO$_2$ cathode material (NMP immersion+ultrasonication 20 min., 450°C, 1 hour thermal treatment)	1.25 M Ascorbic acid (C$_6$H$_8$O$_6$)		70	20	25	98.5% Li; 95% Co	100% Co; 96% Li		Li et al. (2012)
LiCoO$_2$ cathode material (NMP at 100°C ultrasonication 1 hour for Al foil removal+ calcination at 700°C for 5 + 2 hours grinding-planatery mill;	1.5 M Succinic acid	4% (v) H$_2$O$_2$	70	40	1.50	94.65% Li; 99.83% Co			Li et al. (2015)

(Continued)

TABLE 6.2 (Continued)
Previous Organic acid Leach Results

LIB Materials	Leach Reagents	Reducing Agent	T (°C)	t (min)	S/L Ratio (g/L)	Recovery (%)	Separating Method	Additives	Reference
6.76% Li; 57.94% Co; 0.76% Ni; 0.91% Mn NMC cathode	2.0 M formic acid HCOOH	30% (v) H_2O_2	60	120	50	99.93% Li; 99.31 Li&Mn; 95.46% Al		Li_2CO_3 purity: 99.90%	Gao et al. (2017)
(18.32% Ni; 18.65% Co; 17.57% Mn; 6.15% Li; 7.86% Al) NMC cathode (Discharge + dismantle + NaOH leach+calcination, 610 C + grinding (39 minutes) + lactic acid leach+resynthesis of NCM)	1.5 M lactic acid	0.5 M H_2O_2	70	20	20	97.7% Li; 98.2% Ni; 98.9% Co; 98.4% Mn	Sol-gel method resynthesize NMC		Li et al. (2017a)
$LiNi_{1/3}Co_{1/3}Mn_{1/3}O_2$	3.0 M TCA	4% H_2O_2	60	30	20	93% Ni; 91.8% Co; 89.8 %Mn; 99.7% Li	E_a: 44.8 and 28 kJ/mol for Co and Li		Zhang et al. (2015)
6.15% Li; 18.32% Ni; 18.65% Co; 17.57% Mn; 7.9% Al	4.5M H_2SO_4		70	240	10	91.7% Ni; 87.7% Co; 75.9% Mn; 99.6% Al			
$LiNi_{1/3}Co_{1/3}Mn_{1/3}O_2$	15 vol. % TFA		40	180	125				Zhang et al. (2014)

resultant leachate. Leaching temperatures of 90°C, CA concentrations of 0.5 M, H_2O_2 concentrations of 1.5 vol.%, leaching times of 60 minutes, and pulp densities of 20 g/L were found to be the ideal conditions. Li, Co, Ni, and Mn had leaching efficiencies that were greater than 95%. Leaching efficiencies for Li and Co were greater than 99%. A sol-gel approach was utilized to resynthesize fresh $LiCo_{1/3}Ni_{1/3}Mn_{1/3}O_2$ material from the leachate. The initial discharge capacity of the resynthesized material (NCM-spent) at 0.2 C was 152.8 mAh/g, which was more than the 149.8 mAh/g of NCM-syn, according to a comparison of the electrochemical characteristics of the two materials. The NCM-spent and NCM-syn had discharge capacities of 140.7 and 121.2 mAh/g, respectively, after 160 cycles. The NCM-spent material maintained a higher capacity of 113.2 mAh/g than the NCM-syn (78.4 mAh/g) after 300 cycles of discharge at 1 C. Trace Al doping was the cause of the NCM-spent's superior performance. To explain the kinetics of the leaching process, a brand-new formulation built on the shrinking-core concept was put out. Indicating that the leaching was a chemical reaction-controlled process, the activation energies of the Li, Co, Ni, and Mn leaching were calculated to be 66.86, 86.57, 49.46, and 45.23 kJ/mole, respectively.

Chen et al. (2015b) investigated the environmentally friendly recovery of metals from S-LIBs. To discharge the remaining electricity, used batteries were submerged in a 10 w/v% Na_2SO_4 solution for roughly 24 hours. Then, these depleted batteries were cleaned with deionized water before being baked at 80°C for 12 hours. Batteries were then physically disassembled into metallic shells, separators, anodes, and cathodes for manual dismantling. Metal separators and shells were recycled right away. For the peeling process, the anodes and cathodes were chopped into little pieces measuring about 1 cm by 1 cm. The PVDF binder was dissolved and the Al/Cu foils were separated from the cathode/anode materials using a green solvent (*N*-methyl-2-pyrrolidone, NMP) during the recycling of the Cu/Al foils. After evaporation and reclamation, Al/Cu foils might then be reused repeatedly in their metallic forms. The resulting CAMs were then filtered, and the carbon materials were removed by calcining them for 2 hours at 700°C in a muffle furnace as the final thermal and mechanical treatment. The waste CAMs were then ground down into finer fractions with a higher specific surface area, which will help with the subsequent leaching process.

Chen et al. (2015b) investigated three leaching systems (H_3Cit/H_2O_2, H_3Cit/PA, and H_3Cit/TWsystems) for the creative application of various biomass as reductants. In the H_3Cit/H_2O_2 system, leaching was possible under optimal conditions of 80 minutes, 70°C, 2.0 M, 0.6 g/g reductant dosage, and 50 g/L slurry density for roughly 98% Co and 99% Li. The H_3Cit/TW system could achieve similar leaching efficiency (96% Co and 98% Li) under the ideal conditions of 120 minutes, 90°C, 1.5 M, reductant dose of 0.4 g/g, and a slurry density of 30 g/L. For the H_3Cit/PA system, the optimal conditions are 120 minutes, 80°C, 1.5 M, reductant dose 0.4 g/g, and slurry density 40 g/L. This yields inferior leaching results (83% Co and 96% Li). Eqs. 6.1 and 6.2 can be used to succinctly represent the leaching reactions in the H_3Cit/PA and H_3Cit/TW systems (Chen et al., 2015b). The extraction of Co and Li from waste $LiCoO_2$ will be made easier during the leaching processes if there are any degraded products or reducible chemicals present, in accordance with the oxidation mechanisms of PA and TW.

$$LiCoO_2 + HCit + TW \rightarrow Co_3(Cit)_2 + Co(HCit) + Co(H_2Cit)_2$$
$$+ Li_3Cit + Li_2(HCit) + Li(H_2Cit) + OD_1 \tag{6.1}$$

$$LiCoO_2 + HCit + PA \rightarrow Co_3(Cit)_2 + Co(HCit) + Co(H_2Cit)_2 + Li_3Cit$$
$$+ Li_2(HCit) + Li(H_2Cit) + OD_2 \tag{6.2}$$

where OD_1 and OD_2 stand for the corresponding oxidized derivatives.

The leaching reaction (H_3Cit/H_2O_2 system) could be expressed as follows:

$$H_3Cit + LiCoO_2 + H_2O_2 \rightarrow Co_3(Cit)_2 + Co(HCit) + Co(H_2Cit)_2 +$$
$$Li_3Cit + Li_2(HCit) + (H_2Cit) + H_2O + O_2 \tag{6.3}$$

$$LiCoO_2 + H_2O_2 + H_2C_2O_4 + H_3PO_4 \rightarrow CoC_2O_4 + Li_3PO_4$$
$$+ O_2 + H_2O \tag{6.4}$$

Table 6.3 shows the ionization constants for various acids in the aqueous phase at 25°C. K_{a1} and pK_{a1} (determining the acidity) present in the following order: $H_2C_2O_4$ > H_3PO_4 > H_3Cit (K_{a1}:5.4×10^{-2}, 7.5×10^{-3}, and 7.4×10^{-4} and pK_{a1}:1.27, 2.12, and 3.13 for $H_2C_2O_4$, H_3PO_4, and H_3Cit, respectively). Therefore, it is theoretically feasible the addition of stronger acids ($H_2C_2O_4$ and H_3PO_4) to substitute weaker acids (H_3Cit) from the leaching solution by selective precipitation (Chen et al., 2015b).

The selective precipitation method allows for the simultaneous recycling of metals and used CA. After five cycles under the same conditions, recycled CA likewise exhibits comparable leaching performance to fresh acid, recovering around 99% Co and 93% Li as $CoC_2O_4 2H_2O$ and Li_3PO_4, respectively.

TABLE 6.3
Ionization Constants for Different Acids in the Aqueous Phase (25°C)

		Ionization Constant	
Acid	Ionization Equation	K_a	pK_a
$H_2C_2O_4$	$H_2C_2O_4 = HC_2O_4^- + H^+$	$5.4 * 10^{-2}$ (K_1)	1.27 (pK_{a1})
	$HC_2O_4^- = C_2O_4^{2-} + H^+$	$5.4 * 10^{-5}$ (K_2)	4.27 (pK_{a2})
H_3PO_4	$H_3PO_4 = H_2PO_4^- + H^+$	$7.5 * 10^{-3}$ (K_1)	2.12 (pK_{a1})
	$H_3PO_4^- = HPO_4^{2-} + H^+$	$6.3 * 10^{-8}$ (K_2)	7.20 (pK_{a2})
	$HPO_4^{2-} = PO_4^{3-} + H^+$	$4.4 * 10^{-13}$ (K_3)	12.36 (pK_{a3})
H_3Cit	$H_3Cit = H_2Cit^- + H^+$	$7.4 * 10^{-4}$ (K_1)	3.13 (pK_{a1})
	$H_2Cit^- = HCit^{2-} + H^+$	$1.7 * 10^{-5}$ (K_2)	4.76 (pK_{a2})
	$HCit^{2-} = Cit^{3-} + H^+$	$4.0 * 10^{-7}$ (K_3)	6.40 (pK_{a3})

$$Co_3(Cit)_2 + 3H_2C_2O_4 \leftrightarrow 3CoC_2O_4\downarrow + 2H_3Cit \qquad (6.5)$$

$$Co(HCit) + H_2C_2O_4 \leftrightarrow CoC_2O_4\downarrow + H_3Cit \qquad (6.6)$$

$$Co(H_2Cit)_2 + H_2C_2O_4 \leftrightarrow CoC_2O_4\downarrow + 2H_3Cit \qquad (6.7)$$

$$Li_3Cit + H_3PO_4 \leftrightarrow Li_3PO_4\downarrow + H_3Cit \qquad (6.8)$$

$$3Li_2(HCit) + H_3PO_4 \leftrightarrow 2Li_3PO_4\downarrow + 3H_3Cit \qquad (6.9)$$

$$3Li(H_2Cit) + H_3PO_4 \leftrightarrow Li_3PO_4\downarrow + H_3Cit \qquad (6.10)$$

pK_{sp} (solubility product) constants of relevant CoC_2O_4, $Li_2C_2O_4$, and Li_3PO_4 precipitates are

$$CoC_2O_4\downarrow \leftrightarrow Co^{2+} + C_2O_4^{2-} \quad pKsp: 7.2 \qquad (6.11)$$

$$Li_2C_2O_4\downarrow \leftrightarrow 2Li^+ + C_2O_4^{2-} \quad pKsp: 1.9 \qquad (6.12)$$

$$Li_3PO_4\downarrow \leftrightarrow 3Li^+ + PO_4^{3-} \quad pKsp: 3.4 \qquad (6.13)$$

The fact that CoC_2O_4 has a substantially higher pK_{sp} (7.2) than $Li_2C_2O_4$ (1.9) suggests that when $H_2C_2O_4$ is added to the leaching fluid, Co^{2+} will precipitate first, followed by Li^+. Then, Li_3PO_4 solution was used to precipitate the Li ions. At 60°C, 30 minutes, 300 rpm, and $n(H_2C_2O_4):n(Co^2) = 1.05$ or $n(H_3PO_4):n(Li^+) = 0.4$ (molar ratio), respectively, over 99% of Co and about 93% of Li may be recovered as $CoC_2O_4 2H_2O$ and Li_3PO_4 (99.3% and 98.5% in purities). Following the addition of a saturated oxalic acid solution, the XRD pattern of the cobalt oxalate precipitation is shown in Figure 6.2.

The recovery of metals from S-LIB active materials was studied by Li et al. (2013) utilizing citric ($C_6H_8O_7H_2O$), DL-malic ($C_4H_5O_6$), and l-aspartic ($C_4H_7NO_4$) acids as leachants. The aforementioned three organic acids were chosen due to their traits, which include simple natural degradation and the absence of hazardous fumes during the reaction. CA is the least acidic of the three acids, followed by malic acid and aspartic acid. Figure 6.3 displays both the residues recovered from the acid-leaching stage and the XRD patterns of the S-LIB cathodic materials after calcination and grinding (but before leaching). The cathodic components before leaching are shown by the XRD data to primarily consist of $LiCoO_2$ and minor amounts of Co_3O_4, a performance-degrading breakdown byproduct created during battery operation. The lack of carbon peaks shows that the majority of the carbon leftovers are burned off during the calcination process. The majority of the black residue recovered after citric and DL-malic acid leaching was insoluble Co_3O_4, as revealed by the XRD results.

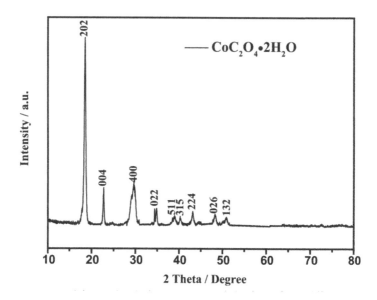

FIGURE 6.2 XRD pattern of formed cobalt oxalate precipitation after adding a saturated oxalic acid solution.

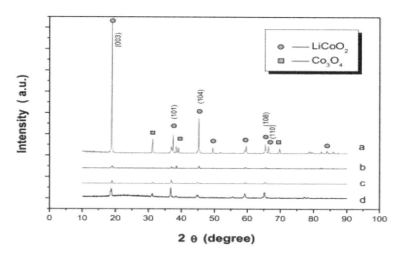

FIGURE 6.3 XRD diffractograms of samples of (a) the cathodic material following dismantling, calcination, and grinding, (b) the residues after leaching with $C_6H_8O_7 \cdot H_2O$, (c) the residues after leaching with $C_4H_5O_6$, and (d) the residues after leaching $C_4H_7NO_4$.

However, due to an incomplete interaction between the spent $LiCoO_2$ and the acid, low-intensity peaks of $LiCoO_2$ were found in the XRD patterns of the residue recovered by leaching with aspartic acid (Li et al., 2013).

The leaching reaction of spent $LiCoO_2$ and the organic acids was described by Li et al. (2013) as occurring in two steps: (i) the reduction of Co^{3+} to Co^{2+} in the presence of H_2O_2 and subsequent dissolution of spent $LiCoO_2$ in the acid solution and

(ii) the chelation of Co^{2+} and Li with citrate, malate, or aspartate. Figure 6.4 depicts the reaction between the used $LiCoO_2$ and citric, DL-malic, and L-aspartic acids, along with some potential schematic products.

Li et al. (2013) used 1.25 M citric acid and 1.5 M malic acid to leach more than 90% of Co and nearly 100% of Li. However, when the aspartic acid concentration was raised from 0.5 to 1.5 M, almost 60% of Co and Li were leached. When the concentration of H_2O_2 approached 2.0 vol.%, more than 90% of Co and 99% of Li were leached from citric and malic acids, respectively. The increase in the leaching efficiency of Co and Li was negligible as H_2O_2 was further raised to 6 vol.%. When H_2O_2 was increased to 4.0 vol.%, Co and Li's aspartic acid leaching efficiency increased to 60%, demonstrating the critical role that H_2O_2 plays. At the optimum S/L ratio of 20 g/L, 91% and 93% Co were leached with citric and malic acids, respectively. Aspartic acid leached only 36% Co and Li at this S/L ratio. CA can be produced from corn or potato farming. Corns or potatoes are wet milled to produce starch, after fermentation and separation CA can be produced.

Chen et al. (2016a) were firstly dissolved S-LIB CAMs with citric acid and D-glucose as leachant and reductant, respectively. About 99%, 91%, 92% and 94% Li, Ni, Co, and Mn could be leached under the following optimized conditions: retention time of 2 hours, leaching temperature of 80°C, CA concentration of 1.5 mol/L, pulp

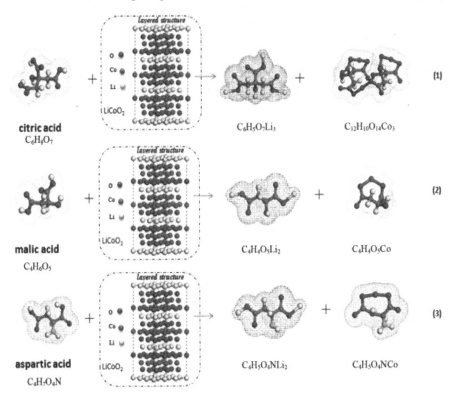

FIGURE 6.4 Possible schematic products of leaching reaction of spent $LiCoO_2$ with citric (1), DL-malic acid (2), and L-aspartic acid (3).

density of 20 g/L and reductant dose of 0.5 g/g. The recovery of high-value-added metals was then accomplished through the selective precipitation method. It was also discovered that the residual leachate after metals recovery can be re-utilized as a leaching reagent with potentially excellent performance as a fresh leachate. Additionally, the leaching and precipitation mechanism was also tentatively investigated in terms of glucose oxidation pathway and materials recovery. Finally, atom utilization efficiency was calculated and the atom utilization efficiency can achieve as high as 98% for the whole recovery process. Both experimental and theoretical results obtained can support a sustainable and desirable process for the comprehensive recovery of metal values from S-LIBs in a closed-loop manner.

After a predetermined leaching time of 30 minutes, the glucose was introduced to the reactor. The leaching response is shown in Eq. 6.14:

$$18LiNi_{1/3}Co_{1/3}Mn_{1/3}O_2 + 18H_3Cit + C_6H_{12}O_6 \rightarrow 6Li_3Cit$$
$$+ 2Ni_3(Cit)_2 + 2Co_3(Cit)_2 + 2Mn_3(Cit)_2 + 33H_2O + 6CO_2 \qquad (6.14)$$

where it is assumed that all D-glucose was completely oxidized and decomposed into CO_2 and H_2O.

In both strong and mild oxidation conditions, D-glucose can be oxidized via two distinct routes (Pagnanelli et al., 2014). The carbon chain of D-glucose can be shortened to aldaric acid with a smaller carbon chain by the synthesis of mono-carboxylic poly-hydroxy acids during the reductive leaching of MnO_2 ores, which is another reason why it has been widely employed as a low-cost and ecologically friendly reductant. To gain a fundamental understanding of the reductive leaching process, the oxidation mechanism was tentatively investigated. According to Chen et al. (2016a), Figure 6.5 shows a potential glucose oxidation pathway during a leaching reaction.

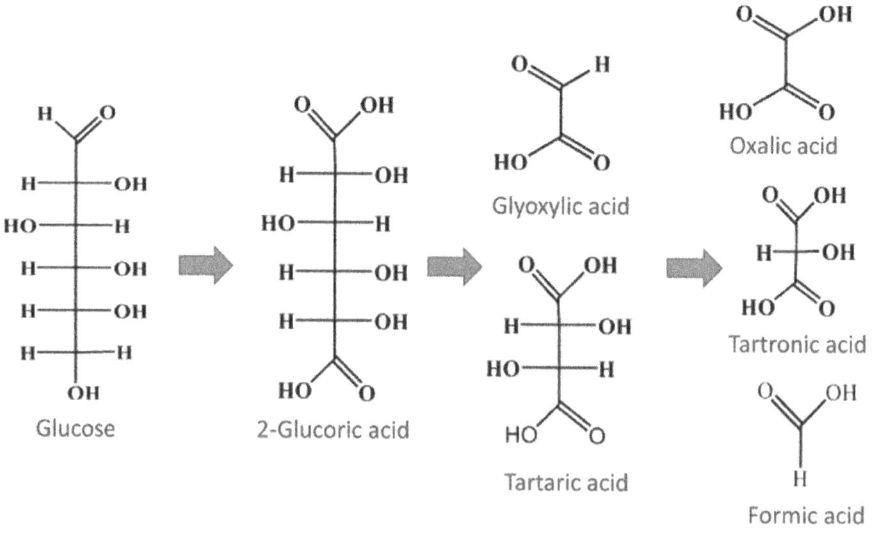

FIGURE 6.5 A probable glucose oxidation route during the leaching reaction.

It may be concluded that the D-glucose utilized in this work is capable of being successively oxidized and degraded into 2-glucoric, glyoxylic, tartaric, tartronic, oxalic, and formic acids in the presence of powerful oxidants (Co^{3+} and Mn^{4+}). Then formic acid may oxidize into CO_2 and water. The results of the research shown above show that all of the decomposed intermediate products are environmentally friendly acids, gases (CO_2), or liquids (H_2O), suggesting that glucose could be utilized as a green reductant with great reducibility.

Following purification and leaching, Ni, Co, and Li ions were successively precipitated by oxalic acid (0.5 M $H_2C_2O_4$), phosphoric acid (0.5 M H_3PO_4), and dimethylglyoxime reagent (0.2 M DMG, $C_4H_8N_2O_2$). Nickel dimethylglyoxime chelate/ $Ni(C_4H_6N_2O_2)_2$, $CoC_2O_4 2H_2O$, and Li_3PO_4 were used to selectively recover Ni, Co, and Li ions as various precipitates. The greatest recovery rates for Ni, Co, and Li are 98.5%, 96.8%, and 92.7%, respectively, and the majority of the Ni, Co, and Li ions could be precipitated within 30 minutes (Chen et al., 2016a). According to Eqs. 6.15– 6.17, all metal values in the current investigation can be selectively recovered using the precipitation reactions listed below:

$$6C_4H_8N_2O_2 + Ni_3(Cit)_2 \rightarrow 3Ni(C_3H_6N_2O_2)\downarrow + 3H_3Cit \qquad (6.15)$$

$$(C4H_6N_2O_2)_2\downarrow \rightarrow Ni^{2+} + 2C_4H_6N_2O^{2-} \; pK_{sp}: 23.4$$

$$Co_3(Cit)_2 + 3H_2C_2O_4 \rightarrow 3CoC_2O_4\downarrow + 2H_3Cit \qquad (6.16)$$

$$CoC_2O_4\downarrow \rightarrow CO_2^+ + C_2O_4^{2-} \; pK_{sp}: 7.2$$

$$Li_3Cit + H_3PO_4 \rightarrow Li_3PO_4\downarrow + H_3Cit$$

$$Li_3PO_4\downarrow \rightarrow 3Li^+ + PO_4^{3-} \; pK_{sp}: 3.4 \qquad (6.17)$$

To recover Co and Li from S-LIBs, Li et al. (2010a) investigated a hydrometallurgical technique based on leaching. As leaching agents, $C_6H_8O_8.2H_2O$ and H_2O_2 were utilized to create the corresponding citrates. Scanning electron microscopy (SEM) and X-ray diffraction (XRD) are used to characterize the leachate. The method that was suggested involved a chemical leaching process and mechanical separation of metal-containing particles. With 1.25 M citric acid, 1.0 vol.% H_2O_2, an S:L of 20 g/L, and agitation at 300 rpm in a batch reactor, the best conditions for recovering more than 90% Co and nearly 100% Li were achieved experimentally. This led to a highly effective recovery of the metals within 30 minutes of the processing time at 90°C. The recovery of valuable metals from used LIBs is proven to be a straightforward, environmentally responsible, and adequate hydrometallurgical method.

For the recovery of Co and Li from S-LIBs, Fan et al. (2016) employed a hydrometallurgical and environmentally benign technique. Pretreatment, CA leaching, selective chemical precipitation, and circulatory leaching are all steps in this procedure.

Cu and Al foils are recycled directly after pretreatment (manual disassembly, NMP immersion, and calcination), and the CAMs are successfully removed from the cathode. Pretreatment circumstances included:

NMP immersion: 100 mL NMP, 90°C, 90 minutes under mixing,

Drying: 2 hours, 80°C in an oven, and

Calcination: 750°C, 2.5 hours to remove PDVF and C.

The resulting CAMs (waste $LiCoO_2$) were then initially leached with a solution of 1.25 M citric acid and 1 vol.% H_2O_2. Then, using $H_2C_2O_4$ and a 1:1.05 molar ratio of Co^{2+}, Co was precipitated. Following filtration, the filtrate—which contained Li^+ and H_2O_2—was used as a leaching agent, and the ideal circumstances were carefully examined. Under the following conditions: leaching temperature of 90°C, 0.9 vol.% H_2O_2, and S/L ratio of 60 ml/g for 35 minutes, the leaching efficiency can reach as high as 98% for Li and 90.2% for Co, respectively. More than 90% Li and 80% Co may be leached under the same leaching conditions after three rounds of circulatory leaching. With this hydrometallurgical method, Li and Co may be efficiently recovered, and waste liquor can be used again, potentially providing advantages for both the economy and the environment.

To recover Co and Li from CAMs of S-LIBs, Li et al. (2013) evaluated three distinct organic acids, namely CA, malic acid, and aspartic acid, in the presence of hydrogen peroxide. More than 90% of Co and Li were recovered through leaching with citric and malic acids, although aspartic acid recovery was much lower. The active component ($LiCoO_2$) is probably first dissolved in the presence of H_2O_2 before Co^{2+} and Li are chelated with citrate, malate, or aspartate. The recovery of Co from S-LIBs may need less energy and emit fewer greenhouse gases than the production of virgin cobalt oxide, according to an environmental analysis of the process.

6.2.2 OXALIC ACID LEACH

The conjugate base of oxalic acid ($H_2C_2O_4$), the most basic dicarboxylic acid, is the oxalate anion ($C_2O_4^{2}$). The dihydrate version of oxalic acid (OA) is offered for sale in the marketplace. In 1776, Scheele produced OA for the first time by oxidizing sugar with HNO_3. Wohle first produced OA in 1824 by hydrolyzing cyanogen. OA can exist in solution as a number of distinct species, including $H_2C_2O_4$, HC_2O_4, and $C_2O_4^{2}$, depending on the pH. Figure 6.6 depicts the oxalate anion's structure. Below pH 1.23, H_2C_2O is the main species, whereas at pH 4.19 or higher, $C_2O_4^{2}$ is the dominant species.

FIGURE 6.6 The structure of oxalate ion.

The mole balance around the pK$_a$ values of OA was used to produce the OA specia- tion curve displayed in Figure 6.7. A bidentate anionic ligand that can give two pairs of electrons to a metal ion is the oxalate ion (IUPAC: ethane dioate ion). Due to its capacity to connect to a metal cation in two locations, this ligand is also known as a chelate. The study of oxalates as a precipitant, chelating agent, and reducing agent has attracted a lot of attention. The application of oxalates as leaching agents is expanded by the possibility of creating other oxalate complexes (such as $Fe(C_2O_4)_2^{2-}$) with excess oxalate. The majority of simple oxalate compounds, like $FeC_2O_4 2H_2O$, are insoluble in water. In contrast to a compound, which is made up of two or more elements joined together by an ionic or covalent connection, a metal complexion con- tains a metal centre with a number of ligands linked by coordinate bonds, generating a net cationic or anionic charge. Oxalate has long been employed in rare earth extrac- tion procedures as a precipitating agent. According to Wikipedia.org, oxalate/OA can also be used as a moderate reducing agent in the following oxidation reaction:

$$C_2O_4^{2-} \leftrightarrow 2CO_2 + 2e^- \quad E^0 = 0.49 \text{ V} \quad (6.18)$$

This reaction's thermodynamic viability is demonstrated by its positive potential, which also raises the prospect that it might decrease a metal ion. According to oxa- late, Co^{3+} and Fe^{3+} are reduced to Co^{2+} and Fe^{2+}, respectively (Verma et al., 2019). The leaching efficiency may be impacted by this reduction feature, especially if the metal's solubility changes as it is reduced. The majority of metals either produce sim- ple oxalate compounds, complexes of oxalate, or both. Depending on the interactions between the metal and ligand, simple oxalate compounds can either be water-solu- ble (like $Li_2C_2O_4$) or insoluble (like $FeC_2O_4 2H_2O$), whereas all oxalate complexes are water-soluble. The soluble and insoluble metal oxalate complexes are shown in Table 6.4. The development of oxalate crystals is caused by an increase in the solubil- ity of simple oxalate compounds when there is an excess of oxalate.

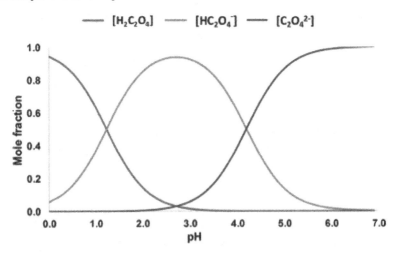

FIGURE 6.7 Oxalic acid speciation as a function of pH (pK = 1.23 and pK$_{a2}$ = 4.19) at room temperature.

TABLE 6.4

Solubility and Insolubility of Metal Oxalate Compounds

Insoluable (<0.1 g/mL at 25°C)	Soluable
$Al_2(C_2O_4).H_2O\downarrow$	$Fe_2(C_2O_4)_3$
$CoC_2O_4.2H_2O\downarrow$ ($2.69*10^{-10}$ g/100 g H_2O	$Li_2C_2O_4$ (8 g/100 g
$CuC_2O_4.0.5H_2O\downarrow$ ($2.16*10^{-10}$ g/100 g H_2O)	H_2O)
$FeC_2O_4.2H_2O\downarrow$	$Na_2C_2O_4$
$NiC_2O_4.2H_2O\downarrow$ ($3.98*10^{-10}$ g/100 g H_2O	$K_2C_2O_4$

Among organic acids, OA is a relatively potent acid. The equilibrium constant (K_a) for the loss of the initial H^+ at ambient temperature is $5.37*10^2$ (pK$_a$ = 1.27). The yields of the oxalate ion with an equilibrium constant of $5.25*10^5$ (pK$_a$ = 4.28) are caused by the loss of the second H^+. The following equations can be used to describe the OA dissociation reaction:

$$C_2O_4H_2 \rightarrow C_2O_4H^- + H^+ \tag{6.19}$$

$$C_2O_4H^- \rightarrow C_2O_4^{2-} + H^+ \tag{6.20}$$

To further enhance the separation and process economics, the S/L ratio can be modified based on the OA content. According to the following reaction, theoretically, a stoichiometric molar ratio of two between OA and lithium cobalt oxide should be sufficient to enable full recovery and precipitation of Li and Co, respectively:

$$4H_2C_2O_{2(aq)} + 2LiCoO_{2(s)} \leftrightarrow Li_2C_2O_{2(aq)} + 2CoC_2O_4.2H_2O_{(s)} + 2CO_{2(aq)} \tag{6.21}$$

Sohn et al. (2006) examined two acidic leaching processes to choose the most efficient recycling method for S-LIBs (derived from $LiCoO_2$). They specifically created two varieties of acidic leaching for powders that have been crushed. One of these involves the reducing agent H_2O_2 and the acid H_2SO_4 leaching. At the conditions of 2.0 M H_2SO_4, 10 vol.% H_2O_2, 75°C, 300 rpm mixing speed, 50 g/L S/L ratio, and 75 minutes reaction time, the leaching rates of Co, Li, and the other metals were above 99%. The leaching of OA is the additional leaching procedure. At the conditions of 3.0 M oxalic acid, 80°C reaction temperature, 300 rpm mixing speed, 50 g/L initial S/L ratio, and 90 minutes extraction time, more than 99% of Li and less than 1% of Co were dissolved in this method. Figure 6.8 shows the dissolution behavior of Li and Co with H_2SO_4, HNO_3, $H_2SO_{4+}H_2O_2$, and OA. There is a very good selectivity between Li and Co with OA leaching. Inorganic acid selectivities between Li and Co are very low.

Each method has benefits and drawbacks. Leaching reagent is relatively inexpensive in H_2SO_4 leaching, and Co can be recovered into $Co(OH)_2$. OA is more expensive than H_2SO_4 acid, on the other hand, but Li may be dissolved selectively. Additionally, Co might be converted into cobalt oxalate and, after being heated, transformed into Co_3O_4. The recovery rate and purity of cobalt hydroxide and cobalt oxalate were

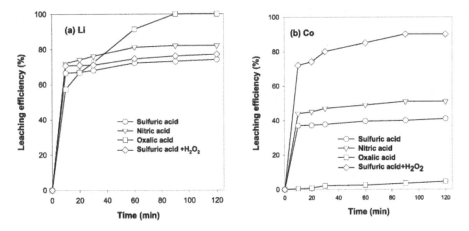

FIGURE 6.8   Dissolution of $LiCoO_2$ with different inorganic and organic acids.

compared to choose the most efficient recycling method. It was also looked into which method was more cost-effective and environmentally beneficial.

$LiCoO_2$ can be dissolved by $C_2O_4H_2$ (Song et al., 2006):

$$LiCoO_2 + 3(C_2O_4H_2) \rightarrow H_3Co(C_2O_4)_3 + LiOH + H_2O \qquad (6.22)$$

$$H_3Co(C_2O_4)_3 + 2H_2O \rightarrow CoC_2O_4.2H_2O + CO_2 + 1.5H_2C_2O_4 \qquad (6.23)$$

From Eqs. 6.22 and 6.23, selective recovery of Co from LCO in the form of CoC_2O_4 is through OA extraction. After Co was precipitated Li-ion in the PLS is converted into Li_2CO_3 by using Na_2CO_3.

Thermal phase change of $CoC_2O_4.2H_2O$ is observed at 196°C, 290°C, and 896°C, respectively. Firstly, the dehydration of precipitate, the decomposition of oxalate, and the phase change of cobalt oxide. The reactions are given below:

$$CoC_2O_4.2H_2O \rightarrow CoC_2C_{4(s)} + 2H_2O_{(g)} \qquad (6.24)$$

$$3CoC_2O_{4(s)} \rightarrow Co_3O_{4(s)} + 4CO_{(g)} + 2\,CO_{2(g)} \qquad (6.25)$$

$$Co_3O_{4(s)} \rightarrow 3CoO_{(s)} + 1/2O_2 \qquad (6.26)$$

Zeng et al. (2015a) used OA for S-LIB ($LiCoO_2$) leaching. The optimal parameters for the leaching process was 1.0 M oxalic acid leaching at 95°C for 150 minutes retention time at a 15 g/L S/L ratio. 98% Li and 97% Co recoveries were achieved.

The leaching reactions using oxalate as leachate and precipitant for $LiCoO_2$ may be represented as follows (Sun and Qiu, 2012):

$$7H_2C_2O_4 + 2LiCoO_{2(s)} \leftrightarrow 2LiHC_2O_4 + 2Co(HC_2O_4) + H_2O + 2CO_{2(g)} \qquad (6.27)$$

$$4H_2C_2O_4 + 2LiCoO_{2(s)} \leftrightarrow Li_2C_2O_4 + 2CoC_2O_{4(s)} + 4H_2O + 2CO_{2(g)} \qquad (6.28)$$

The Co^{3+} can be converted to Co^{2+} by the evolved $CO_{2(g)}$ from oxalate during the oxalate leaching process, which aids in the dissolution. According to Ferreira et al. (2009), Lee and Rhee (2003), and Swain et al. (2007), H_2O_2 is frequently used during reaction processes to increase the leaching efficiency of Co, which can convert Co^{3+} to Co^{2+} and so facilitate the forward reaction. Since the conversion of Co^{3+} to Co^{2+} is accelerated by the addition of H_2O_2 during oxalate leaching, the following Eqs. (29) and (30) illustrate the potential reactions:

$$3H_2C_2O_4 + LiCoO_{2(s)} + 1.5H_2O_2 \leftrightarrow LiHC_2O_4 + Co(HC_2O_4)_2 + 3H_2O + O_{(g)} \qquad (6.29)$$

$$3H_2C_2O_4 + 2LiCoO_{(s)} + H_2O_2 \leftrightarrow Li_2C_2O_4 + 2CoC_2O_{4(s)} + 4H_2O + O_{2(g)} \qquad (6.30)$$

Due to the absence of redox reactions, the leaching reactions of CoO are significantly simpler than those of $LiCoO_2$ and can be illustrated as follows:

$$2H_2C_2O_4 + CoO_{(s)} \leftrightarrow Co(HC_2O_4)_2 + H_2O \qquad (6.31)$$

$$H_2C_2O + CoO_{(s)} \leftrightarrow CoC_2O_{4(s)} + H_2O \qquad (6.32)$$

The majority of Li could be leached into the solution, whereas Co was left in the residue after filtration because Li and Al's oxalates are substantially more soluble than cobalt oxalate. The residue's components were CoC_2O_4 produced during reactions, unreacted CoO_2, unreacted CoO, and carbon that did not react with oxalate. According to Sun and Qiu (2012), CoC_2O_4 is the most often utilized raw material to manufacture various metal Co powders and cobalt oxides under various heat treatment procedures.

Different types of organic and inorganic acids (2.0 M citric ($C_6H_8O_7$), 1.0 M oxalic ($C_2H_2O_4$), 2.0 M H_2SO_4, 4.0 M HCl, and 1.0 M HNO_3 acid)) and reducing agents (hydrogen peroxide (H_2O_2), glucose ($C_6H_{12}O_6$), and ascorbic acid ($C_6H_8O_6$)) for the recovery of valuable metals from S-LIBs were tested by Aaltonen et al. (2017). The average amounts of Co, Li, and Ni, Cu, Mn, Al, and Fe in the crushed and sieved material were 23% (w/w), 3% (w/w), 1–5% (w/w), respectively. At 25°C with a slurry density of 5% (w/v), the results showed that mineral acids (4.0 M HCl and 2.0 M H_2SO_4 with 1% (v/v) H_2O_2) produced typically better yields than organic acids. Li, Co, and Ni nearly completely dissolved. When utilizing $C_6H_8O_6$ as a reducing agent (10% g/g_{scraps}) at 80°C, additional leaching studies using H_2SO_4 medium and various reducing agents with a slurry density of 10% (w/v) reveal that virtually all of the Co and Li may be leached out in sulfuric acid (2.0 M). As the later divalent metals are known to precipitate as oxalates, it was also demonstrated that $C_2H_2O_4$ was the most selective leaching medium between Li and Ni/Co. Ascorbic acid $C_6H_8O_6$, D-glucose, and H_2O_2 were shown to be the most effective reduction agents in LIBs H_2SO_4 acid (2.0 M) leaching, with the parameter range evaluated ($C_6H_8O_6 = 0$–12% g/g$_{scraps}$, D-glucose $= 0$–16% g/g$_{scraps}$, and $H_2O_2 = 0$–5% (v/v)). By utilizing 10% (g/g$_{scraps}$) $C_6H_8O_6$ as a reducing agent, the greatest metal extraction into H_2SO_4 was accomplished. Based on solution analysis, the industrially crushed LIBs' heterogeneous nature led to the highest metal yields of 100.7% for Co, 95.1% for Li, and 105.9% for Ni. The remaining levels of Co, Li, and Ni in the leaching residue were thus also examined, and it was discovered that they were, respectively, as low as 0.11%, 0.03%, and 0.17%. The depletion of natural resources and potential pollution from used batteries make the recycling of valuable metals from S-LIBs more crucial.

Zhang et al. (2018d) studied how OA leaching and calcination, two straightforward and innovative processes, can regenerate LIBs' used NCM cathodes. A thorough inquiry was done to learn more about the regeneration process initially. In contrast to conventional acid leaching, OA leaching causes the transition metals in NCM cathodes to form precipitates, and the extent of the transformation may be managed by the leaching period. The layer-structured NCM cathodes are regenerable after Li_2CO_3 calcination, and those with a quick leaching time of 10–30 minutes perform better electrochemically. The highest initial specific discharge capacity of 168 mAh/g at 0.2 C is delivered by the regenerated NCM cathode with a 10-min leaching time, and this capacity is maintained at 153.7 mAh/g after 150 cycles with a high capacity retention of 91.5%. After 300 cycles of cycling at 1 C and an efficiency of 86.7%, a high reversible discharge capacity of 137.3 mAh/g may be produced. The submicron particles and voids produced during the regeneration process, along with the retention of the optimal elemental proportions, are responsible for the remarkable electrochemical performances. This novel regeneration method has been successfully applied to recycle spent NCM cathodes and it is believed that it has great potential to regenerate other similar spent cathodes as well in the future. The simple and novel feature provides a new perspective on recycling spent cathodes of LIBs.

Utilizing OA's minimal natural effects, Li et al. (2018b) used it as a leaching reagent to recover Li as a resource and remove phosphorus from $LiFePO_4$ batteries. Energy-dispersive X-ray spectroscopy (EDS) with XRD and SEM are used to determine the physical characteristics of spent cathode materials (before leaching) and residues (after leaching). ICP-AES analyzes the total quantities of Li and Fe. OA concentration of 0.3 M, temperature of 80°C, reaction time of 60 minutes, and S/L ratio of 60 g/L are the required parameters for the procedure to take place. $LiFePO_4$ can be efficiently precipitated into $FeC_2O_4 2H_2O$ at a rate of 92%, while Li can be leached at a rate of up to 98% This procedure provides a novel, low-cost, and environmentally friendly technique for disposing of used LFP batteries.

The leaching reaction mechanism of waste $LiFePO_4$ with $H_2C_2O_4 \cdot 2H_2O$ solution is proposed as follows:

$$12LiFePO_{4(s)} + 6H_2C_2O_4 \cdot 2H_2O_{(aq)} \rightarrow 3Li_2C_2O_{4(aq)} + 3FeC_2O_{4(s)}$$
$$+ 4H_3PO_{4(aq)} + 3Fe(PO_4)_{2(aq)} + 2Li_3PO_{4(aq)} + 12H_2O_{(l)} \qquad (6.33)$$

6.2.3 ASCORBIC ACID ($C_6H_8O_6$) LEACHING

The electrons in the double bond, hydroxyl group lone pair, and carbonyl double bond form a conjugated system, which causes ascorbic acid (AA) to act as a vinylogous carboxylic acid. The hydroxyl group in AA is substantially more acidic than usual hydroxyl groups because the two primary resonance structures stabilize the deprotonated conjugate base. In other words, AA can be viewed as an enolin whose stabilized enolate form is the deprotonated form. The pKa values for AA are 4.10 for pK_{a1} and 11.6 for pK_{a2}. Additionally, AA is a moderate reducer that can be double oxidized to generate the stable form of dehydroascorbic acid ($C_6H_6O_6$) or oxidized by one electron to a radical state.

The waste $LiCoO_2$ was first dissolved with AA to create a soluble $C_6H_6O_6Li_2$ during the leaching process, and the AA further reduced the Co^{3+} in the $LiCoO_2$ to a soluble Co^{2+}. The AA was also converted into dehydroascorbic acid ($C_6H_6O_6$) at the same time. The leaching products containing Co^{2+} have several theoretically feasible configurations, which are shown in Figure 6.9. However, a quick calculation of the thermodynamics revealed that only product (b), namely $C_6H_6O_6Co$, is thermodynamically advantageous during the leaching. According to Li et al. (2012), the AA leaching reaction can be shown as follows:

$$4C_6H_8O_6 + 2LiCoO_2 \leftrightarrow C_6H_6O_6 + C_6H_6O_6Li_2 + 2C_6H_6O6Co + 4H_2O \quad (6.34)$$

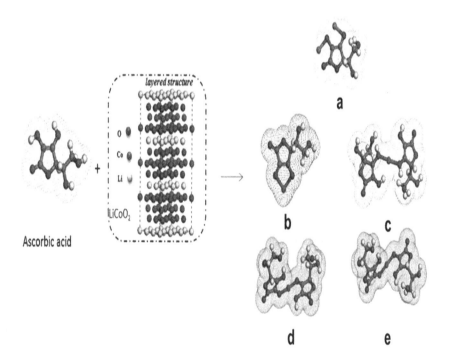

FIGURE 6.9 Possible schematic products of the leaching reaction of waste $LiCoO_2$ with ascorbic acid.

Li et al. (2012) investigated a novel method that combined ultrasonic washing, calcination, and organic acid leaching to extract Co and Li from CAMs (including $LiCoO_2$ and Al). After being physically separated from the cathode, the anode materials can also be used to recover Cu. To increase the Co recovery efficiency, AA is used as a reducing agent as well as a leaching reagent. With a 1.25 M AA solution, leaching temperature of 70°C, leaching period of 20 minutes, and S/L ratio of 25 g/L, leaching efficiencies as high as 94.8% for Co and 98.5% for Li are attained. On the basis of the structure of AA, the mechanism of the acid leaching reaction has been preliminary examined. This technique was proven to be an effective way to reuse priceless components from S-LIBs and it may be scaled up for commercial use.

6.2.4 MALIC ACID LEACH

Co and Li were recovered from the CAMs of S-LIBs using a leaching procedure that was environmentally friendly, according to Li et al. (2010). As a leaching agent, DL-malic acid ($C_4H_5O_6$), an easily broken-down organic acid, was employed. XRD and SEM were used to describe the structural and morphology of the CAMs both before and after leaching. Atomic absorption spectrophotometry (AAS) was used to calculate the concentrations of Co and Li in the leachate. By adjusting the conditions for achieving a recovery of more than 90 wt.% Co and nearly 100 wt.% Li was determined experimentally by varying the concentrations of leachant, time and temperature of the reaction as well as the initial solid-to-liquid (S/L) ratio. They discovered that H_2O_2 in a DL-malic acid solution is an effective reducant because it enhances the leaching efficiency. Leaching with 1.5 M DL-malic acid, 2.0 vol.% H_2O_2 and a S:L of 20 g/L in a batch extractor results in a highly efficient recovery of the metals within 40 minutes at 90°C.

Nayaka et al. (2016) researched new organic acid mixtures to recover the valuable metal ions from the CAM of S-LIBs. The $LiCoO_2$ CAMs collected from LIBs are dissolved in mild organic acids, iminodiacetic acid (IDA), and maleic acid (MA), to recover the metals. Almost complete dissolution occurred in slightly excess (than the stoichiometric requirement) of IDA or MA at 80°C for 6 hours, based on the Co and Li released. The reducing agent, ascorbic acid (AA), changes the dissolved Co^{3+}- to Co^{2+}-L (L = IDA or MA) allowing for the selective recovery of Co as Co^{2+}-oxalate. The UV-Vis spectra of the dissolved solution as a function of dissolution time show that Co^{3+}- and Co^{2+}-Li are formed. Thus, the proposed reductive-complexing dissolution mechanism. Unlike mineral acids, these mild organic acids are not harmful to the environment.

6.2.5 SUCCINIC ACID LEACH

Succinic acid (SA) is an organic weak acid that occurs naturally. Its $C_4H_6O_4$ molecule has two carboxyls. SA has pK_a values of 4.102 and 5.408. With the help of H_2O_2, the cathodic material's cobalt content is converted from Co^{3+} to Co^{2+}, making it more stable and soluble in the acid-aqueous solution. This is the beginning of the leaching process. The used $LiCoO_2$ is then dissolved in SA, allowing the H^+ ion to extract the Li and Co ions, potentially producing three different products: $C_4H_4O_4Li_2$,

$C_4H_4O_4Co$, and $C_8H_{10}O_8Co$. Figure 6.10 shows the acid-leaching reaction using SA as the leaching agent. Theoretically, Figure 6.10 lists $C_4H_4O_4Co$ and $C_8H_{10}O_8Co$ as two potential leaching products containing Co^{2+}. However, a quick calculation of the thermodynamics revealed that only the product $C_4H_4O_4Co$ is thermodynamically advantageous during the leaching. Notable is the closed-loop structure of $C_4H_4O_4Co$, which is symmetrical of the atomic species on both sides of the Co atom. This structure may be more stable and simpler to create, which would increase Co's leaching efficiency relative to Li's.

To recover Li and Co from the CAMs in S-LIBs, Li et al. (2015) also developed a hydrometallurgical technique involving natural organic acid leaching. H_2O_2 is used as a reductant and SA is used as a leaching agent. Inductively coupled plasma-optical emission spectroscopy (ICP) is used to determine the Co and Li contents from the succinic acid-based treatment of S-LIBs to determine the leaching efficiency. By using XRD and SEM, the spent $LiCoO_2$ samples after calcination and the residues after leaching are characterized. The findings demonstrate that under the ideal conditions of SA concentration of 1.5 M, H_2O_2 content of 4 vol.%, S/L ratio of 15 g/L, temperature of 70°C, and reaction time of 40 minutes, nearly 100% of Co and more than 96% of Li are leached. Results for fitting the experimental data to acid-leaching kinetic models are also provided.

6.2.6 FORMIC ACID LEACH

Formic acid's acid ionization equation is described by

$$HCOOH_{(aq)} \rightarrow HCOO_{(aq)} + H_{(aq)} \text{ (pK}_a\text{: 3.77, 25°C, H}_2\text{O)} \tag{6.35}$$

Formic acid is not only a weak acid, but due to its aldehyde, it is also reductive to metal ions with high oxidative potentials. While most of the Ni, Co, and Mn can be precipitated out as hydroxides, formic acid, unlike other acids, can electively leach

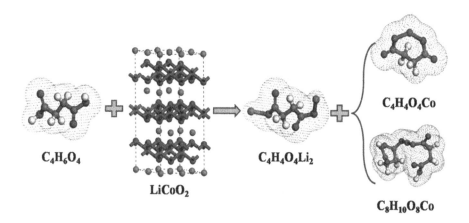

FIGURE 6.10 Leaching reaction and possible reaction products in the succinic acid leaching process.

Li with a lower leaching rate of the Al to obtain high-purity Al foil. According to Gao et al. (2017), the chemical reactions that occur during the leaching process are as follows:

$$2Al_{(s)} + 6HCOOH_{(aq)} \rightarrow 2CHAlO_{(aq)} + 3H_{(g)} \tag{6.36}$$

$$6LiNi_{1/3}Co_{1/3}Mn_{1/3}O_{2(s)} + 18HCOOH_{(aq)} \rightarrow 2C_2H_2NiO_{4(aq)} + 2C_2H_2CoO_{4(aq)}$$
$$+ 2C_2H_2MnO_{4(aq)} + 6CHLiO_{2(aq)} + 3CO_{2(g)} + 12H_2O_{(aq)} \tag{6.37}$$

with H_2O_2:

$$6LiNi_{1/3}Co_{1/3}Mn_{1/3}O_{2(s)} + 21HCOOH_{(aq)} + 3H_2O_2 \rightarrow 2C_2H_2NiO_{4(aq)}$$
$$+ 2C_2H_2CoO_{4(aq)} + 2C_2H_2MnO_{4(aq)} + 6CHLiO_{2(aq)} + 3CO_{2(g)} + 12H_2O_{(aq)} \tag{6.38}$$

Gao et al. (2017) achieved 99.93% Li using 2.0 M formic acid + 30 vol.% H_2O_2 at 60°C, 50 g/L solid ratio and 120 minutes leaching time.

6.2.7 LACTIC ACID LEACH

Li et al. (2017a) resynthesized $LiNi_{1/3}Co_{1/3}Mn_{1/3}O_2$ from the leachate using a sol-gel technique. The leaching and chelating agent of choice is lactic acid (LA). By using ICP-OES to measure the amounts of metal elements like Li, Ni, Co, and Mn in the leachate, the leaching efficiency is investigated. XRD and SEM are used to examine the used CAMs for the pretreatment procedure as well as the newly generated and newly synthesized materials. According to the findings, Li, Ni, Co, and Mn had respective leaching efficiencies of 97.7%, 98.2%, 98.9%, and 98.4%. The ideal conditions are 1.5 M lactic acid concentration, 20 g/L S/L ratio, 70°C leaching temperature, 0.5 vol.% H_2O_2 content, and a 20-minute reaction time. The leaching kinetics of cathode scrap in LA fit well with the Avrami equation. Electrochemical analysis indicates that the regenerated $LiNi_{1/3}Co_{1/3}Mn_{1/3}O_2$ cathode materials deliver a highly reversible discharge capacity, 138.2 mAh/g, at 0.5 C after 100 cycles, with a capacity retention of 96%, comparable to those of freshly synthesized $LiNi_{1/3}Co_{1/3}Mn_{1/3}O_2$ cathodes.

The plastic and steel cases of the S-LIBs were manually disassembled in order to remove and recycle them. The cathode foils were then divided into small pieces that measured 1 cm by 1 cm after the anodes and cathodes had been separated. After the used carbon materials have been removed, the Cu foil in the anode can be recycled. The cathode foils were submerged in a simple stirring NaOH solution at room temperature. One way to describe the procedure is as follows:

$$2Al + 2NaOH + 2H_2O \rightarrow 2NaAlO_2 + 3H_2 \tag{6.39}$$

The CAM and Al foil were successfully separated after being cleaned with distilled water and going through filtration. The filtrate's Al ions were recycled and used again in the subsequent experiment, and the residue was dried at 60°C for 24 hours. By calcining the cathode material at 610°C for 5 hours in a muffle furnace and then cooling to room temperature, impurities like carbon and PVDF binder were burned off. A system that consists of a cooler, a condensation chamber, activated carbon filters, and bag filters can be used to collect and purify the waste gas. After roasting and cooling, the powder was ground for 30 minutes in a planetary ball mill to create smaller particles with more surface area, which would speed up the dissolution process and leaching efficiency.

A thermostatic 100 mL Pyrex reactor with three necks and a round bottom was used for all of the metal-leaching experiments. To keep the reaction temperature constant, the reactor was submerged in water. The reactor had a condenser pipe to prevent water loss and a stirrer to quicken the reaction. There was measured powder added to the reactor. The reactor was filled with a known concentration of LA and H_2O_2 solution, and the mixture was stirred at 300 rpm. LA concentration (0.25–2 M), S/L ratio (10–40 g/L), temperature (40°C–90°C), H_2O_2 volume percentage (0–3 vol.%), and reaction time (10–60 minutes) were examined as variables in the leaching process.

6.2.8   TCA (CCl_3COOH) + $H_2O_2/Na_2SO_3/Na_2S_2O_3$ (ORGANIC) LEACH

Zhang et al. (2015) used three types of organic acids (HAc, TFA, and TCA) and a commonly used inorganic acid (H_2SO_4) as the leachants and H_2O_2, Na_2SO_3, and $Na_2S_2O_3$ as reductant. 3.0 M organic acid TCA and 4.5 M mineral acid H_2SO_4 displayed the best leaching ability. TCA ionizes in the following way:

$$CCl_3COOH_{(aq)} \rightarrow CCl_3COO^-_{(aq)} + H^+_{(aq)} \tag{6.40}$$

The reaction between TCA solution and $LiNi_{1/3}Co_{1/3}Mn_{1/3}O_2$ in the presence of H_2O_2 may be presented as follows:

$$3LiNi_{1/3}Co_{1/3}Mn_{1/3}O_2 + 9CCl_3COO^-_{(aq)} + 9H^+_{(aq)} + H_2O_{2(aq)} \rightarrow Ni(CCl_3COO)_{2(aq)}$$
$$+ Co(CCl_3COO)_{2(aq)} + Mn(CCl_3COO)_{2(aq)} + 3Li(CCl_3COO)_{(aq)}$$
$$+ 5.5H_2O_{(aq)} + 1{:}25O_{2(g)} \tag{6.41}$$

When the acid concentration reached 4.5 M, H_2SO_4 demonstrated the best leaching abilities for Ni, Co, Mn, Li, and Al in the cathode scraps, with the corresponding leaching rates being 91.7%, 87.7%, 75.9%, 99.7%, and 99.6%.

The following reactions take place, and SO_2 and HSO^{3-}, which were acting as reductants, accelerated the leaching of Ni, Co, Mn, and Li.

$$2H^+_{(aq)} + SO_2^{3-}_{(aq)} \leftrightarrow H_2O_{(aq)} + SO_{2(g)} \tag{6.42}$$

$$SO_{2(g)} + H_2O_{(aq)} \leftrightarrow HSO_3^-_{(aq)} + H^+_{(aq)} \tag{6.43}$$

Similarly, the leaching rates of Ni, Co, and Mn can be increased by $Na_2S_2O_3$ reductant:

$$2H^+_{(aq)} + S_2O_3^{2-}_{(aq)} \rightarrow SO_{2(aq)} + S_{(s)} + H_2O_{(aq)} \tag{6.44}$$

4% H_2O_2 in the TCA solution was sufficient to leach metals from CAM. More Na_2SO_3 and $Na_2S_2O_2$ were needed to achieve the same results (Zhang et al., 2015).

6.2.9 TFA LEACH

Trifluoroacetic acid (TFA) was used to dissolve the organic binder PTFE and then separate the cathode material from Al foil in Zhang et al.'s (2014c) attempt to close the materials loop for recycling and resynthesizing $LiNi_{1/3}Co_{1/3}Mn_{1/3}O_2$ from the LIBs cathode scrap. However, it was discovered that a sizeable amount of aluminum could be leached while the cathode material was being separated from the aluminum foil. Before further recycling of Ni, Co, Mn, and Li from the leachate, the Al must be removed. Furthermore, if the purification and recirculation of TFA cannot be accomplished, the cost-effectiveness of scaling up this process further may become uncertain. In this study, we present a method for recycling cathode scrap that fully recycles the cathode scrap while minimizing Al loss and environmental impact and improving the materials efficiency of the recycling process. We introduce formic acid, which can act as a leachant and a reductant.

In this study, a novel process for recycling and resynthesizing $LiNi_{1/3}Co_{1/3}Mn_{1/3}O_2$ from the cathode scraps generated during manufacturing is proposed to address the recycling challenge for aqueous binder-based LIBs. The cathode material is separated from the Al foil using TFA. Systematically examined are the effects of TFA concentration, L/S ratio, reaction temperature, and time on the separation efficiencies of the CAM and Al foil. Under the ideal experimental conditions, which include 15 vol.% TFA solution, L/S ratio of 8.0 mL/g, reacting at 40°C for 180 minutes, along with suitable agitation, the CAM can be completely separated. By using a solid-state reaction technique, it is possible to successfully synthesize $LiNi_{1/3}Co_{1/3}Mn_{1/3}O_2$ from the separated cathode material. To confirm the typical properties of the resynthesized $LiNi_{1/3}Co_{1/3}Mn_{1/3}O_2$ powder, various characterizations are carried out. Electrochemical tests reveal that the resynthesized $LiNi_{1/3}Co_{1/3}Mn_{1/3}O_2$ has initial charge and discharge capacities of 201 and 155.4 mAh/g (2.8–4.5 V, 0.1 C), respectively. Even after 30 cycles, the discharge capacity holds steady at 129 mAh/g, with a capacity retention ratio of 83.01%.

6.3 BIOLEACHING

Another eco-friendly technique is bioleaching, which uses the acid that is reduced during the microorganism's metabolism to remove the material from used batteries. In general, fungi produce organic acid while bacteria typically produce inorganic acid. Due to its lower cost, environmental friendliness, and lower demand in industrial implementations, bioleaching is a desirable substitute for conventional acid leaching. However, the disadvantages of bioleaching include the lengthy culturing time (i.e., time-consuming) required to incubate microbes, the low percentage of solid usage, and the susceptibility to contamination. These shortcomings limit its applications. Dobo et al. (2023) have extensively reviewed the recycling of S-LIBs and have covered bioleaching in detail.

Moreover, high concentrations of metal ions in the leaching solution are deemed to be hazardous for cells; therefore, only the low metal concentration solution is produced by bioleaching. It was found that the leaching rates of Li and Co significantly declined from 72% to 37% and from 89% to 10%, respectively, with the rise of pulp densities (2%–4%) (Niu et al., 2014). Poor metal concentrations of leaching solutions lead to arduousness for subsequent processing.

In bioleaching, fungi and bacteria are frequently used. While bacteria secrete inorganic acids, fungi produce organic acids. In a published study (Zeng et al., 2012), Acidithio-bacillus ferrooxidans (bacteria) were used with the addition of Cu-ions (0.75 g/L concentration) to effectively leach S-LIBs. In the leaching solution after 6 days, 99.9% of the Co present in the batteries had been dissolved. Similar to this, according to Zeng et al. (2013), Acidithiobacillusferrooxidans with 0.02 g/L Ag-ions for 7 days could leach 98.4% Co from S-LIBs. In comparison to bacteria, fungi have a higher toxic tolerance, a faster rate of leaching, and a shorter lag phase. To recover heavy metals from different types of solid waste, fungus bioleaching has frequently been used (Madrigal-Arias et al., 2015; Zeng et al., 2015).

Using chemolithotrophic and acidophilic bacteria, acidithiobacillus ferrooxidans, which used elemental S and ferrous ion as the energy source to produce metabolites like H_2SO_4 acids and ferric ions in the leaching medium, Mishra et al. (2008) conducted bioleaching tests for the extraction of Co and Li from S-LIBs containing $LiCoO_2$. These metabolites assisted in the dissolution of S-LIB metals. According to research findings, acidophilic bacteria can be used to dissolve metals from CAMs of S-LIBs. These cells can develop in an environment that contains the elements S and Fe as a source of energy. Results showed that a ferrooxidan culture could produce H_2SO_4 acid to indirectly leach metals from the LIBs. Li was dissolved more slowly than Co.

Horeh et al. (2016) suggested a method for leaching S-LIBs that makes use of the fungus Aspergillus niger. During the bioleaching procedure, the used fungi could release CA, OA, gluconic acid, and malic acid (MA). CA is the acid that the leaching process mostly depends on to be effective. The optimal conditions (30°C, 16 days, and 1% (w/t) pulp density resulted in the leaching of 95% of Li, 45% of Co, 38% of Ni, 65% of Al, 70% of Mn, and 100% of Cu.

Bahaloo-Horeh and Mousavi (2017) used a bio-hydrometallurgical method to leach used LIBs, which were cultivated with Aspergillus niger. 100% Cu and Li, 77% Mn, and 75% Al were recovered with a pulp density of 2% (w/v), whereas 64% Co and 54% Ni could be recycled with a pulp density of 1% (w/v). This team added that compared to other organic acids made by Aspergillus niger, CA was more important in bioleaching. Reducing agents improve the efficiency of the leaching procedure because lower valence metals dissolve more easily. There are both inorganic and organic species as well as metallic current collectors (Cu, Al) that can be used as reductants; H_2O_2 is the most popular one. Following leaching, solvent extraction, chemical precipitation, electrolysis, and ion exchange are treatment options to separate metals or get rid of impurities. The HPLC results showed that CA played a significant role in the efficiency of bioleaching with A. niger, especially when compared to other detected organic acids (gluconic, oxalic, and MA). The battery powder was subjected to FTIR, XRD, and FE-SEM analyses before and after the bioleaching process, and the results showed that the fungi were quite effective. Additionally, compared to chemical leaching, bioleaching had a higher removal efficiency for heavy metals. This study (Horeh et al., 2016) showed the enormous potential of the bio-hydrometallurgical route to recover heavy metals from used lithium-ion mobile phone batteries. Following leaching, solvent extraction, chemical precipitation, electrolysis, and ion exchange are treatment options to separate metals or get rid of impurities. The HPLC results showed that CA played a significant role in the efficiency of bioleaching with A. niger, especially when compared to other detected organic acids (gluconic, oxalic, and MA). The battery powder was subjected to FTIR, XRD, and FE-SEM analyses before and after the bioleaching process, and the results showed that the fungi were quite effective. Additionally, compared to chemical leaching, bioleaching had a higher removal efficiency for heavy metals. This study (Horeh et al., 2016) showed the enormous potential of the bio-hydrometallurgical route to recover heavy metals from used lithium-ion mobile phone batteries.

7 Industrial Lithium-Ion Battery Recycling

ABBREVIATIONS

ABTC	American Battery Technology Company
CAM	Cathode active material
CAPEX	Capital expenditure
DMC	Dimethyl carbonate
DOE	Department of Energy
ESS	Energy storage system
EU	European Union
EV	Electrical vehicle
EW	Electrowinning
IT	Information technology
LCO	Lithium cobalt oxide
NCA	Nickel cobalt aluminum
NMC	$LiNi_{0.33}Co_{0.33}Mn_{0.33}O_2$
S-LIB	Spent lithium-ion battery
ReLIB	Recycling of lithium-ion battery
OEM	Original equipment manufacturers
PFAS	Polyfluoroalkyl substances
PVDF	Poly(vinylidene fluoride)
UHT	Ultra-high temperature
USBM	United States Geological Survey
VOC	Volatile organic compounds

7.1 INTRODUCTION

Asia, specifically China, South Korea, Japan, and other Asian nations presently dominate the battery value chain due to an increase in demand brought on by the global growth of electrical cars (EVs). The demand for essential metals will put pressure on the world's battery cell markets and supplies. Recycling will take precedence as a strategy for the European Union (EU), which has a limited domestic raw material supply and aims to achieve carbon neutrality by 2050, to lessen reliance on outside sources and reduce environmental effects. Although it may appear straightforward, recycling has a variety of difficulties. More energy is used in recycling than in primary production. With less expensive primary production, it must contend. New commercial and economic concepts are required for recycling. Even though the market for recycled spent lithium-ion batteries (S-LIB) won't fully develop until 2025–2030 when electric vehicles will substantially increase the amounts of garbage

 DOI: 10.1201/9781003384557-7

collected, research and investment decisions for processing and infrastructures must be made now.

Typical LCO and NCA lithium-ion batteries have 31% cathode active material (CAM), 22% anode, 17% Cu, 8% Al, 4% carbon black & binder, 15% electrolyte solution, and 3% separator plastics. CAM contains about 49% Ni, 9% Co, 7% Li, and 1.5% Al metals. Since EV cost decreases, LIB demand increases globally. In 2025 and 2030, the dead LIB cell amount will be expected to be 0.7 and 5.9 Mt, respectively. Targeting >90% recovery of key battery constituents is very important. LIB recycling should environmentally friendly and safe process. Reducing life cycle CO_2 emissions and eliminating hazardous waste products should be one of the main objectives. The recycling process should treat multiple battery chemistry and formats. It has the flexibility to process. Cost studies should confirm operating cost efficiencies. The recycling process should recover all battery materials and produce high-purity chemicals. The EU enforces battery producers to take responsibility for recycling waste batteries. Developing environmentally sound, safe, and economically attractive solutions is globally significant.

7.2   LIB RECYCLING

S-LIBs can be recycled in direct and indirect ways after physical treatment (Figure 7.1). After discharging and pretreatment (dismantling and size reduction), black electrode powder/mass can be produced. Indirect recycling can use either hydrometallurgical leaching processes (Figure 7.2) or pyrometallurgical treatment. Table 7.1 presents and compares the recycling steps, products, and their following steps for the electrode and battery production stages. Figure 7.3 shows the direct recycling cycle details.

In direct recycling, the battery is again shredded and the black mass—a mix of cathode and anode powders—is recovered. Powders are coated in polymers—the glue that attaches them to the metal foils. **ReCell** process uses those powders after careful purification without damaging the active powder surface—for example, removing the glue using a solvent or depolymerizing with heat.

Although in theory the maximum value product can be recovered by direct recycling, recyclers must take into account the initial necessity for recycling the battery. The active cathode material has frequently decayed, and in order to make it functional once more, additional Li must be added. You may estimate the amount of Li that is absent using spectroscopy, and additional Li can be added by combining the cathode with LiOH and heating it to 220°C. Manufacturers constantly switch to newer chemistries with higher capabilities due to the market's quick change. It may be necessary to improve recovered cathodes in order to use them again. For instance, the amount of Ni in older nickel magnesium cobalt batteries was significantly smaller than in more recent models. Increasing the Ni concentration of the recovered material is important to recover something that the market will actually want to purchase. Making the Ni uniform throughout the cathode is one challenge. Something's charge affects how quickly it diffuses; the more highly charged something is, the slower it diffuses. In this instance, researchers are working hard to homogenize 3+ cations in the solid by repeatedly moving them into it.

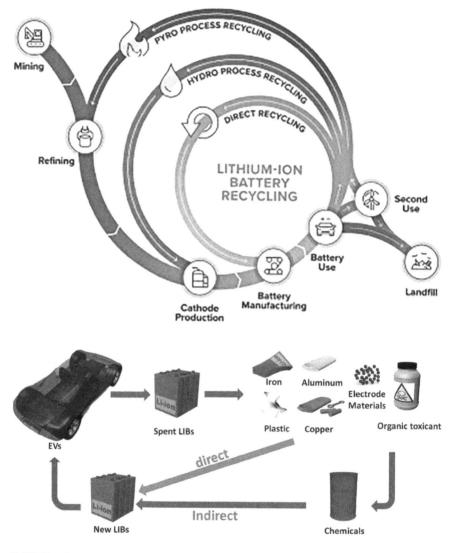

FIGURE 7.1   Direct and indirect S-LIB recycling cycles. (compiled from Gaines et al., 2021.)

However, not everyone destroys batteries. To create a purer stream of material, the Faraday Institution's Recycling of Lithium-Ion Batteries (**ReLiB**) initiative in the UK is doing what some businesses are trying to do in the far east by taking the pack and separating it into the anode, cathode, and separator. To eliminate the risk of electric shocks for technicians performing manual disassembly, ReLiB is creating an automated solution. The metal foils are separated from the active material surface. To that aim, the team at ReLiB has developed an ultrasound-based delamination technique that can separate the cathodes from the metal current collectors 100 times more quickly than using acids in just a few seconds as opposed to hours.

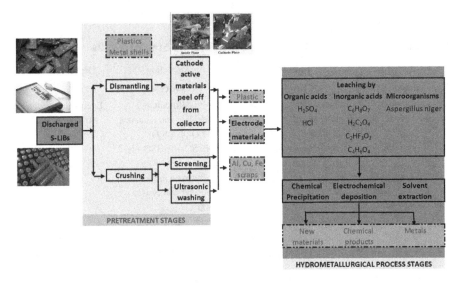

FIGURE 7.2 Diagram of typical indirect recycling S-LIBs processes. (compiled from Huang et al., 2018.)

TABLE 7.1
Comparison of the Direct and Indirect Recycling Step of S-LIBs

Physical Process	Indirect Chemical	Recycling Process	Direct Recycling
Discharge	Pyrometallurgy	Hydrometallurgy	
Pretreatment	Heat treatment	Leaching	Black mass
Sorting	Involves three stages,	Separation	purification by a
Dismantling	including pyrolysis (the	Extraction	solvent or
Crushing	breakdown of organic	Chemical precipitation	depolymerization
Screening	components), metals	Electrochemical	with heat
Magnetic separation	reduction at 1,500°C, and	precipitation	All types of cathodes
Washing	gas incineration and		are economically
Thermal treatment	quenching (to minimize		handled
	dioxins) at about 1,000°C.		Can be lithiated
Product	Ni + Co + Cu mixed alloy	Focuses on metal	Directly used as
	Li + Al + Mn is lost in slag	recovery.	cathode material
	Metal separation is difficult.	All metals can be	
		selectively extracted.	
Loss	Anode material, organic		
	materials, separators,		
	polymer binders, and		
	electrolytes are burned off.		
Product goes to	Refining + Cathode		
	production + Battery	Cathode	Battery
	manufacturing	production + Battery	manufacturing
		manufacturing	

(Continued)

TABLE 7.1 (*Continued*)

Comparison of the Direct and Indirect Recycling Step of S-LIBs

	Indirect	Recycling	Direct Recycling
Physical Process	**Chemical**	**Process**	
Extra investment costs	To handle HF, expensive gas treatment equipment are needed. High energy consumption. Energy demand: 53.8 MJ/ kg.	Co^{3+} is difficult to dissolve, and reductant (H_2O_2, $Na_2S_2O_5$, Na_2SO_3, $NaHSO_3$) is required. Co^{3+} can be reduced to Co^{2+} by pyrolysis at 600°C. A high amount of wastewater production. Purity problem for the final product. Energy demand: 41.6 MJ/kg.	Energy demand: 6 MJ/kg
Advantage		Higher metal recovery and selectivity than pyrometallurgy	Cheapest. Not energy intensive. Lower environmental impact. Most sustainable. Patented process.
Cathode chemistry and suitability	Not economical for $LiMn_2O_4$ or $LiFePO_4$ cathodes because of the low value of metals.	Not economical for $LiMn_2O_4$ or $LiFePO_4$ cathodes because of the low value of metals.	

Note: Economy, eco-friendliness, and sustainability increases

In their endeavors, ReCell and ReLiB are not acting alone. Around the world, more and more recycling research networks are developing their own inventions and methods. It is yet unclear which of these approaches, if any, would be the preferred one once recycling infrastructure is put in place to recycle at scale. You must be aware of the worth of your waste and weigh it against the cost of your operation (https://www.chemistryworld.com/features/the-drive-to-recycle-lithium-ion-batteries/401222-2.article).

The development of a more effective procedure is complicated by a number of problems. The fact that these batteries are made for high performance and longevity rather than for recycling presents a fundamental difficulty because these features could be compromised in favor of creating a battery that is more recyclable. Each cell in a LIB pack contains a cathode, anode, separator, and electrolyte. A LIB pack is made up of several thousand cells arranged in modules. In most cases, cathodes are made of a powdered active transition metal oxide combined with carbon black and

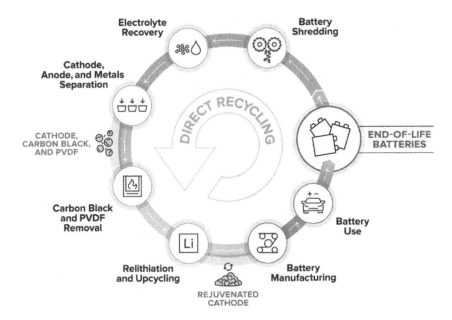

FIGURE 7.3 Direct recycling flowsheet (https://www.chemistryworld.com/features/the-drive-to-recycle-lithium-ion-batteries/4012222.article).

adhered to an Al-foil current collector with a substance like poly(vinylidene fluoride) (PVDF). The electrolyte is typically a solution of $LiPF_6$ salts, and the anodes are made of graphite that has been PVDF-glued to a Cu foil. In Figure 7.4, a used pouch cell is shredded to produce a variety of goods, including powdered black anode carbon.

FIGURE 7.4 A spent pouch cell and its shredded products, including anode carbon in a powdered black form.

7.3   A BATTERY RECYCLING BOOM

The EV market is only now starting to take off. In comparison to just one facility a few years ago, the USGS survey highlighted that roughly 20 companies in North America and Europe are recycling LIBs or have plans to. Traditional techniques couldn't effectively recover high-grade Li for use in rebuilding batteries at the few facilities that can recover materials from LIBs. The pyrometallurgy method, for instance, is simple to scale and compatible with any battery format, but it requires a process that uses a lot of energy to burn the battery at a high temperature. Pyrometallurgy can emit hazardous gases and restrict the recovery of other important components, even though the ash will contain beneficial elements. According to the battery technology being employed, other techniques require shredding the battery and then extracting the contents using drawn-out, difficult chemical processes.

An alternative is direct recycling, which involves disassembling the battery and recovering the cathode and anode components. Although this technique is still in its infancy, it has the potential to be less expensive, safer, and more effective. The technique is complicated by the requirement to manually disassemble a broad range of battery forms. Modules of a Li battery pack comprise cells, and it is in these cells that precious metals are located. Automation is required to process large volumes because reaching these cells manually is time-consuming.

The Swedish battery manufacturer has a number of programs through an initiative it's calling Revolt, including a pilot recycling plant that has been operating since late 2020. A full-scale recycling facility called Revolt Ett—Swedish for "one"—is also being built. It will be able to recycle 125,000 tpy of batteries starting in 2023.

Like most businesses, **Northvolt** does not directly recycle anything. Prior to any crushing, shredding, or chemical operations, the batteries are first broken down to the level of modules. The first battery from Northvolt was made entirely from recycled materials. In its pilot plant, Northvolt is fine-tuning a robot in anticipation of extensively automating the majority of the disassembly procedure in the future.

7.4   THE STATE OF RECYCLING TODAY

Researchers are examining techniques to maximize recovery despite the lack of a recycling-friendly design. It would be more effective to use a pure hydrometallurgical method, in which the valuable metals would be extracted from the crushed cathode material using aqueous solutions. To do that, materials that are insoluble in Co^{3+} are often converted into soluble Co^{2+} using a reducing agent consisting of a mixture of H_2SO_4 and H_2O_2. By altering the pH of the solution, the Co and Li can be recovered as salts through precipitation after leaching. You are left with highly pure beginning ingredients after that process, which you can use to make fresh batteries. The value that comes from the cathode material itself is lost, although hydrometallurgical methods recover more valuable materials than purely pyrometallurgical ones. Because most of the value is found in the manufactured cathode oxides rather than the raw materials, it is less useful for cathodes that are less Co-rich, such as lithium cobalt oxide (LCO), even though hydrometallurgical processes can recover up to 70% of these cathodes' value from them.

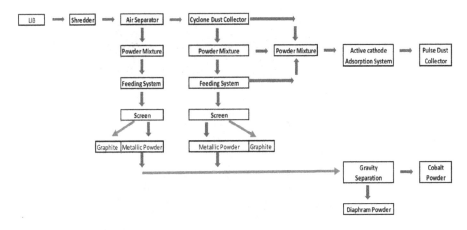

FIGURE 7.5 LIB recycling plant flowsheet.

7.5 LITHIUM BATTERY RECYCLING PLANT

The anode and cathode structures as well as the characteristics of the component materials (Cu, Al, Co, Li powder, etc.) led to the design of the S-LIB recycling facility, which consists of hammer milling, screening, and air-separating. To ensure the highest recovery rate and zero pollution, the entire process is carried out in a sealed vacuum environment. With the increase of EVs and digital products, the S-LIB recycling business will be better and better. Figure 7.5 shows the S-LIB recycling plant flowsheet with air separation. Gravity separation is used for Co powder and discharge powder generation.

7.6 WORKING PROCESS OF LITHIUM BATTERY RECYCLING PLANT

Lithium batteries include important metal resources, and improper handling of them could result in significant environmental harm and resource waste. So how can waste be turned into treasure? Hydrometallurgy, pyrometallurgy, and physical recycling are the methods that are widely employed. The physical recycling method has no problem with pollutant emissions because it doesn't require any chemical reagents, unlike hydrometallurgy and pyrometallurgy. Figure 7.6 shows the Li battery recycling plant for black mass production in China. The capacity of such plants changes from 500 kg/h up to 2,500 kg/h (https://www.alibaba.com).

7.7 WHICH KIND OF RAW MATERIAL WE CAN OBTAIN FROM THE LITHIUM BATTERY RECYCLING PLANT?

The scope of the raw materials are cylindrical LIBs, pouch LIBs, cellphone LIBs, 18,650 Li batteries, anode & cathode of Li batteries, ternary batteries, Li batteries of abandoned vehicles, LIBs, etc. Figure 7.7 shows different types of LIBs, anode and cathode plates, graphite powder, Cu granule, Al granule, and LCOs. LIB recycling

FIGURE 7.6   Li battery recycling plant for black mass production in China (500–2,500 kg/h).

plant is used for dismantling and recycling soft package batteries, cell phone batteries, Shell batteries, cylindrical batteries, etc. Different types of LIBs have different recycling procedures. The price changes from 50 to 100,000 $/t. Figure 7.8 shows manual EV battery dismantling.

Cell Lithium Battery   EV Lithium Battery   Graphite Powder   Copper Granule

Anode Plate   Cathode Plate   Aluminum Granule   Lithium Cobalt Oxides

FIGURE 7.7 LIB types and products.

7.8 HOW DOES THE RECYCLING FACILITY FOR LITHIUM BATTERIES OPERATE?

- The raw material is crushed by the primary crushing machine to 30–100 mm in size.
- Coarsely crushed materials are delivered to the secondary crushing machine to 2–5 mm in size by the sealed conveyor.
- A high-pressure air conveyor feeds the finely crushed material into the air separator, where it is divided into coarse/rough powder, fine powder, and septum. While air pressure and speed are carefully controlled by transducers, the fine powder enters the cyclone-dispersing device and the coarse powder enters the vibrating screen.
- The material that enters the cyclone-discharging system is transported to the vibrating screen by an air conveyor, where it is separated into Cu, Al, lithium cobalt combination powder, and septum powder.

FIGURE 7.8   Manuel EV dismantling study.

- The high-pressure air conveying system will deliver the coarse powder, which includes more metal, to the vibrating screen, where the lithium cobalt combination powder and Cu/Al powder will be separated.
- The air gravity separator, which can completely separate Li, Co, and Cu-Al, is the last separating system.
- To prevent dust pollution and Li-Co leakage, the entire plant operates in a vacuum.

7.9 ELECTROLYTE RECOVERY

The recovered graphite, electrolyte, and cathode material were evaluated chemically and electrochemically on a lab scale before being used again in LIBs. By performing a static extraction in an autoclave setup with various electrolytes and separators, the proof principle of electrolyte recovery was demonstrated. Following that, commercial 18,650 cells were extracted using organic solvents successfully. The conducting salt, however, could only be recovered in limited amounts (Grützke et al., 2014). A flow-through configuration was hence chosen (Grützke et al., 2015). It was shown that over 90% of the electrolyte, including the conducting salt and aging products, could be recovered from commercial $LiNi_{0.33}Co_{0.33}Mn_{0.33}O_2$ (NCM)/graphite 18,650 cells by using either subcritical or supercritical CO_2 with additional solvents. Commercial S-LIBs pouch-bag cells with an NCM cathode, a graphite anode, and a $LiPF_6$/organic carbonate solvent-based electrolyte were used as a source for the recycling process, along with production rejects from the NCM electrode fabrication (Krüger et al., 2014). The CAM was separated from the cells and dissolved in

10% H_2SO_4. The transition metal oxides were then separated under alkaline conditions by precipitation as hardly soluble carbonate salts. The actual resynthesis used a hydrometallurgical precursor synthesis to create NCM active material with electrochemical properties that were comparable to those of materials made from pure solutions. Finally, the recycling of graphitic anode material was integrated with the two aforementioned techniques (Rothermel et al., 2016). In the best situation, the electrochemical performance of recycled graphite outperformed the benchmark consisting of a newly manufactured graphite anode, including a 90% recovery of the electrolyte, by using subcritical CO_2 for electrolyte extraction.

Overall, it is possible to reuse almost all of the S-LIB's components, which is quite advantageous in terms of recycling effectiveness. The recycled material is also on par with brand-new material. The elimination of the electrolyte is also advantageous for environmental concerns and pilot plant equipment since the fluorinated substances inside an electrolyte pose a risk for both. These benefits extend beyond the recycling efficiency.

7.10   RECYCLING PROCESS RESTORES CATHODE'S ATOMIC STRUCTURE AND LITHIUM CONCENTRATION

The cathode material loses some of its Li-atoms as a LIB age. Additionally, the cathode's atomic composition changes, making it less effective at moving ions in and out. The cathode's Li concentration and atomic structure are both returned to their former states by the recycling procedure that was created. In total, 5.9 MJs of energy, or about three-quarters of a cup of gasoline, are used in the recycling process to restore 1 kg of cathode material. At least twice as much energy is used by a number of different LIB cathode recycling techniques that are currently being explored.

An energy-efficient recycling method that repairs used cathodes from S-LIBs and renders them functionally equivalent to brand-new ones has been devised by nanotechnologists. The degraded cathode particles from a used battery are extracted, boiled, and heat-treated in this procedure. Using regenerating cathodes, researchers created new batteries. Battery longevity, charge storage capacity, and charging time were all returned to their prior values.

To begin, cathode particles from S-LIBs must be collected. The cathode particles are then pressurized in a hot, alkaline solution that contains Li salt; this solution may be recycled and used to produce more batches. The particles next undergo a brief annealing procedure in which they are heated to 800°C and then progressively cooled. The regenerated particles were used to create fresh cathodes, which were subsequently tested in lab-built batteries. The energy storage capability, charging time, and longevity of the replacement cathodes were identical to those of the originals.

7.11   INDUSTRY DEMONSTRATIONS

7.11.1   GIGAFACTORY (RENO, NEVADA-USA)

The American Battery Technology Company (ABTC) is an advanced technology, first-mover LIB recycling, and primary battery metal extraction company that uses in-house developed proprietary technologies to produce domestically sourced battery-grade critical and strategic metals at significant cost savings and with less environmental impact than currently available conventionally sourced battery met als. For more information, visit https://americanbatterytechnology.com.

This disassembly is the first step of their two-step recycling procedure, which also includes a chemical or hydrometallurgical step. In northern Nevada, ABTC is now constructing its first facility, which it claims has the capacity to recover battery-grade materials in less than three hours. The facility was anticipated to be finished by the end of 2022 and have the capacity to accept 20,000 tpy of recyclables. That would equal around 5% of the weight of raw Li generated in 2021 if it were accomplished.

The battery metal supply chain has traditionally operated more linearly, with primary metals like Li being initially harvested or extracted from ores and brines underground. Large chemical refiners refine these metals before turning them into high-energy-density CAMs. The producers of battery cells purchase these synthetic cathode materials to create batteries. The original equipment manufacturers (OEMs), such as significant automobile vehicle firms, purchase the materials from the battery cell producers (Figure 7.9).

To purify these battery metals to the same, or even better, quality criteria than conventional materials obtained from virgin mining operations, ABTC developed

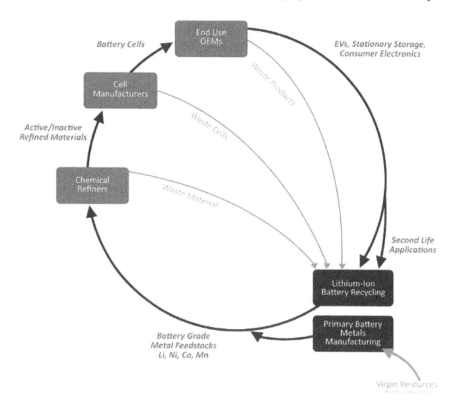

FIGURE 7.9   Closed-loop LIB recycling chain.

a closed-loop battery recycling process. This process separates and recovers crucial elements from EoL batteries. This method has the capacity to process materials in a very short amount of time because it is very streamlined and effective.

This integrated battery recycling system was created by the ABTC team using a tactical de-manufacturing strategy. ABTC created a "de-manufacturing" approach to remove metals and recover materials from S-LIBs as opposed to the brute force techniques used today, such as placing batteries in high-temperature furnaces for smelting or size reduction in shredding or grinding systems. The ABTC recycling process uses an automated deconstruction process along with a focused hydrometallurgical, non-smelting process to break down battery packs into modules, modules into cells, and cells into sub-cell components, which are then carefully sorted and separated. When compared to traditional virgin-sourced materials, the ABTC recycling process reportedly reduces waste, conserves water and natural resources, reduces smelting-related air pollution, increases operational efficiencies, increases material recovery rates, and captures battery-grade materials without compromising quality (https://americanbatterytechnology.com/solutions/lithium-ion-battery-recycling):

- Strategic design prevents the release of air and liquid pollutants; no high-temperature operations are performed.
- Early separation of low-value components during processing steps promotes high product recovery and purity.
- The supply chain can be locked off so that metal items made to battery cathode standards can re-enter it.
- Recycling facilities in each region have the same throughput as manufacturing facilities.
- Low CAPEX is achieved by avoiding high-temperature procedures and producing less trash.
- Short processing residence times are achieved through material handling and high-speed strategic disassembly.

7.11.2   TOXCO PROCESS AND RETRIEV TECHNOLOGIES (USA/CANADA) (OHIO, USA; B.C., CANADA)

Toxco process patent was filed in 1994 in the USA. It uses a preliminary cryogenic milling treatment with liquid N_2 at $-195°C$ which reduces the material reactivity. Toxco processes Li and Ni-based batteries in Canada and the USA. The batteries are then crushed and treated with an alkaline solution to recover Li salt. Li is processed by cryo-milling in Trail, Canada and Ni is processed by pyrometallurgy in Baltimore-OH in the USA.

Battery recycling business Retriev Technologies (formerly "Toxco Inc.") was established in the United States in the 1990s. All types of LIB compositions, including LAB, NiCd & NiMH, Li-ion, Li-primary, and alkaline batteries, can be recycled using Retriev Technologies' method, which has a 4,500 tpy capacity. To recycle LIBs and recover all cathode metals, this procedure combines cryogenic, mechanical, and hydrometallurgical treatments. To prevent violent reactions, the larger battery packs are disassembled and shredded in a Li brine solution or a cryogenic liquid nitrogen atmosphere. This nitrogen is typically utilized when there are a lot of batteries. Making the exterior plastic casing brittle and rigid will help it break off easily during the shredding process. Smaller batteries, such as those found in cell phones, are introduced into the process without being disassembled beforehand. The crushed material is divided into three stages when it enters a shaker table that is sprayed with water. The Cu-Co product is in the initial stage. The Li-ion fluff, which is the second phase, is subjected to a Na_2CO_3 treatment to precipitate Li_2CO_3. After that, it is cleaned, washed, and sold. The slurry, which is the third step, is filtered to create a sludge that contains Co, Cu, Ni, Mn, and Fe. The Co-Mg product is then produced and sold to smelters.

The following steps make up the majority of Retriev's Li-ion recycling process: (i) Manual disassembly (of large-size batteries); (ii) Crushing in an aqueous solution, which results in the production of three different materials: metal solids, metal-enriched liquid, and plastic fluff. (iii) Additional processing (off-site): The metal solids can be further processed to recover the metals for use in new products. The metal solids contain varying proportions of Cu, Al, and Co, depending on the kind of LIB processed. Utilizing filtering technology, the metal-enriched liquid solidifies and is sent somewhere for more metal purification.

Downstream smelters (Glencore) who are interested in the Co or Ni content, the filter cake and metal solid are sold. Plastics can be discarded or recycled right away. Li_2CO_3 can be created here to recycle Li.

Retriev received 9.5 million dollars from the US Department of Energy in 2009 to construct a new facility in Lancaster, Ohio to assist recycling of US Li-containing batteries, particularly hybrid and electric vehicle (EV) batteries. Li-primary and LIBs are both subjected to mechanical processing by Retriev. Early in the 1990s, the Retriev Li battery recycling method was developed to handle primary Li metal batteries, mostly from military applications. Cryogenic refrigeration with liquid N_2 is used in this process for safety reasons. Retriev (Toxco) presented a newly developed Li-ion recycling method in patent US 8616475 (B1) in 2013, which included the primary phases of crushing, screening, filtration, froth flotation of carbon, filtration, and thermally regenerating LIB cathode material. The company announced to commercialization the described process. However, the situation as it is now does not indicate an implementation.

7.11.3 Sony & Sumitomo Process (Tokyo, Japan)

A team from both Sony Electronics and the Sumitomo Metal Mining Company created the Sony-Sumitomo process in 1996, which has a capacity of about 150 tons annually. Hydrometallurgy and pyrometallurgy are both used in the process. At around 1,000°C, the battery undergoes calcination, during which electrolytes and organic components are burnt off. The leftover mixture is then mechanically separated, allowing minerals like Al, Cu, and Fe to be magnetically separated and reused (Lupi et al., 2005). The resulting powder is presently put through a hydrometallurgical process in order to obtain Li and Co, which are then utilized to make batteries.

7.11.4   UMICO Process (Hoboken, Belgium)

The umicore process uses a combined pyro-hydro treatment along with electrowinning (EW) for all types of batteries on an industrial scale. The plant-permitted LIB capacity is 7,000 t/y. They are burned yielding two kinds of products; an alloy containing valuable metals like Co, Ni, and Cu, designed for further hydro-processes a slag containing Li which can be used either in the construction industry or processes to recover Li (Umicore, 2019). Figure 7.10 shows the Umicore battery recycling process flowsheet (Elwert et al., 2016). This process recovers 95% of Co, Ni, and Cu just as additional quantities of Li. Umicore and Johnson Mattey (UK) produced 15,100 tons (NMC-LCO) and 2,650 tons LFP of cathode materials in 2015 (Danino-Perraud, 2020).

The Umicore process is primarily composed of smelting (in a shaft furnace) and subsequent hydrometallurgical treatment stages (Figure 7.11) (Tytgat, 2015). Li-ion and NiMH batteries as well as production waste are put straight into the smelter without going through a lot of processing. Large industrial batteries can only be prepared after being broken down to a tiny size (like cell level). In addition to batteries, the smelting furnace is fed with additives including coke, sand, and limestone (slag formers).

A schematic of the smelting process in the shaft furnace is shown in Figure 7.12. Based on the reactions that take place as a result of the process' gradual temperature increase,

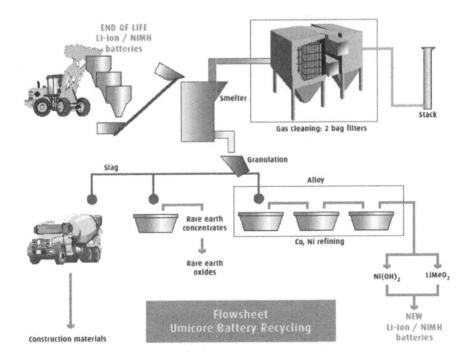

FIGURE 7.10   Umicore battery recycling flowsheet.

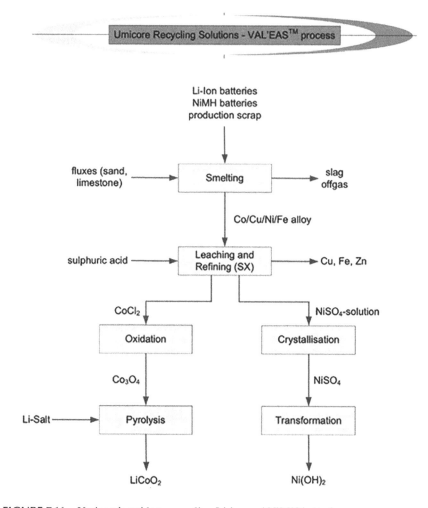

FIGURE 7.11   Umicore's guide to recycling Li-ion and NiMH batteries.

the smelter can be divided into three zones: (i) the upper zone (300°C), where electrolytes are evaporated; (ii) the middle zone (700°C), where plastics are pyrolyzed; and (iii) the bottom zone (1,200°C–1,450°C), where smelting and reduction happen.

Electrolyte solvents and plastic housings types of organics and graphite, which account for about 25–50 wt% (depending on battery type) of batteries, are employed as combustible compounds and reducing agents for metal oxides. The smelter is stated to be heated enough by the energy released during these reduction events. To completely breakdown the organic components and make sure that no dangerous dioxins or volatile organic compounds (VOCs) are created, a gas cleaning system has been built. The flue dust is a collection of fluorine (F).

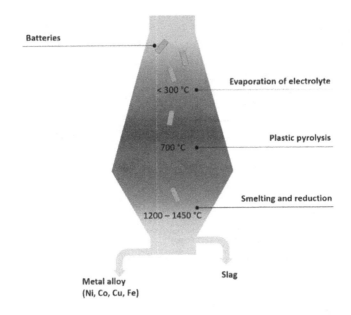

FIGURE 7.12 Schematic diagram of Umicore's shaft furnace for pyrometallurgical operation.

The smelter generates the following outputs:

- Metal alloy: Contains Co, Ni, Cu, and Fe and is capable of being sent to a subsequent hydrometallurgical process to recover Co, Ni, and Cu. It should be noted that the alloy must be granulated before being sent through the hydrometallurgical process's leaching step because the hydrometallurgical process needs fine particles as input material (particularly in the leaching step).
- Slag: This material, which contains Al, Li, and Mn, can be used in the building sector or processed further to extract metal. Li-recovery flowsheets are stated to incorporate LIB slag through collaboration with outside parties.
- Cleaned gas: UHT technology cleans off-gas. The cleaning system's halogen-containing flue dust, which primarily contains F, is landfilled.

7.11.5   AMERICAN MANGANESE INC./RECYCLICO BATTERY MATERIALS INC.

American Manganese Inc/RecycLiCo Battery Materials, is a battery materials company that specializes in recycling and upcycling S-LIBs. Kemetco Research Inc. is its R&D partner. The proprietary, closed-loop hydrometallurgical method provides excellent LIB materials for direct integration into the re-manufacturing of new LIBs with only a few processing stages and up to 99% extraction of Li, Co, Ni, and Mn.

The LithoRec method (Figure 7.13) is one modern strategy. By using the majority of the materials from a LIB, which is a fundamental component of this process,

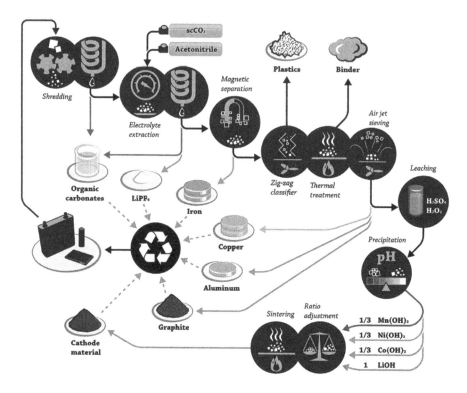

FIGURE 7.13 Mechanical—Hydrometallurgical LithoRec recycling flowsheet (https://analyticalscience.wiley.com/do/10.1002/gitlab.15680/full/).

this mechanical-hydrometallurgical method intends to satisfy the requirements of the EU rule. Batteries that have been deeply discharged by external resistance or power are broken down, and the individual cells are then destroyed in an inert environment. The electrolyte, which evaporates during this stage and typically consists of a combination of linear and cyclic carbonates and a conducting salt, is condensed and collected. The leftover electrolyte is then recovered using one of many potential techniques in the next phase. The conducting salt $LiPF_6$, which is the most expensive component of the electrolyte, might be lost during a thermal drying stage, which is a drawback. Dimethyl carbonate (DMC) can also be used as a liquid extractant to recover electrolytes like $LiPF_6$. Utilizing subcritical or supercritical carbon dioxide (sc CO_2) to effectively recover the organic carbonate solutions is another extraction technique. Adding more co-solvents to the extractant will also increase the yield of the conductive salt. Fe components are then extracted via magnetic separation and sent to a facility that recycles scrap metal. A zigzag air classifier is fed the remaining non-magnetic material. Here, the shred material is further divided into two fractions: one with the separator and plastic foils, and the other with the current collectors and active ingredients. Heating the material to between 400°C and 600°C removes the binder while also causing the current collectors to separate from the active material particles. The active materials are separated from the current collectors by the use of extra air jet sieves. Graphite is also removed from the recycling process at this point. While the active material dissolves in an acidic mixture and is further refined by a hydrometallurgical process, Li is leached out of the cathode material.

The RecycLiCo patented process was created in 2016 by the battery recycling and upcycling business RecycLiCo Battery Materials, with headquarters in Vancouver, Canada. A closed-loop hydrometallurgical procedure called RecycLiCo can extract Li, Co, Ni, and Mn from S-LIBs up to 100% of the time. Figure 7.14 shows the demo hydrometallurgical plant in Vancouver, Canada. The company claims zero loss in battery production with 62% less CO_2 than competing hydrometallurgical processes (Figure 7.15). From waste to the cathode, RecycLiCo creates valuable LIB materials,

FIGURE 7.14 RecycLiCo Battery Materials Inc. demo hydrometallurgical plant in Vancouver.

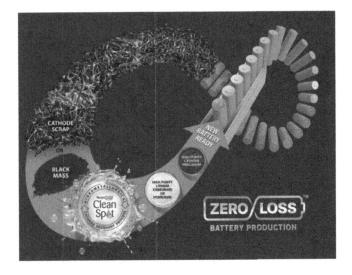

FIGURE 7.15 Zero loss recycling.

with low environmental impact, for direct integration into the re-manufacturing of new LIBs (https://recyclico.com/). Using RecycLiCo's patented closed-loop process, the Demo Plant has verified a 163% increase in actual leach processing capacity over the 500 kg/day planned capacity since it was first put into operation earlier in 2022 (https://americanmanganeseinc.com/recyclico-battery-materials-advances-to-lithium-recovery-stage-of-demonstration-plant-project/). At the Demo Plant, the company is conducting tests using feedstock materials like various cathode production scrap and black mass. RecycLiCo will also be enhancing its current line-up of distinctive products, which is still being shipped to potential partners for advanced product qualification and assembly in battery cells, by adding next-generation, battery-ready cathode precursor and lithium products.

Nanoramic labs and RecycLiCo create a strategic partnership for LIB recycling. For testing with RecycLiCo's patented LIB recycling process, which has demonstrated up to 99% extraction of Li, Ni, Mn, and Co from conventional waste cathode materials, Nanoramic will provide its discarded Neocarbonix® LIB cathode electrodes, notable for their high-performance, low-cost nanocarbon mesh binding structure. Together, Nanoramic and RecycLiCo will be integrated into commercial operations in the future, according to the goals of the collaboration. The novel NMP- and PVDF-free electrodes from Nanoramic provide an answer to prospective restrictions on per- and polyfluoroalkyl substances (PFAS) in LIBs. Both parties are attempting to "close the loop" by sending the recycled product back to Nanoramic for cell manufacture and additional electrochemical analysis by utilizing RecycLiCo's expertise in converting waste cathodes into valuable battery-grade materials.

7.11.6 KEMETCO RESEARCH INC. (BC-CANADA)

A private company that integrates science, technology, and innovation is Kemetco Research. Their Contract Sciences division offers consulting services, applied research and development, bench scale studies, fieldwork, laboratory analysis and testing, and pilot plant investigations to both the private sector and the public sector. Their customers range from young start-ups creating novel technology to established huge global organizations. In the areas of Specialty Analytical Chemistry, Chemical Process, and Extractive Metallurgy, Kemetco offers scientific competence. Kemetco can offer a wider range of backgrounds and experience than other laboratories because it conducts research in numerous sectors. Kemetco creates and assesses cathode materials as well as batteries (https://www.kemetco.com/battery-research.html).

Kemetco developed a proprietary hydrometallurgical recycling process for the recovery & purification of battery-grade Li from waste batteries. This environmentally sound and energy-efficient technology was fully patented by the USPTO in 2018 and advanced to pilot-scale demonstrations in 2019. Figure 7.16 presents pilot plant for LIB recycling and LIB CAM and recycled Li products.

FIGURE 7.16   Pilot plant for LIB recycling and LIB CAM and recycled Li products.

7.11.7   C4V LiSER Technology (Binghamton, NY, USA)

A Binghamton, New York-based company called C4V specializes on intellectual property and has made patented advancements in the creation and composition of LIBs. In order to produce next-generation storage materials that can be smoothly incorporated into existing cell manufacturing lines, C4V makes use of its knowledge in electrode design and process development.

Lithium Slim Energy Reserve is referred to as LiSER. The LiSER technology from C4V has a proprietary, patent-pending battery cell architecture that enables OEMs to omit modules and construct the pack directly. This platform features designs with long, slender cells that can charge and discharge extremely quickly without sacrificing the advantages of energy density. LiSER additionally enables OEMs to achieve the highest possible cell-to-pack performance translation (Figure 7.17) (*source:* https://www.chargecccv.com/technology/innovation). The Co and Ni-free LIB cell technology developed by LiSER is a first of its type, and it promises a power density of up to 2,000 KW/kg and an energy density of up to 228 Wh/kg. Even if the present covid-driven supply-chain issues subside, the high costs of Co and Ni will ultimately jeopardize EV makers' ability to maintain cost leadership.

Four key elements—the anode, cathode, separator, and electrolyte—make up about 80% of the cost of producing LIBs. C4V has discovered, patented, and commercially produced processing technologies for next-generation anode and cathode materials in collaboration with current industry titans in the electrolyte and separator sectors. The business is also working to optimize these materials at the same time. C4V has quantified and validated more than 15 additional components that are present in lithium-ion battery cells in addition to the four crucial ones mentioned above. The development of cells in prismatic, cylindrical, and pouch form factors for a range of applications is also a crucial step for C4V. The P, N, and S series of LIBs were created by C4V (Figure 7.18).

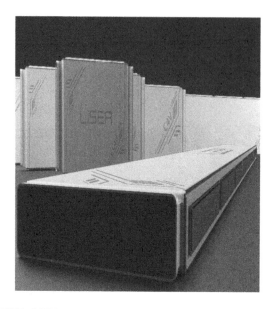

FIGURE 7.17   LiSER C4V batteries.

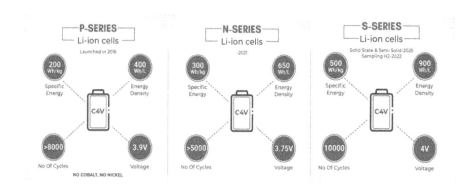

FIGURE 7.18   Summary of the properties of C4V's P, N, and S series of LIBs.

The Co- and Ni-free cathode chemistry of the *P-Series batteries*, which will be commercially available in 2019, boasts the best voltage and cycle lifetime of any material used in current commercial products. In addition, compositionally patented changes at the crystal level guarantee unparalleled safety in the case of thermal runaway or fire exposure in addition to an increase in energy density and high-rate capabilities. Graphite made with in-house processing techniques and combined with its composite silicon will be used in C4V anodes. Graphite, which makes up the majority of current anodes, frequently needs costly thermal and chemical treatments to boost purities beyond 99.95% for authorized usage. By modifying commercially available machinery that can produce yields of more than 70%, C4V's innovative processing technology for purification and spheronization has completely done away with the necessity for either phase. With processing methods that are more than 20 years old, most modern commercial technologies can only achieve about 40% yield. Its integration with composite silicon is guaranteed to satisfy the specified performance parameters thanks to proven control over morphology, surface area, and particle size distributions.

The production technique for C4V's composite silicon, a game-changer for increased energy density and volumetric capacity, is also affordable, scalable, and green. Given that silicon has a specific energy capacity of more than 3,000 mAh/g, it has long been coveted for use in the anode. Commercial viability has been affected by futile attempts to control the dramatic volume growth that occurs during lithiation. Through nano-structuring, C4V's composite silicon technology prevents catastrophic volume expansion. Improvements in performance are made possible by allowing the primary particles to internally expand without cracking. Energy densities and volumetric capacities of 200 Wh/kg and 500 Wh/L will be demonstrated by P-Series batteries. Currently, cylindrical and prismatic cells have been the focus of design and optimization work, with the manufacture of the 2,170 and 3,270 types with increased heat dissipation. Production at a nominal 1 GWhs scale began in the second half of 2022 with the aid of strategic partners and a significant investment in cell manufacturing equipment (https://www.chargecccv.com/technology/portfolio).

The N-Series Battery will be improved by enhancing the anode and cathode compositions and adding a more thermally stable electrolyte. While the anode and cathode materials determine the theoretical limits for a battery's storage capacity, the electrolyte—which acts as an interfacial Li transport medium for either electrode—determines the power density, imparts electrochemical stability within the cell through passivation of both anode and cathode surface layers, and to a large extent determines a battery's chemical response to non-equilibrium conditions like overheating, which can cause fires.

Li-ion batteries can only be used in a limited range of temperatures due to the current commercial electrolytes' sensitivity to environmental temperatures. To maintain the ideal thermal conditions for cycle life and performance, an electronic battery management system and other forms of insulation must be used. C4V has found and is experimenting with an electrolyte that might reduce the amount of upkeep necessary to keep the batteries cold. This electrolyte has so far shown stable up to 65°C. With advancements in energy density and volumetric capacity to 300 Wh/kg and 650 Wh/L, respectively, implementation for volume manufacturing began in 2021.

In the upcoming years, S-Series, C4V intends to keep developing battery technologies and processing methods that are better, more affordable, and safer. This commitment includes creating a solid-state battery, which is a sizable portion of it. There are currently no specific plans or projections for bringing solid-state batteries to market, although they are theoretically conceivable. This technology might need about further 20 years to fully commercialize and establish a reliable production and supply chain infrastructure. To address some of these issues, C4V is adopting a different and more useful strategy. To create semi-solid-state technology that combines the benefits of both systems, they were able to replace more than 50% of the liquid electrolyte with a solid-state electrolyte. They also show energy densities and volumetric capacities of 400 Wh/kg and 800 Wh/L in their preliminary testing using S-series batteries. They firmly forecast that a reliable supply chain and these special batteries will be prepared for mass production by 2025.

7.11.8 BATREC INDUSTRIE AG (WIMMIS, SWITZERLAND)

Since more than 35 years ago, battery collecting in Switzerland has operated effectively; close to 85% of all batteries sold are collected and given to Batrec. In Switzerland, the INOBAT battery recycling technology is employed. To recycle LIBs, Batrec Industrie AG uses a hydrometallurgical method. In a CO_2 atmosphere, the spent batteries are sorted and crushed. The released Li is subsequently neutralized by wet air. The protective CO_2 gas is crushed, neutralized, and then scrubbed in a gas scrubber before being exhausted. The leftover scrap components are treated in an aqueous solution that has been acidified, and the resulting leaching liquor and solid fraction are separated for further purification (https://batrec.ch/en/).

7.11.9 TES-AMM (FORMERLY RECUPYL) PROCESS (DOMENE, FRANCE)

A specialized battery recycling business called Recupyl was established in the Grenoble, France, region in 1993. Using mechanical separation technique, Recupyl recycles batteries, including alkaline, Zn-carbon, and LIBs. A pilot test with a yearly capacity of several hundred tons was constructed at the Grenoble factory. Recupyl has established joint ventures or sold a number of patent licenses since 2007. Due to Recupyl's bankruptcy in 2019, some of its subsidiaries either became independent (Recupyl, Poland) or were acquired by other businesses, such as TES-AMM, a global recycler of e-waste with headquarters in Singapore.

TES-AMM is operating in Li-ion activities since 2006 in two sites (France and Singapore). French site (partner in this project) has been operating since 2008 an industrial installation dedicated to power pack batteries from e-mobility and allowed by permit from environment authorities. Through the first company RECUPYL SAS, TES-RECUPYL as cumulated experience within four European projects as well as five French National projects all dedicated to recycling advanced batteries.

Besides mechanical and electrical assets, TES-REC has in France a full pilot plant for the treatment of mixed cathode-anodes powder using a sustainable hydrometallurgy process with free-water release. Upsizing the technology led to the first industrial facility in Asia using hydrometallurgy at 6,000 t/y.

After discharge, Recupyl uses a pin mill to grind the LIBs into powder, followed by sieving the materials to separate them into fractions of plastics, aluminum, copper, steel, and active mass powder. Although their capacity is unknown, it is thought to be modest. Before sending the active mass produced to the hydrometallurgical process, it must first undergo pyrometallurgy pre-refinement (Sojka et al., 2020). Valibat, a hydrometallurgical method developed by Recupyl for the recycling of LIBs, involves mechanically treating used batteries while they are being surrounded by a CO_2 inert

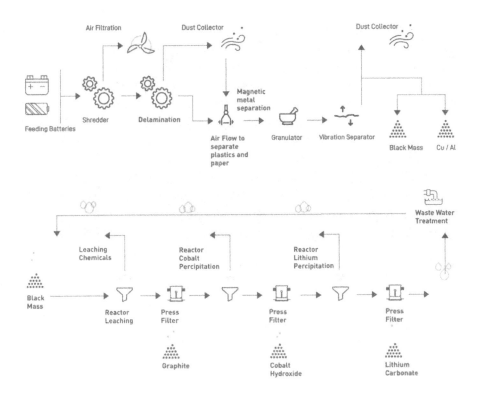

FIGURE 7.19 TES-AMM battery recycling—black mass process and battery recycling—chemical refinement process.

gas combination and physically separating the metals steel, copper, and polymers. Li, mixed metal oxides, and C are produced in an alkali solution by further leaching the fine particles. Metal oxides go through a second leaching after Li precipitates out as Li_2CO_3 or Li_3PO_4. NaClO is then added in order to precipitate Co as cobalt (III) hydroxide after the Cu and other impurities have been eliminated. Figure 7.19 depicts the TES-AMM battery recycling process using black mass and the battery recycling process using chemical refining.

The commercial battery recycling facility in Singapore operated by TES-AMM has the capacity to process 14 tons of LIBs per day, which is equal to 280,000 smartphones. About 90% of materials are recovered throughout the recycling process. Products like Co and Li are 99% pure. standards that are so high they can be repurified and used to make new batteries. Li_2CO_3, $Co(OH)_2$, Na_2SO_4, graphite, Cu, and Al are obtained (https://www.tes-amm.com/it-services/commercial-battery-recycling#!/). Figure 7.20 shows a picture of the TES-AMM plant in Singapore.

7.11.10 LITHO REC

The German Federal Ministry of Environment supported the LithoRec project, which sought to create a high recycling rate, commercially feasible, and environmentally

FIGURE 7.20   TES-AMM plant in Singapore.

advantageous recycling method. The hydrometallurgical treatment of active materials extracted from the recycling stream is performed by Albemarle Germany GmbH
(Rockwood Lithium GmbH), and the recovered lithium and transition metal salts can
be used to create new cathode materials.

7.11.11   BRUNP RECYCLING TECHNOLOGY CO. LTD. (GUANDONG, CHINA)

In Fushan, China, Brunp Recycling Technology Co. Ltd. was established in 2005.
Recycling S-LIBs and automobiles are the two primary commercial sectors. In the
provinces of Hunan and Guangdong, Brunp maintains facilities. From 2013 to 2015,
LIB producer CATL continuously bought out Brunp's stock, increasing its ownership
to 69.02%, and Brunp became a CATL subsidiary. According to a report from 2018,
Brunp's yearly waste battery treatment capacity (including S-LIB and other batteries)
exceeds 30,000 tons. As a result, it can be said that Brunp mostly handles LIB trash
generated from production scrap. The actual proportion of EoL LIBs to industrial
scrap and other trash is uncertain, though. The annual production of nickel, cobalt,
and manganese hydroxide, according to Brunp, is 10,000 tons. The primary steps in
Brunp's LIBs recycling process are hydrometallurgy, mechanical treatment, thermal
pretreatment, and discharge. The plant in the province of Hunan is where the company does its hydrometallurgical processing. The hydrometallurgical process entails
crystallization, solvent extraction, acid/basic leaching, etc. China is where the hydrometallurgical method is most frequently used. As one of the top battery recycling
companies, Brunp processes S-LIBs using acid leaching (H_2SO_4 acid and H_2O_2),
and the metal hydroxides that are produced can be used to make cathodes. The corporation provided details on the LIB recycling technologies in a number of patents,
however, it is unclear which technology is currently in use. Ni-Mn-Co hydroxide,
cobalt sulfate, nickel sulfate, cobalt chloride, etc., which are typical by-products of
hydrometallurgical operations, are the principal end products from the Brunp battery

recycling process. According to speculation, the high recovery rate achieved during the hydrometallurgical process accounts for the high recycling rate of Ni, Mn, and Co. However, it is uncertain how often Li is recycled. As of right now, Brunp is the second company in the world and the first in Asia when it comes to recycling used batteries and using resources (https://en.brunp.com).

7.11.12 GEM Ltd. (Green Eco-Manufacturer Hi-tech Co., Ltd) (Shenzhen, China)

Recycling business GEM Ltd. was established in Shenzhen, China, in 2001. GEM processes "Urban Mines" such as Ni/Co/W metal scraps, scrapped automobiles, battery wastes, and electronic wastes. GEM removes Co and Ni from Ni/Co metal scraps and S-LIBs using mechanical and hydrometallurgical methods and then creates Co/Ni salts or ultrafine Co/Ni powder. A total of 5,000 tons of NCM and NCA ternary precursor material can be produced annually by GEM. Figure 7.21 from the company's recycling process flowchart (Chen et al., 2019).

As seen in Figure 7.21 above, S-LIBs and wastes containing Ni/Co are largely comminuted. The mechanically separated, ground-up Ni/Co-containing material is next subjected to hydrometallurgical processing. Metal ions like Cu^{2+}, Al^{3+}, Fe^{2+}, Co^{2+}, and Ni are divided into distinct solutions once the material is dissolved in a solution. To create Co/Ni salts for battery manufacture, the resultant Co/Ni-containing solutions are subsequently treated using liquid phase synthesis and high-temperature treatment. The business also said that it produces ultrafine metal powders from used LIBs. According to business patents, the atomizing-hydrolysis method, the ball milling + H_2 reduction method, or the airflow crushing technology can all be used to manufacture ultrafine Co/Ni products. The manufacturing of hard alloys can utilize ultrafine metal particles. More information on the mechanical pretreatment of S-LIB's sorting and comminution can be found in a patent. However, a lack of information prevents the risk assessment for mechanical separation. Additionally, there is limited knowledge regarding the removal of other pollutants such as electrolytes and PVDF binders. The precise amount of batteries

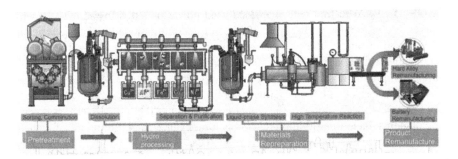

FIGURE 7.21 Flowsheet of Ni/Co waste and LIB recycling process at GEM plant in China.

recycled is unknown because Co/Ni products are made from a mixture of batteries and other metal trash. Nevertheless, it is known that approximately 11,000 t/a of mixed S-LIBs are collected via the GEM battery collection program. One can calculate that the treatment of LIB is roughly 2,000 t/a (Danino-Perraud, 2020) assuming that 20% of them are LIB.

Major Chinese Co refiner and Li cathode precursor manufacturer Green Eco-Manufacture (GEM) is building a production complex to recycle decommissioned power and energy storage batteries and used battery materials in Yibin in southwest China's Sichuan province.

The complex, with a total investment of 239 M$, will have a 100,000 t/y capacity for decommissioned power batteries and scrap, 50,000 t/y for used lithium iron phosphate (LFP) material, and 3 GWh/y for used energy storage battery packs. The project will be developed by GEM subsidiary Wuhan Power Battery Renewable Technology in partnership with domestic chemical company Yibin Tianyuan Group. The first phase of the project, with 50,000 t/y capacity for decommissioned power batteries and scrap, 20,000 t/y for used LFP material and 1 GWh/y for used energy storage battery packs, will be completed and put into production within a year. The whole project aims to be competed no more than 3 years from the start of the first phase of construction. More details, including the start date, were undisclosed.

GEM has been accelerating its development of power battery recycling business in the past few years. It signed an agreement in August 2021 with major Chinese LIB manufacturer EVE Energy for supplies of 10,000 t/y for recycled Ni-containing products, including $NiSO_4$, ternary precursors and ternary cathode active materials. EVE Energy will receive these products from GEM for 10 years from 2024.

It also signed an agreement in August 2021 with Chinese power and energy storage manufacturer Farasis Energy (Ganzhou) to develop a recycling project for used power batteries and scrap. GEM will receive used power batteries and scrap from Farasis' global supply chain and process them into recycled battery-grade Ni, Co, and Mn sulfates and ternary precursors and ternary CAMs. These products will then be returned as feedstock supplies to Farasis for its battery production. More details including the capacity and supply volumes were undisclosed. GEM in May this year told its investors that it aims to increase its power battery recycling output to above 300,000 t by 2026 in line with increased used power batteries in the market, with a target to disassemble and recycle 35,000 t of decommissioned power batteries in 2023, doubling that of the previous year. It is also on target to have 150,000 t/y metal equivalent of Ni output capacity in Indonesia by 2026, as well as 50,000 t/y of recycled Ni production by the same year.

Argus forecasts that LIB recycling from EV and gigafactory scrap will add up to over 1 Mt of cathode material to the global supply mix by 2033. GEM's ternary precursor shipments, including lithium nickel-cobalt-manganese and lithium nickel-cobalt-aluminum precursors, increased by 67% from a year earlier to 152,300 t in 2022, driven by rapid demand growth from the Li battery and EV sectors (https://www.argusmedia.com/en/news/2458422-chinas-gem-expands-battery-recycling-production).

7.11.13 GANZHOU HIGHPOWER (GUANGDONG, CHINA)

An integrated "clean energy" supplier, Highpower International Inc. was established in Guangdong province, China, in 2002. Energy storage systems and R&D, production, and sales of NiMH and Li-ion rechargeable batteries are the major businesses. Highpower produces Li-polymer, cylindrical Li-ion, Li-polymer coin, high energy density, high voltage (4.4–4.5 V, 650–850 Wh/L), fast charge, high-low temperature (−40, +70°C), and high-rate batteries (https://www.highpowertech.com/li-ion-li-polymer-rechargeable-batteries).

The company established a facility in Ganzhou City, Jiangxi province, in 2012 and began recycling batteries and electronic garbage there. In 2016, Ganzhou Highpower made the announcement that it would build a production facility with a 10,000 tpy capacity for processing production scraps and used (NiMH & Li-ion) batteries. Unknown is the capacity for therapy at the moment. The business claimed to be involved in echelon usage of spent EV batteries in addition to recycling. The business claims that the LIB recycling technologies use physical deconstruction (mechanical treatment), pyrometallurgical, and then hydrometallurgical processes. There are no more specifics available.

7.11.14 BATTERY RESOURCES/ASCENT ELEMENTS (MASSACHUSETTS, USA)

The Worcester, Massachusetts-based company Battery Resources has created a closed-loop method to recycle LIBs. Here, the cathode powder is dissolved using a hydrometallurgical procedure. Different $LiNi_xMn_yCo_zO_2$ (NMC) are synthesized using the leaching solution.

Ascend Elements is an independent producer of cutting-edge battery components that utilize priceless elements recovered from S-LIBs. The industry as a whole has made a significant improvement in sustainability because of the patented Hydro-to-Cathode® direct precursor synthesis process, which converts waste from today into high-value components for tomorrow's EV batteries. Reduced prices and CO_2 Shredding, leaching, impurity extraction, cathode manufacture, and battery manufacturing are all processes in Ascend's hydro-to-cathode direct precursor synthesis process (Figure 7.22). Sustainable battery materials produced by the Hydro-to-CathodeTM direct precursor synthesis technique can outperform conventional materials created with recently mined and refined metals. According to a recent study, battery cells built using recycled cathode material had an 88% higher power capacity and over 50% longer cycle life than conventionally made cells (https://ascendelements.com/).

Figure 7.23 compares hydro-to-cathode, hydrometallurgy, and pyrometallurgy LIB recycling methods. Ascend claims the hydro-to-cathode method is the most efficient process for LIB recycling (https://ascendelements.com/innovation/). Ascend Elements is delivering both precursors (pCAM) and finished cathode materials (CAM) for the LIB market that help promote the growth of demanding EV market applications.

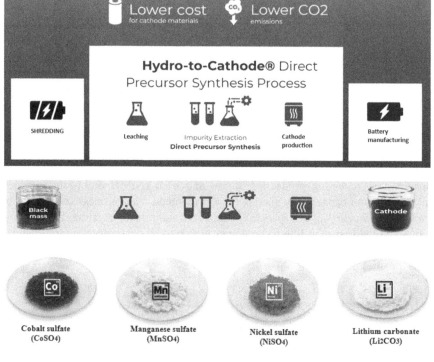

FIGURE 7.22 Lower cost and CO_2 Hydro-to-Cathode direct precursor synthesis process of Ascend.

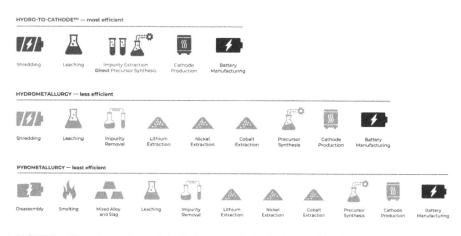

FIGURE 7.23 Comparison of the hydro-to-cathode, hydrometallurgical, and pyrometallurgical steps for LIB recycling.

- NMC hydroxide and sintered NMC cathode materials
- Various metal ratios including 111, 532, 622, 811
- Particle d_{50} is available in both 5–6 and 10–12 μm.

7.11.15 ONTO TECHNOLOGY (BEND, OR-USA)

OnTo processes Li-ion-based batteries with liquid-liquid technology in the USA. OnTo removes hazards for safe and inexpensive transport. This eliminates Li-ion logistics hazards and reduces the overall cost of recycling by ~50%. OnTo recovers clean, valuable cathode precursor from any LIBs and adds flexibility to direct recycling.

An approach for manufacturing sustainability that shows promise is the direct recycling of Li-ion batteries. Because it recovers the functioning cathode particle without converting it into substitute elements or dissolving and precipitating the entire particle, it is more effective than traditional approaches. This application of cathode-healing™ to a battery recall case study illustrates an industrial strategy for recycling Li-ion, whether they are used in consumer electronics or EV batteries. The whole procedure entails recycling the cathode and anode before extracting electrolytes with CO_2, shredding industrial materials, harvesting electrodes, froth flotation, and cathode-healing™. Using direct recycled cathodes and anodes from an industrial source, the finished products showed usable capability in the first complete cells (Sloop et al., 2020; https://onto-technology.com/).

The company has received the following four safety patents for LIB recycling since 2009:

- Sloop, S.E.; Crandon L.E. US 62/934,446 "Battery Deactivation" (2019),
- Sloop, S.E. US#10,014,562 "Distinguishing Batteries in a Recycling Stream" (2014)
- US10014562B2,
- Sloop, S. E.; Parker, R. US Patent #8,823,329 "Discharging of batteries" (2011)
- US8823329B2, and
- Sloop, S. E. US Patent # 8,497,030 B2 "Recycling Batteries Having Basic Electrolytes" (2009), US8497030B2.

7.11.16 AQUA METALS (USA) AND YULHO MATERIALS (S. KOREA)

A cooperation existed between the South Korean manufacturer of storage solutions and battery components Yulho and the US-based battery recycling Aqua Metals. Yulho collects and turns battery manufacturing scrap and S-LIB into black matter through its subsidiary, Yulho Materials. It is constructing an 8,000-ton-per-year high-purity black mass facility. According to Aqua Metals, it will be the biggest in South Korea. Production is anticipated to begin later this year, with plans to increase capacity to 24,000 tons later (Figure 7.24) (https://www.bestmag.co.uk/aqua-metals-in-tie-up-with-south-korean-storage-solution-and-battery-materials-company-yulho/).

FIGURE 7.24   Aqua metals plant in S. Korea.

7.11.17   SungEel HiTech (Gunsan, Jeollabuk-Do, South Korea)

The company (SHA and SHT) was established in 2000 in Korea. To focus on LIB recycling. Pretreatment plant has been used and operated all kinds of S-LIBs since 2008. The hydrometallurgical plant has started to operate since 2011 to recover metals (Co, Li, Mn, Ni, Cu, and Al) with a more than 95% recovery rate based on metal oxide powder. Use mechanical + (thermal) + hydrometallurgical route. The process is an eco-friendly, closed-loop supply chain only in Korea. The annual capacity of an industrial plant is 8,000 t of S-LIBs and scraps. Figure 7.25 shows the process steps of the plant. EVs, mobile phones and IT devices, e-mobility, energy storage systems (ESS), power tools, and manufacturing scraps are used as waste in the plant. High-purity battery materials such as $CoSO_4$, $NiSO_4$, Li_2CO_3, $MnSO_4$, electrolytic Ni, and Cu can be produced from black mass. Technology reduces 40%–67% carbon footprint compared to metals mined in traditional mines and smelters (https://www.sungeelht.com/en)

To prevent fire, the LIBs (mostly production waste) are pulverized into water. The ground material is then mechanically separated into three different groups: the ferrous metal fraction, which is sent to a processing plant for ferrous metals; the Cu and Al foils, which are sent to a processing plant for non-ferrous metals; and the battery active mass, which is sent to an internal hydro-process (leaching, precipitation, etc.) for the metal-salts' final recovery. The recovery of lithium phosphate is reportedly possible but has not yet been verified. SHA produced 1,000 tons of Co, 720 tons each of Mn, Ni, and Li (Li_2CO_3 equivalent) in 2018. The installation of a new hydrometallurgical system at SHT's headquarters factory in Gunsan was announced for 2020. The site's capacity for LIB was subsequently increased by a

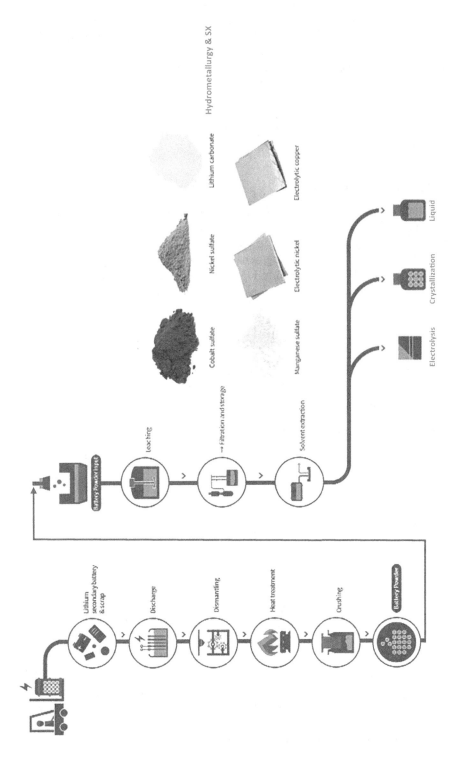

FIGURE 7.25 SunEel HighTech process flowsheet.

factor of three to 24,000 tpy. Outside of Korea, SungEel has bases. SMCC, a joint venture between SungEel HiTech and Metallica Commodities Corp. LLC (MCC, with US headquarters), was announced as being established in New York, US, in 2018. According to the announcement, the joint venture will initially recycle more than 3,000 tons of S-LIBs per year. It was planned to open this joint venture plant in New York in 2020. According to the permit, the joint venture facility's primary processes are: (1) drying/heating in a rotary dryer (T < 600°C), (2) grinding, and (3) hydrometallurgical process. As a result, the method includes a phase for thermal pretreatment.

7.11.18 KOBAR LTD. (S. KOREA)

Kobar Ltd, a specialized recycler of rechargeable batteries such as NiCd, NiMH, and LIBs, is based in Seoul, Korea. Kobar uses mechanical crushing and subsequent pyrometallurgical procedures to recycle NiCd and NiMH batteries, yielding Fe-Ni alloys and Cd metal. Crushing, screening, and hydrometallurgical processing are used to recycle LIBs, as shown in Figure 7.26. The major outcome of recycling LIBs is co-containing powders. Two fractions are created following H_2SO_4 leaching:

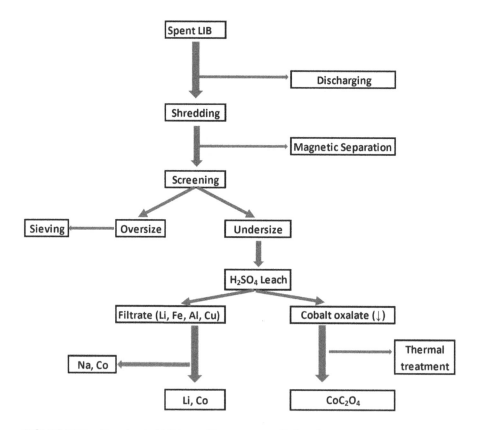

FIGURE 7.26 Flowsheet of LIB recycling process at Kobar plant.

(i) cobalt oxalate, which was most likely synthesized by precipitation with ammonium oxalate; and (ii) a filtrate that contained Li, Fe, Al, and Cu. The filtrate's post-treatment, however, is unclear. According to reports, Kobar has a yearly recycling capacity of 900 t for LIBs and 1,200 t for NiCd and NiMH batteries (http://www.kobar.co.kr/).

7.11.19   ReCell

ReCell, a Lithium-Ion Recycling Center that would host a multi-institution initiative led by Argonne National Laboratory, was recently announced by the US Department of Energy (DOE). The center's goal is to create a closed-loop recycling R&D process that focuses on new materials and techniques in order to increase the battery recycling industry's commercial viability.

7.11.20   Duesenfeld Process (Germany)

Duesenfeld claims eco-friendly recycling of LIBs process. Duesenfeld combines mechanical, thermodynamic, and hydrometallurgical processes in a patented process. The process claimed to achieve the greatest material recovery rates with little energy consumption. This is made possible by a process control with low temperatures, in which toxic HF is not produced. Exhaust gas scrubbing is not necessary for the mechanical processing step. The fluorides are removed in a targeted and safe manner in hydrometallurgy. Duesenfeld process capacity is about 3,000 t/y and can be considered a pilot scale. Products are Co, Ni, and Mn as active materials (https://www.duesenfeld.com/recycling_en.html)

Duesenfeld runs the only recycling process that provides graphite, electrolytes, and Li for material recycling in addition to conventional metals. In the process of recycling materials, all metals are recovered at high recovery rates in the form of premium secondary raw materials that can be used to make batteries. Compared to the primary extraction of raw materials, the Duesenfeld recycling process reduces CO_2 emissions by 8.1 tons for every ton of recycled batteries. The Duesenfeld process saves 4.8 tons of CO_2 per ton of recovered batteries compared to conventional pyrometallurgical processes, and 1 ton of CO_2 per ton of recycled batteries compared to mechanical processes with exhaust gas scrubbing.

For LIBs, the Duesenfeld process utilizes mechanical processing to obtain a material recovery rate that is more than twice as high as traditional recycling techniques. It is feasible to recycle materials almost completely when combined with hydrometallurgical techniques. EoL batteries are typically carried in battery transport containers and are categorized as hazardous commodities. By mechanically processing the materials with on-site modular recycling devices to separate the electrolyte from the other components, the need for specialized hazardous goods transport containers is eliminated. Since these intermediates are transported in regular containers, a truck may carry seven times as much of them. A significant portion of the total cost of recycling batteries is saved because of this decrease in the transit of dangerous items. Figure 7.27 depicts the Duesenfeld LIB recycling process flowsheet.

Duesenfeld Recycling

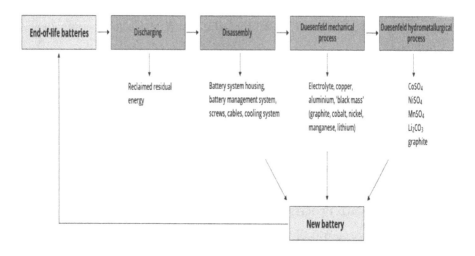

FIGURE 7.27   LIB recovery flowsheet at Duesenfeld Recycling.

The efficiency of modern state-of-the-art LIB recycling is currently 32%. The comparison of recycling rates is shown in Figure 7.28 without the use of battery housings, fastening mechanisms, screw connections, cabling, or electronics. With the processing of the black mass in Duesenfeld hydrometallurgy, the material recycling rate rises to 91%. Duesenfeld obtains a LIB recycling rate of 72% in mechanical recycling. The only parts of the electrolyte that are not currently recovered are the separator film and the high boiler portion. As a result, Duesenfeld goes considerably beyond what is now required under EU Battery Directive 2006/66/EC.

Discharging: The recycling of LIBs starts with the patented deep discharge of the batteries and the recovery of the energy. The deep discharge of the batteries using the Duesenfeld process ensures the safe and efficient discharging of cells, modules, or packs connected in series. The intelligent control software enables an automated deep discharge of the batteries independent of various factors such as the state of charge, the voltage, the age of the battery, or the manufacturer. Due to the high linear discharge power, a large throughput and the highest possible efficiency is achieved. The recovered electricity can be used to operate the recycling plant or fed into the grid.

The Duesenfeld discharging systems guarantee maximum safety and protection against incorrect operation through permanent software monitoring during the discharging process. For example, the connected LIBs can be safely and easily replaced during operation through quick contact. There is no further electrical risk involved in the subsequent process steps. At the same time, there is no need for high-voltage authorized personnel for the subsequent optional disassembly of LIB packs. Duesenfeld discharging of LIBs ensures employee protection, process safety, and high efficiency. Figure 7.29 shows the industrial scale Duesenfeld discharge equipment.

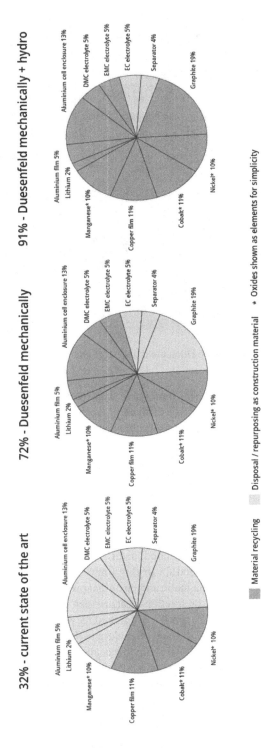

FIGURE 7.28 Comparison of the current state-of-the-art and Duessenfeld recycling efficiency of a LIB cell.

FIGURE 7.29   Industrial scale Duesenfeld discharge equipment.

Mechanical Processing: The flammable electrolyte and hazardous components make mechanically treating LIBs a difficult process. Duesenfeld has created and patented a method for secure reprocessing that gets rid of the particular risks involved. Following discharge and disassembly, the batteries are ground up under the atmosphere of an inert gas, and the electrolyte solvent is then extracted from the ground-up material by vacuum distillation. Because poisonous gas generation is prevented by a low process temperature, there is no need to scrub exhaust emissions. The separated solvent has a very high purity level because the production of HF from reaction products is avoided. The solvent is sent to the chemical industry for further processing. The plant is also able to process dry electrode scraps from battery cell production. No retooling is necessary and the shredded material can be taken to the next stage of the sorting process without the need for drying time. Based on physical characteristics such as grain size, density, magnetic and electrical properties, the shredding material is divided into various material fractions that are then processed metallurgically. The fractions of Fe, Cu, and Al are sent to recognized recycling channels. Duesenfeld has created a patented hydrometallurgical procedure to prepare the so-called black mass, which contains the electrode active components and the conductive salt. This method allows for the extraction from the black mass of the metals Co, Li, Ni, and Mn as well as graphite. Figure 7.30 shows the recovered electrolyte in the collection tank.

Hydrometallurgy: Only Co and Ni are recovered in the majority of the present industrial hydrometallurgical techniques for processing the black matter. These activities result in the loss of Li, Mn, and graphite, which removes them from the material cycle. Duesenfeld has created and patented a technology that enables the

FIGURE 7.30   Recovered electrolytes in the collection tank.

complete recycling of electrode active components through the manufacturing of battery-quality raw materials.

The fluorine-containing conductive salt, which can cause HF to develop during wet chemical processing, presents a unique problem in the hydrometallurgical processing of the black material. Duesenfeld thoroughly eliminates the fluoride before leaching using a proprietary, particular pretreatment procedure, effectively preventing the creation of HF. Following the removal of the fluoride, the metals are leached and subsequently removed from the graphite. Li, Co, Ni, and Mn are then separated from one another using various extraction techniques, purified, and recovered as salts. The salts are used as raw ingredients to create fresh CAMs. The recovered graphite is depicted in Figure 7.31.

7.11.21   ERLOS GmbH (Germany)

In recent years, the 2002-founded and Zwickau, Germany-based recycler of plastic waste (particularly from the automobile industry), ERLOS GmbH, has pursued this direct recycling approach for recycling of LIB. In the direct recycling process, cathode or anode materials are separated by mechanical separation, reconditioned and then directly reused for manufacturing LIBs, which has also drawn many people's interest. The mechanical separation of electrodes is the first step in recycling, followed by washing, filtering, and drying (Figure 7.32).

FIGURE 7.31   Recovered graphite with Duesenfeld process.

The following is a description of the process in detail: (i) The battery packs are entirely disassembled to battery cell level after discharge (excess energy used within the organization), primarily by hand. (ii) The pouch cells then move into an enclosed chamber that is filled with inert gas to avoid fire or explosion. Here, with the aid of robots, the pouch cells are automatically opened and split into anodes—an exhaust system with an activated carbon filter that needs to be changed regularly to collect the electrolyte vapor. In addition to the electrodes, the separator foils are gathered for thermal use in incineration plants or plastic recycling. (iii) In the subsequent cleaning procedure, the cathode and anode electrodes are handled separately. Al-foil and the coating (active mass containing Ni, Mn, Co, and C) are separated from one another by treating the cathode with water and/or a sodium hydrogen carbonate solution at 20°C–30°C and a maximum of 90 bar air pressure. (iv) Marketable recycled cathode materials are obtained from the liquid fraction containing the cathode materials through filtering, pressing, washing, and drying in a gas oven. These materials can be used in part to create new cells and cathodes. In comparison to manual processing, this automation provides an additional safety measure. However, it is necessary to combine primary cathode materials with recycled cathode materials. The cathode's collected aluminum foils are all delivered to a separate metal recycler. (v) The anode is handled in a similar manner, with the Cu-foil and the active mass of the graphite being separated and recovered. However, alternative uses for the recycled graphite material exist in addition to making battery anodes. Cu-foils from the anode are all collected, dried, and sold for recycling to third parties. (vi) Sewage treatment facilities purify and clean washing water because it contains parts of the electrolyte and other impurities (Sojka et al., 2020).

    Erlos claims that while this type of recycling method can be used with pouch and prismatic cells, it is less appropriate for cylindrical cells. Additionally, direct

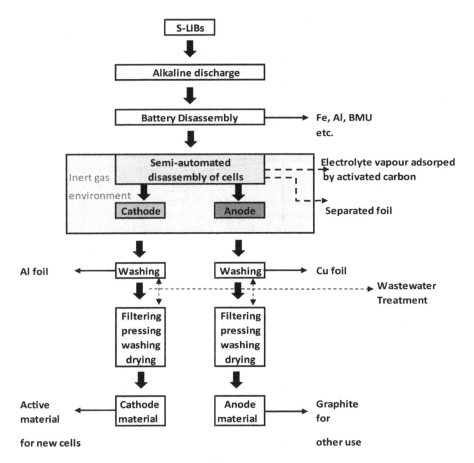

FIGURE 7.32 Erlos uses a direct recycling process that includes mechanical separation and reconditioning.

recycling may have trouble processing feedstock with unclear or ill-characterized provenances due to cross-contamination, and there may be commercial resistance to material reuse if product quality is compromised. Therefore, treating production waste or a pure type of material rather than battery mixes may be more pertinent. Additionally, the product's quality needs to be demonstrated.

7.11.22   NEOMETALS (NMT) (AUSTRIA) & SMS GROUP (GERMANY) J.V.

Neometals use mechanical and hydrometallurgical processes to produce Co, Ni, Cu, Li, and graphite from LCO and NMC spent and scrap LIBs recycling. The process flowsheet and EV battery value chain are given in Figure 7.33. In the first stage, shredding is performed to separate black mass from plastics and metal foils. In stage 2, carbon and metal refining is achieved by hydrometallurgical methods. A recycling commercialization JV ("Primobius GmbH") with the German company SMS group (a leading supplier and constructor of metallurgical plant) has resulted from 3 years of R&D research, including bench and pilot experiments, feasibility studies, and

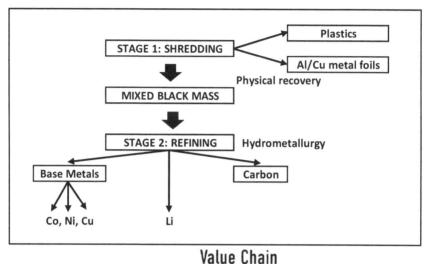

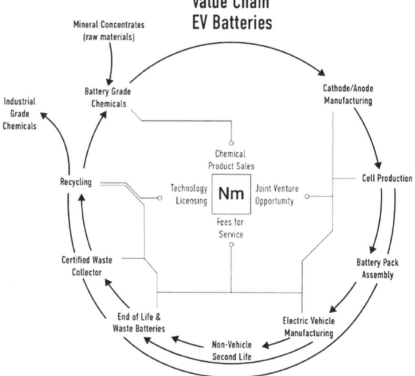

FIGURE 7.33 The process flowsheet and EV battery value chain for Neometals/Primobius.

engineering. Prior to a decision on commercial deployment, the JV will jointly fund the development of a showpiece LIB recycling demonstration plant in Germany and finish the feasibility and market analysis. Demonstrating plant capacity was increased from 1 to 10 t/d operations. The advantages of NMT technology are: (https://www.neometals.com.au/)

- A transparent and reliable route to commercialization and cash flow through the Primobius JV;
- Innovative, pending-patent technology;
- Eco flowsheets allow for a true closed loop with a reduced CO_2 footprint, which helps to cut down on greenhouse gas emissions;
- Recognition of a variety of battery chemistries;
- There is no necessity to release S-LIBs;
- Flexible business models;
- High purity finished cathode production eliminates the requirement to sell mixed metals to refiners as an intermediary product; and
- The hub and spoke design eliminates the need for co-located shredding and refining, which lessens the difficulties associated with transporting hazardous trash.

Figure 7.34 shows the Neometal LCO and NMC LIB recycling plant photos.

FIGURE 7.34 Neometal LCO and NMC LIB recycling plant photos.

7.11.23 ACCUREC RECYCLING GMBH (GERMANY)

Accurec Recycling GmbH, founded in 1995, is a dedicated battery recycling company based in Germany for all kinds of modern rechargeable batteries. Concerning LIBs, both portable and industrial applications, including automotive batteries, can be recycled at Accurec, which started R&D for portable LIBs in 2011 and EV LIBs in 2013. Krefeld facility construction started in 2015 and plant opening was accomplished in 2016 with a capacity of 4,300 tpy LIB. In 2019, Accurec reached a treatment capacity of 3,000 tpy of LIBs. The Accurec process includes thermal treatment + mechanical + pyrometallurgical + hydrometallurgical steps. The company uses mechanical and electric furnace processes. EcoBatRec process was developed in 2016. Capacity is 6,000 tpy. Co-alloys and Li_2CO_3 are produced. The recycling process is flexible and continuous; there is no need for the battery discharging; the process is robust; the process is energy neutral; and HF emission is 0%. Figure 7.35 shows the Accurec LIB recycling technology steps. Batch-wise processing of each Li sub-chemistries; highest safety at processing, no emissions of hazardous electrolytes, low energy consumption, and highest occupational worker safety are some characteristics of the process.

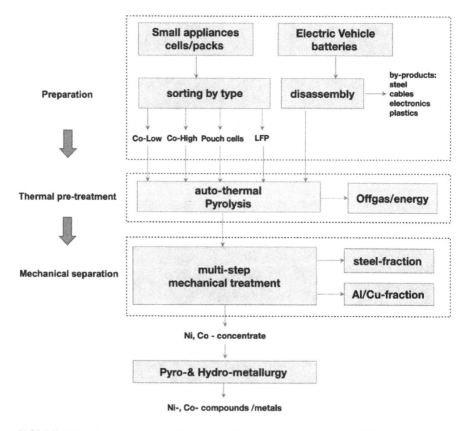

FIGURE 7.35   Current Li-ion batteries recycling process at Accurec (2020).

Through vacuum thermal pretreatment, mechanical separation, which uses sieves, air separation, vibrating screens, and/or zigzag separators, Accurec recycles LIBs. Batteries must first be ready, which includes sorting, disassembling, and discharging (according to needs). After that, Accurec uses thermal treatment (currently provided by an outside provider in a rotary kiln) to pyrolyze and shatter the organic components. To prevent the oxidation of certain non-noble metals like aluminum, the temperature in this rotary kiln treatment is restricted to 600°C. An afterburner and quenching system purify the off-gas, and extra energy from burning organic materials is converted into high-pressure steam and utilized in industrial operations. After that, Accurec's internal multi-step mechanical separation plant screens, crushes, and classifies/sorts the pyrolysis battery cells.

While the Co-Ni concentrate can be sent to pyrometallurgy and subsequent hydrometallurgy plants for final recovery of Co- and Ni-salts or -metals, this treatment results in the formation of metallic fractions, producing a Fe-Ni/steel fraction, an Al fraction, and some Al-Cu fraction. The remaining portion is mixed with a binder before being squeezed and agglomerated into briquettes. These briquettes are now reduced in two stages using pyrometallurgy. The briquettes are processed in two stages: first, at 800°C in a rotating furnace, and then, at 1,200°C in an electric arc furnace. Afterward, pure Co is obtained, but Mn and Li are lost in the process. However, by using hydrometallurgy to leach the slag, Li may be recovered from it. As technology advances, Accurec is now creating and installing new techniques, which will be operational by March 2021 at the company's Krefeld Li-ion battery recycling facility. The recovery of Li and graphite is a goal for a later phase in addition to Ni, Co, and Cu.

The performance of batteries and accumulators has changed significantly as a result of battery evolution, but components and risks have also altered. While uncontrolled fire propagation is the significant risk of today's battery systems, leaching has historically been the most dangerous concern of storage with prior battery chemistries. A very secure LIB storage and safety approach has been established by Accurec at its Li battery recycling facility in Krefeld. All bins are opened before storage following the unloading of customers' deliveries in order to inspect the batteries' packaging and insulation. Containers that don't adhere to the rules are repackaged or sand-filled (Figure 7.36). After that, the storage of battery trash for sorting and treatment is constantly watched over 24 hours per day, 7 days per week. Every half hour, day and night, stationary infrared cameras scan each bin to look for hot areas and stop any heat from spreading.

7.11.24 NICKELHÜTTE AUE (GERMANY)

In Aue, Germany, where it was established in 1635, Nickelhütte Aue GmbH (NHA) is a pyrometallurgical works and recycling business. NHA specializes in recycling non-ferrous metals from waste products. The facility is particularly interested in used catalysts that contain valuable metals like Ni, Cu, Co, V, and Mo from the chemical and pharmaceutical industries as well as the petrochemical industry and the hydrogenation of oils and fats. However, as long as they contain one or a combination of the aforementioned metals, they can also handle various types of wastes such ashes, dusts, grindings, liquids, and slurries. NHA began participating in the recycling of

FIGURE 7.36   Accurec storage facility.

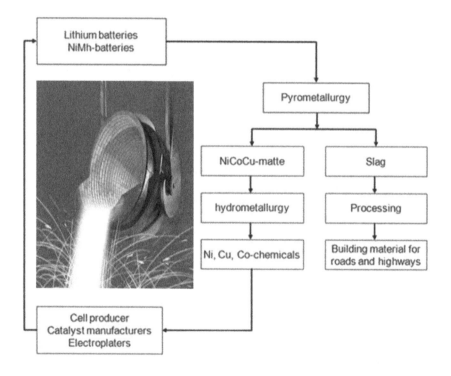

FIGURE 7.37   Li-ion & NiMH batteries recycling process flowsheet of NHA.

LIBs in 2011 (https://nickelhuette-aue.de/en/about). Pyrometallurgy, hydrometallurgy, and thermal pretreatment make up the bulk of the recycling process used. Figure 7.37 depicts the flowsheet for NHA's battery recycling procedure. The batteries can be melted in a batch-wise short drum furnace during the pyro-treatment stage to create NiCoCu-Matte. The business also partially pretreats the batteries thermally using a rotary kiln. There is presently no material available that goes into greater detail on the final hydrometallurgy process. Thermal rotary kiln pretreatment is supposed to increase processing capacity to 7,000 t/a.

7.11.25   EUROPE'S TOP EV AND BATTERY CHAIN COMPANIES

Figure 7.38 shows the top EV and battery chain companies in Europe. They are Britishvolt, ACC, VW, Tesla, Mercedes, BYD Co., LG Energy Solutions, SK Innovation, Samsung SDI, and Stellantis.

7.12   CHALLENGES, AND FUTURE OUTLOOK OF EVs

By 2026, the EV LIB market is anticipated to reach a value of about 90 B$. As indicated by the fluctuating price of raw materials, particularly Li and Co, the rapid, continuous adoption rates of EVs represent a threat to the materials supply chain. There is little question that LIB recycling will be crucial to the supply of key materials. To ensure long-term material availability and to stabilize the LIB supply chain, the US Department of Energy (DOE) has recognized LIB recycling as a crucial necessity. Recycling is actually thought to be a lever with the power to lower future battery costs and energy consumption, lower the cost of pristine resources, and lessen dependency on imported materials (Chen et al., 2019).

As previously said, each of the processes—pyrometallurgical, hydrometallurgical, and direct recycling—has pros and cons of its own. The high Co concentration of LIBs for portable electronics is a key component of the business models for both pyrometallurgical and hydrometallurgical technologies, both of which have been commercialized.

FIGURE 7.38   Europe's top EV and battery chain companies.

However, as they move toward having less and less Co content, these revenue models may become more difficult for EV batteries. To have a significant impact, direct recycling techniques must be developed beyond the lab scale. As a result, the sustainability of EVs depends on the development of adaptable processing techniques that can extract the maximum amount of tangible value from both current and future battery generations. Better recycling technologies are unfortunately required because none of the aforementioned recycling methods currently offers an affordable solution to the extremely dynamic input streams of present and future LIBs. It is commonly acknowledged that the key to effective collection and, ultimately, sustainable closed-loop recycling is a value-based "pull system" in which value extracted outweighs recycling expenses.

Although there are many battery recycling technologies available today, none of them yet provide the ideal solution, necessitating further work. Research on LIBs recycling must keep up with LIB materials research, which is developing quickly and introducing new products to the market. This necessitates LIB recycling systems that are adaptable, practical from an economic standpoint, reliable, and capable of high recycling efficiency.

7.13   THE LIMITS AND OPPORTUNITIES OF THE URBAN MINES

Urban mining supplies secondary raw materials from wastes for sustainable development and circular economy. Recovery means "reuse" or recycling. Overall recycling efficiency is the product of single-step efficiencies (collection, dismantling & pre-processing, and smelting & refining) (Figure 7.39). There is a series loss for collection (about 50%). Thirty percent loss is usual for dismantling and pre-processing. Smelting and refining efficiencies are high (about 95%), and the overall recycling efficiency is about 33% (Hagelüken and Grehl, 2012).

Globally, battery collection rates fluctuate between 5% and 15% for small electronic equipment and 80% for EVs (Danino-Perraud, 2020). Compared to small electronics,

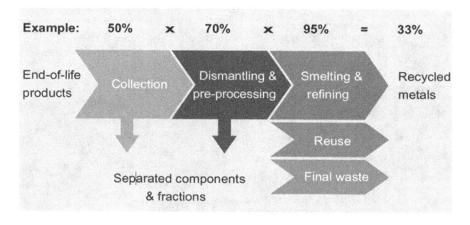

FIGURE 7.39   The main steps and efficiencies of the recycling process chain.

EV batteries have much better rates of collection, recycling, and recovery. Additionally, EVs make up the largest and fastest-growing segment of the battery market. When EV sales reach their EoL in 10–15 years, they will emerge as a significant supplier of energy.

The number of secondary resources in the European urban mine for 12 metals was calculated by the prospecting secondary raw materials in the urban mine and mine waste (ProSUM) project. China recycled about 40% of battery materials. 70% of Li, 67% of Co, 77% of Li, and 95% of graphite were lost in 2016 (Danino-Perraud, 2020).

7.14 SUMMARY

In the near future, there will be a large number of S-LIBs due to the significant increase in EV production. It is important to recycle them properly because they contain a variety of valuable metals and toxic materials. Many recycling techniques have been devised with the goal of recovering the metal values from S-LIBs, and some of them have even been industrialized. All of the S-LIBs' components should be recoverable using a recycling procedure that uses less energy and doesn't harm the environment. There have been advancements in technology research that can currently meet these needs. Due to the variety of materials in LIBs, practically all of them need time-consuming processing processes or expensive reagents. Therefore, there is still much work to be done to develop more potent recycling systems (Huang et al, 2018).

There are other difficulties that the recycling of LIBs must overcome. First off, unlike lead-acid batteries (LABs), LIBs' chemistry and technology are constantly developing, leaving recycling techniques in the dust. For instance, if the cathodes are made of Co-free materials, the pyrometallurgical method might not be economically viable. Second, the issue of cross-contamination of battery type in the recycling stream may become an issue as the LIBs recycling business develops. For use in situations where they can be interchanged, such as on electric bicycles or reduced EVs, LIBs and LABs, for instance, may be made to be geometrically equivalent. This could result in the addition of LIBs to the Pb smelters' input stream, which could cause contamination, fires, and explosions. Thirdly, there aren't any ideal laws, rules, or standard systems for recycling LIBs that can regulate the process and guarantee the secure collecting, transport, and management of used LIBs throughout the recycling procedures.

Finally, additional work may focus on, but not be limited to, the following in order to hasten the construction of an effective recycling system for S-LIBs. (i) Recognizing and classifying S-LIBs for various recycling methods. (ii) Easy-to-use marking or labeling of LIBs throughout production and recycling, which could aid in organizing and monitoring the recycling. (iii) Planning LIBs for recycling and avoiding complex or irreversible assembly. (iv) Quickening the pace of legislation to impose uniformity on recycling procedures.

8 Solution Purification

ABBREVIATIONS

AAS	Atomic absorption spectrophotometry
CAM	Cathode active material
CYANEX 272	Bis(2,4,4-trimethylpentyl) phosphinic acid
DEHPA	Diethylhexyl phosphoric acid
D2EHPA	Di-(2-ethylhexyl) phosphoric acid
ICP	Inductively coupled plasma
NMC	Nickel mangan cobalt
PC-88A	Polytetrafluoroethylene 2-ethylhexylphosphonicacid mono-2-ethylhexylester)
PLS	Pregnant leach solution
PVDF	Polyvinylidene fluoride
O/A	Oil/Aqua
SEM	Scanning electron microscope
S-LIB	Spent lithium-ion battery
SX	Solvent extraction
TOA	Trioctylamine
XRD	X-ray diffraction

8.1 INTRODUCTION

The purification phase is where valuable elements are first separated and recovered from leaching solutions. Some S-LIB components, including Fe, Al, Cu, Mn, Ni, Co, and Li, dissolve and then end up in leaching solutions. The high-value components in the leachate must then be extracted and recovered. On the other hand, removing impurities like Fe, Cu, and Al is crucial for creating high-quality goods. Numerous strategies have been introduced to achieve this, including precipitation, sol-gel method, electrochemical deposition, and solvent extraction (SX). Due to the complexity of the leaching solution composition, it is difficult to separate all the high-value elements using a single method. To separate and recover high-value metallic elements from the leachate, two or more techniques are typically used. A practical method for separating and recovering metals from leaching solutions is the integration of chemical precipitation and SX (Yao et al., 2018).

8.2 LEACHATE PURIFICATION

By using SX and precipitation, leach liquor can be cleaned. Since Co, Ni, and Li losses were kept to a minimum (0.2%, 2%, and 4%, respectively), only partial removal of Mn, Cu, and Al (16%, 12%, and 6%, respectively) was possible during

DOI: 10.1201/9781003384557-8

the first purification of PLS. Al, Mn, and Cu were all removed in up to 93% of the two-stage D2EHPA extraction. Ni is left in the aqueous phase after Co-Ni separation with Cyanex 272, but if Al, Mn, and Cu are also present, these metals are also co-extracted in the organic phase. To remove such metals, D2EHPA extraction was introduced before Cyanex 272 (Granata et al., 2012; Pagnanelli et al., 2016).

8.3   SEPARATING IRON (Fe)

The only impurity in the work described in this article is iron. Iron ions can be completely separated from the solution by adjusting pH. According to the following equations (Zou et al., 2013), the solubility of a metal hydroxide, $M(OH)_n$, depends on the solubility equilibrium and pH:

$$K_{sp} = [M^{n+}] \, [OH^-]^n \tag{8.1}$$

$$K_w = [H^+] \, [OH^-] = 10^{-14} \tag{8.2}$$

$$pH = -\log 10(H^+) \tag{8.3}$$

Table 8.1 shows the pH number of the start and end of precipitating various ions. Figure 8.1 shows schematically various ions precipitating pH areas. Firstly, $Fe(OH)_3$ precipitates between 1.149 and 2.815 pH values. Then $Al(OH)_3$ precipitation finishes at a pH of 4.49. Then $Ni(OH)_2$ starts to precipitate at a pH of 5.156. $Fe(OH)_2$ starts to precipitate at pH 5.844. $Co(OH)_2$ starts to precipitate at a pH of 6.673. $Mn(OH)_2$ precipitates at 7.389. $Cu(OH)_2$ precipitation finishes at a pH of 6.65.

8.4   COPPER REMOVAL

To recover Co and prevent the co-precipitation of CuC_2O_4, Cu^{2+} must be removed from the solution before oxalic acid precipitation. In a wide range of solution pHs, Cu^{2+} can be selectively separated from other metal ions. Due to the low solubility of

TABLE 8.1
The pH Value at the Beginning and End of Different Ions Precipitating

Substance	K_{sp}	C_{Max} (M)	C_{Min} (M)	pH Start	pH End
$Fe(OH)_3$	$2.79*10^{-39}$	1.00	$1*10^{-5}$	1.149	2.815
$Al(OH)_3$	$3.00*10^{-30}$		$1*10^{-5}$		4.49
$Cu(OH)_2$	$2.00*10^{-20}$		$1*10^{-5}$		6.65
$Ni(OH)_2$	$5.48*10^{-15}$	2.667	$1*10^{-5}$	5.156	8.869
$Co(OH)_2$	$5.92*10^{-15}$	2.667	$1*10^{-5}$	6.673	9.386
$Mn(OH)_2$	$2.00*10^{-13}$	3.200	$1*10^{-5}$	7.389	10.151
$Fe(OH)_2$	$4.87*10^{-17}$	1.00	$1*10^{-5}$	5.844	8.344

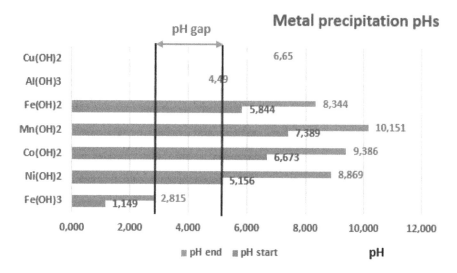

FIGURE 8.1   Various metal ions precipitation pH ranges. (Compiled from Zou et al., 2013.)

CuS in acid solutions, the sulfide precipitation method is more frequently used than the hydroxide and carbonate precipitation methods. Below are some possible reactions at 25°C and 1 atm. pressure (Kang et al., 2010):

$$H_2S \leftrightarrow 2^{H+} + S^{2-}$$
(8.4)

$$CuS_{(s)} \leftrightarrow Cu^{2+} + S^{2-}$$
(8.5)

$$Cu^{2+} + H_2S \leftrightarrow CuS_{(s)} + 2^{H+}$$
(8.6)

Barik et al. (2017) removed Cu at a pH of 5.5 as $Cu(OH)_2$ precipitate after Mn and Al removal from HCl leach PLS.

8.5   PRECIPITATION OF Mn

By performing a series of tests using an exact stoichiometric amount of sodium hypochlorite (NaOCl) solution at different pH values (1.0–2.5) of the feed solution at 30°C for 1 hour, the precipitation behavior of Mn was investigated. The pH increase from 1.0 to 1.5 initially increased the precipitation efficiency of Mn from 42.5% to 88.4%, and the pH increase to 2.5 further decreased it to 48.3%. With an increase in pH from 1 to 1.5 in the case of Co, the precipitation efficiency increased from 8.7% to 31.1% and remained stable after 1.5. Due to Mn's lower solubility (Eq. 8.7), the pH of 1.5 produced the highest precipitation efficiency of Mn. The re-dissolution of precipitated Mn as per Eq. 8.8 may be the cause of the decrease in precipitation efficiency of Mn beyond pH 1.5 (Barik et al., 2017):

$$MnCl_{2(aq)} + 2NaOCl_{(aq)} \rightarrow MnO_2 + 2NaCl_{(aq)} + Cl_{2(g)} \tag{8.7}$$

$$2MnO_{2(s)} + 3NaOCl_{(aq)} + 2NaOH_{(aq)} \rightarrow 2NaMnO_{4(aq)} + 3NaCl + H_2O \tag{8.8}$$

8.6   AL REMOVAL

Barik et al. (2017) removed Al at a pH of 4.5 as $Al(OH)_3$ precipitated after Mn removal from HCl leach PLS.

8.7   PRIMARY PURIFICATION: CHEMICAL PRECIPITATION

Another method frequently used for extracting and recovering high-value elements from leaching solutions is chemical precipitation. With this technique, it is possible to precipitate out precious metals and get rid of tiny amounts of impurities like Fe, Cu, and Al. The solubility of the metal compound under particular temperatures and pH values determines the separation mechanism of chemical precipitation, which must be carefully adjusted during the precipitating process (Dobo et al., 2023).

The commonly used precipitants are carbonates (CO_3^{2-}) containing salts like $(NH_4)_2CO_3$ and Na_2CO_3 as they can form insoluble compounds with almost all high-value metals, comprising Mn^{2+} (Chen et al., 2020; Meshram et al., 2015), Ni^{2+} (Nayl et al., 2015, 2017), Co^{2+} (Pagnanelli et al., 2016), and Li^+ (Zhu et al., 2012; Peng et al., 2019; Zhao et al., 2019; Zhu et al., 2012). Numerous other precipitants have also been documented, including ammonium oxalate ($(NH_4)_2C_2O_4$) (Nan et al., 2005), phosphoric acid (H_3PO_4), sodium phosphate (Na_3PO_4), and oxalic acid ($H_2C_2O_4$) (Sohn et al., 2006a). In chemical precipitation operations, determining the optimal condition and choosing the appropriate precipitant is crucial to preventing precipitate dissolution. It is possible to 99% precipitate Co with oxalic acid/ammonium oxalate. H_3PO_4 can precipitate Li between 83% and 93% and Na_3PO_4 can precipitate between 92% and 96%. Mn can be precipitated around 94% (Dobo et al., 2023).

50 g/L of Co, 10 g/L of Li, 7 g/L of Al, 5 g/L of Ni, 2 g/L of Fe, 3 g/L of Cu, and 1.5 g/L of Mn make up the leach liquor composition obtained under optimal conditions using acid-reducing leaching (90°C, 2 g of H_2SO_4 for every gram of solid, and 50% of stoichiometric excess of reducing agent). The removal of metal impurities, specifically the complete removal of Fe and Al, and 60% removal of Cu, is determined by primary purification (precipitation at pH 5 for NaOH addition) (Granata et al., 2012). At this pH, precipitation was carried out to achieve the best balance between impurity removal and target metal loss. Increased pH at 6.5 would have caused the target metals to be significantly lost, whereas pH 5 is a compromise that gets rid of impurities while preserving Co and Li.

In a mixture leaching solution of H_3PO_4 and H_2O_2, Li and Co were dissolved by Pinna et al. (2017). NaOH and $H_2C_2O_4$ were added to the solution for precipitation after the leaching process. As a result, 99% of Co was precipitated as 97.8% pure CoC_2O_4, and 88% of Li was recovered as 98.3% pure Li_3PO_4. Due to the similar properties of some elements, such as Co, Ni, and Mn, complex procedures are needed to separate these metals. Co-precipitation of all the target metals and direct sintering of precursors into cathodic materials are effective ways to reduce the precipitation steps and recover valuable elements from complex solutions.

8.8   CO-PRECIPITATION

8.8.1   PRECIPITATION OF CO AS CoC_2O_4

Co can be precipitated as Co_3O_4, $CoCO_3$, $CoC_2O_4 \cdot 2H_2O$, or $Co(OH)_2$. Barik et al. (2017) removed Co using Na_2CO_3 after Mn, Al, and Cu removal from HCl leach PLS. After filtration of the leaching solution to remove impurities, $CoC_2O_4 \cdot 2H_2O$ can be precipitated by adding ammonium oxalate or oxalic acid (pK_{a1}: 1.27 and pK_{a2}: 4.28) in the filtrate. The precipitating process could be expressed by Zhu et al. (2012) and Kang et al. (2010b):

$$CoSO_{4(aq)} + (NH_4)_2C_2O_{4(aq)} \rightarrow CoC_2O_{4(s)}\downarrow + (NH_4)_2SO_{4(aq)} \qquad (8.9)$$

$$Co_2^+ + H_2C_2O_4 \leftrightarrow CoC_2O_{4(s)} + 2H^+ \qquad (8.10)$$

The precipitation of CoC_2O_4 is an endothermic process, the increase in temperature facilitated the formation of CoC_2O_4, resulting in the increase in the recovery rate of Co. Zhu et al. (2012) found that the optimum precipitation temperature was 50°C, pH was 2.0, and $n(C_2O_4^{2-}/n(CO_2^+)$ was 1.2 at 300 rpm impeller speed and 60 min reaction time. Figure 8.2 shows the precipitated $CoC_2O_4 \cdot 2H_2O$ picture and its XRD pattern. It was discovered that less than 0.32% of impurities and 94.7% of the Co was deposited as oxalate. In samples, Li could not be found (Zhu et al., 2012). Filtration can be used to separate the precipitate. Oxalic acid that was still present in the solid residue could be removed by washing with distilled water at 50°C and drying the residue for 24 hours in an oven. AAS is capable of analyzing the chemical analysis of the finished product. With only 0.2% Ni and 0.3% Mn as impurities, the residue was 35% Co. Figure 8.3 shows a pink-colored precipitate and an XRD analysis of its crystalline $CoC_2O_42H_2O$ structure. Figure 8.4 shows that at 150°C, the crystalline form changes to CoC_2O_4, and at higher temperatures, to Co_3O_4. Crystalline Co_3O_4 was the end result, and it contained 79% Co along with the following impurities: 0.1% Fe, 0.4% Ni, and 0.6% Mn (Kang et al., 2010b).

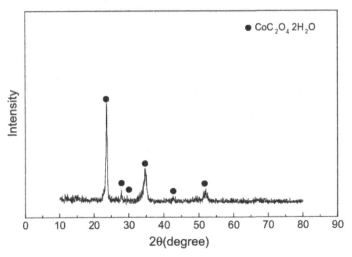

FIGURE 8.2 Crystaine Co-oxalate SEM image and XRD spectra.

8.8.2   PRECIPITATION OF Co(OH)$_2$ FROM CoSO$_4$ SOLUTION

Co(OH)$_2$ can be precipitated using NaOH according to Eq. 8.11. Na$_2$SO$_4$ is produced as a by-product (Zou et al., 2013):

$$CoSO_4 + 2NaOH \rightarrow Co(OH)_2 + Na_2SO_4 \qquad (8.11)$$

By adding one equivalent volume of a 4.0 M NaOH solution, Contestabile et al. (2001) were able to recover the Co that had been dissolved in the HCl solution as Co(OH)$_2$. Co(OH)$_2$ precipitation starts at a pH of 6 and is deemed to be finished at a pH of 8. An ideal method for achieving Co(OH)$_2$ precipitation would be to use a weak base like ammonia solution, which forms a buffer solution at pH 9. Unfortunately, ammonia and Co combine to form stable complexes, which partially dissolve the hydroxide and prevent a quantitative recovery. NaOH is still the best option because it has a strong base and allows for working with small volumes of solution. An adequate pH sensor can be used to regulate this step on an

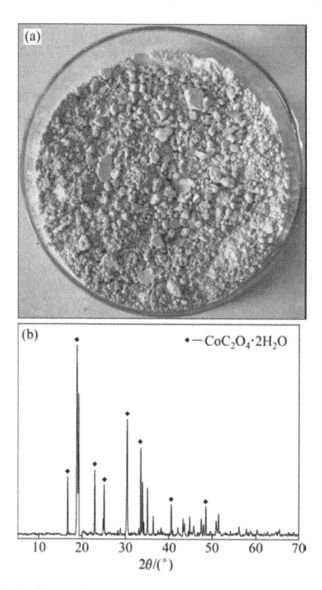

FIGURE 8.3   Precipitated pink-colored $CoC_2O_4 \cdot 2H_2O$ and its XRD pattern.

industrial scale. Filtration makes it simple to separate the $Co(OH)_2$ precipitate from the solution, which can then be recycled.

8.8.3 PRECIPITATION OF $Co(OH)_2$ FROM $CoCl_2$ SOLUTION

These metals could be gradually extracted with the addition of a 40% NaOH solution in accordance with the variations in the solubility product constant between various $M^{n+}(OH)_n$ (Li et al., 2009). At a pH of 7.2, for instance, all Al, Fe, and Cu could theoretically be removed with essentially no loss of Co. To investigate how metal

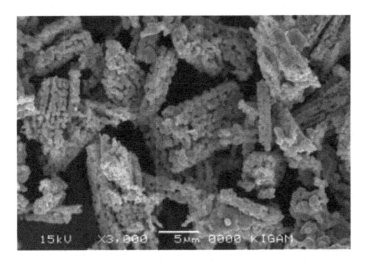

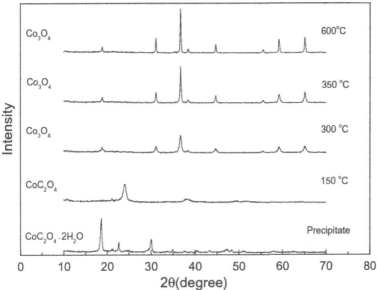

FIGURE 8.4 Conversion of Co-oxalate to crystalline 340 nm Co-oxide with calcination at different temperatures for 2 hours. SEM image and XRD spectra of Co-oxide at different temperatures.

concentrations, change in an acid-leaching solution at various pH levels, experiments were created. Figure 8.5 displays the chemical precipitation results that were determined by experimenting with various metal ion concentrations in solution. They show that as pH was raised, both the Co loss ratio and the impurity removal ratio increased. The pH must be maintained in a roughly range of 4.5–6.0 in order to achieve a low loss ratio of Co and a high removal ratio of impurities (Fe, Cu, and Al).

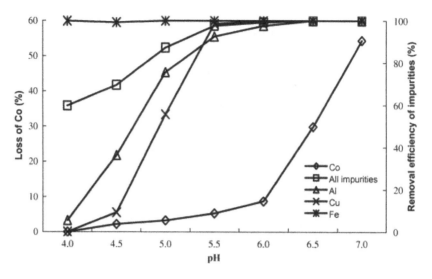

FIGURE 8.5   Removal efficiency of impurities and loss of Co during $Co(OH)_2$ precipitation from $CoCl_2$ solution.

8.9   PRECIPITATION OF Li

8.9.1 Li_2CO_3 PRECIPITATION

Lithium carbonate can be precipitated after recovering CoC_2O_4 by adding an excessive amount of Na_2CO_3 to the leftover filtrate. This system's reaction is illustrated as follows:

$$2Li + Na_2CO_3 \rightarrow Li_2CO_3\downarrow + 2Na^+ \qquad (8.12)$$

The precipitation of Li_2CO_3 is an endothermic reaction, much like the formation of CoC_2O_4, which makes it advantageous to obtain Li_2CO_3 and advantageous for the dissolution of precipitated Li_2CO_3 in the solution. The following are the ideal circumstances for recovering Li_2CO_3: The equilibrium pH is 10, the temperature is 50°C, the Li-ion concentration is 20 g/L, the reaction time is 1 hour, and the agitation speed is 300 r/min. The molar ratio of Na_2CO_3 to Li^+ is 1.1:1.0. Figure 8.6's XRD analysis clearly identifies the crystalline Li_2CO_3 phase. According to the experimental findings, 71.0% of Li was deposited as carbonate with an impurity content of less than 0.52%. In samples, Co was not found (Zhu et al., 2012).

8.9.2   PRECIPITATION OF LiOH FROM Li_2CO_3

Conversion is performed using milk of lime in an agitated tank with the following reaction:

$$Li_2CO_{3(s)} + Ca(OH)_{2(s)} \rightarrow 2Li^+ + 2OH^- + CaCO_{3(s)} \qquad (8.13)$$

FIGURE 8.6 Precipitated white Li_2CO_3 and its XRD pattern.

Crystalline $Li_2CO_{3(s)}$ requires high temperature and intensive washing to achieve > 90% conversion yield. However, intermediate $Li_2CO_{3(s)}$ from hydrothermal leaching requires 20°C–50°C in only 60–90 min residence time to give up to 95% conversion yield.

LiOH solution after conversion is polished by ion exchange for the removal of multivalent cations (such as Ca). Then the solution is vacuum crystallized to LiOH. H_2O with the following reaction:

$$Li^+ + OH^- + H_2O \rightarrow LiOH.H_2O_{(s)} \qquad (8.14)$$

FIGURE 8.7   LiOH.H$_2$O crystals.

In two-stage crystallization, >56.5% LiOH product quality need for LIBs is achieved (Tiilhonen et al., 2020). Figure 8.7 shows LiOH.H$_2$O crystals.

8.9.3   PRECIPITATION OF LiOH FROM Li$_2$SO$_4$ SOLUTION

LiOH can be precipitated using NaOH. Na$_2$SO$_4$ is produced as a by-product.

$$Li_2SO_4 + 2NaOH \rightarrow 2LiOH + Na_2SO_4 \tag{8.15}$$

8.9.4   PRECIPITATION OF Li WITH Na$_3$PO$_3$

Li$_2$SO$_4$ solution can be precipitated by Na$_3$PO$_3$. The precipitation experiment can be carried out at 65°C for 2 hours with a 250 rpm stirring speed. A higher temperature can cause more Li to precipitate because Li$_3$PO$_4$'s solubility decreases as temperature rises. The excess Na$_3$PO$_4$ was then removed from the precipitated Li$_3$PO$_4$ by filtering it and washing it in deionized water heated to 80°C (Li et al., 2017b):

$$3Li_2SO_4 + 2Na_3PO_3 \rightarrow Na_2SO_4 + 2Li_3PO_4\downarrow \tag{8.16}$$

8.10   SECONDARY PURIFICATION: SX

The common method for separating metals from leachate is SX. Target metals and other elements are concentrated and divided into two distinct immiscible liquids during the liquid-liquid extraction process. wherein associated liquids referred to as solvents or extractants are used to present the desired metals. Due to the solvent's excellent selectivity for various metallic ions, this method is a successful method for metal separations. Some benefits of SX include the high yield of pure products and the speedy (30 min) reaction time. The difficulty of the process and the high cost of the solvent, however, restrict its use. Solvents play the most important role in the liquid–liquid extraction process (Dobo et al., 2023).

Table 8.2 presents some SX reagents used for LIB metal extractions. For Co extraction D2EHPA, CYANEX 272, TOA, DEHPA, and PC-88A were tested previously from LIB leach PLSs.

After primary purification, SX is used to test the selectivity of two extractants, D2EHPA and CYANEX 272. To recover high-purity Co products, SX operations aim to separate Co from Ni. According to the following mechanism, metal extraction takes place (Granata et al., 2012b):

$$M_{Aq}^{2+} + A_{org}^{-} + 2(HA)_{2org} \rightarrow MA_2 \cdot 3HA_{org} + H_{Aq}^{+} \tag{8.17}$$

$$Co^{2+}_{(aq)} + 2(RH)_{2(org)} \leftrightarrow Co(R.RH)_{2(org)} + 2H^{+}_{(aq)} \tag{8.18}$$

where the extractant saponified by the following reaction is represented by $A_{org}^{+} + 2(HA)_{2org}$:

TABLE 8.2
SX Reagents That Were Used for LIB Extraction

Chemical Name		Extract	Removal	Separation	Reference(s)
di-(2-ethylhexyl) phosphoric acid	D2EHPA	CoO_2H_4 Li_2CO_3	Cu and Mn Mn, Co, Ni		Chen et al. (2015a), Yang et al. (2017) Chen et al. (2015), Yang et al. (2017)
bis-(2,4,4-tri-methyl-pentyl) phosphinic acid	CYANEX 272	Co Co		Cu-Ni Cu-Ni	Pagnanelli et al. (2016), Kang et al. (2010a), Olushola et al. (2013)
trioctylamine	TOA				
diethylhexyl phosphoric acid	DEHPA				
2-ethylhexyl phosphonic acid mono-2-ethylhexyl ester in kerosene/hexane	PC-88A CYANEX 272 + PC-88A	Co Co		Cu-Ni	Virolainen et al. (2017) Virolainen et al. (2017), Sattar et al. (2019)
5 nonylsalicylaldoxime/ Acorga	Mextral 5640H Mextral 272P	Cu Ni			Pranolo et al. (2010), Nan et al. (2005), Suzuki et al. (2012)
2-ethylhexyl phosphonic acid mono-2-ethylhexyl ester	P507	Co			Chen et al. (2011a,2011b)

$$Na_{Aq}^+ + 1/2(HA)_{2Org} \rightarrow NaA_{Org} + H_{Aq}^+ \tag{8.19}$$

The distribution coefficient of metal in the aqueous and organic phases was used to assess the effectiveness of metal extraction.

$$DCo = Co_{org}/Co_{aq} \tag{8.20}$$

$$DNi = Ni_{org}/Ni_{aq} \tag{8.21}$$

where the residual concentration in the aqueous phase is taken into account in a mass balance to determine the metal concentration in the organic phase.

Then, a measure of the extraction selectivity (β) can be calculated, with Ni serving as Co's primary rival in SX.

$$\beta = D_{Co}/D_{Ni} \tag{8.22}$$

According to experimental findings, the minimum stoichiometric ratio (moles of extractant per moles of Co) required to extract Co quantitatively was four for both SX extractants (D2HEPA and CYANEX 272) (Figure 8.8a and b). All the investigated metals are extracted simultaneously when the amount of D2HEPA is increased above this value, whereas when the amount of CYANEX 272 is increased, Ni begins to be significantly extracted only after Co extraction is complete. According to the trend of the selectivity coefficient, this result indicates that CYANEX 272 has a higher selectivity for Co when compared to D2HEPA. It is clear that CYANEX 272 performs significantly better than D2HEPA for Co-Ni separation at the ideal pH of 6. Using a 4.0 M H_2SO_4 solution (O/A = 1), stripping tests can be carried out to re-extract Co in the aqueous phase (Granata et al., 2012a, 2012b).

Several effective solvents were used, including CYANEX 272 (Kang et al., 2010a; Pagnanelli et al., 2016), di-(2ethylhexyl) phosphoric acid (D2EHPA) (Chen et al., 2015b), and 5-nonylsalicylaldoxime (Acorga M5640) (Pranolo et al., 2010). For the extraction of various metal ions, different solvents are used (e.g., Acorga M5640 for Cu (Nan et al., 2005; Suzuki et al., 2012), D2EHPA for Mn, Co, and Ni

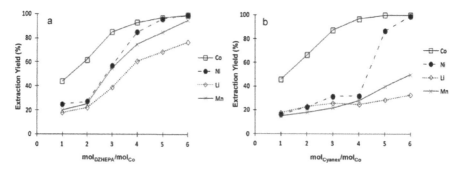

FIGURE 8.8 Effect of stoichiometry on the extraction with D2EHPA (a) and Cyanex 272 (b) at pH 5.5.

(Chen et al., 2015b; Yang et al., 2017), PC-88A and CYANEX 272 for Co (Virolainen et al., 2017; da S. CYANEX 272 has drawn the most interest globally among the aforementioned solvents due to its superior stability and increased effectiveness in extracting Co (Olushola et al., 2013). For instance, using 50% saponified 0.4 M CYANEX 272 at an equilibrium pH range of 5.5–6.0, Co was selectively extracted from the leachate (Kang et al., 2010). It was noted that while Ni is only extracted at a rate of 1%, Co is extracted at a rate of 95%–98%. Olushola et al. (2013) compared different CYANEX solvents for extracting and separating Ni and Co in a different study. CYANEX 923, CYANEX 921, CYANEX 421X, CYANEX 301, and CYANEX 272 are some of the extractants that were examined. According to the findings, CYANEX 272 is the best solvent for separating Co and Ni.

For $LiCoO_2$ wastes, Zhang et al. (1998) used HCl leach and SX. The SX tested 0.29 M D2EHPA and 0.90 M PC-88A. The final PLS had a pH of roughly 0.6 and Co and Li concentrations of roughly 17 and 1.7 g/L, respectively. The Co in the PLS was extracted in a single stage with 0.90 M PC-88A in kerosene at equilibrium pH = 6.7 and an O/A ratio of 0.85:1 with a higher selectivity and nearly completely (at 12.6% Li). After Li was scrubbed with a weak solution of HCl acid containing 30 g/L of Co at a 10:1 O/A phase ratio, the Co in the loaded organic phase was then recovered as cobalt sulfate with high purity (Li/Co $5*10^{-5}$) and the O/A phase ratio was then increased. The displacement reaction shown below can be used to represent the scrubbing process:

$$2\overline{LiA} + Co^{2+} \rightarrow \overline{CoA_2} + 2Li^+ \qquad (8.23)$$

where A represents the monomeric PC-88A anion and the bars over the symbols indicate the species present in the organic phase.

After that, stripping was done using a 2.0 M H_2SO_4 solution with a 5:1 O/A ratio. By adding a saturated Na_2CO_3 solution at just under 100°C, the raffinate was concentrated, and the remaining Li in the aqueous solution was easily recovered as Li_2CO_3 precipitate. Less than 0.07% of Co was found to be present in the Li precipitate. Li recovered at about 80%.

The different solubilities of different metal ions in an organic solvent versus an aqueous liquid are what drives SX. This method has the advantage of a quick reaction time (about 30 min) and a high yield of pure products, but the high cost of solvents and process complexity restrict its use. D2EHPA (di-(2-ethylhexyl) phosphoric acid) and PC-88A (2-ethylhexylphosphonic-acidmono-2-ethylhexylester) were used in the extraction by Wang et al. (2016) to recover the Co from S-LIBs. After removing Cu and Mn with D2EHPA, PC-88A was used to separate Co and Ni. Co was eventually recycled into CoO_2H_4, which had a purity of 99.5%. Successful Li, Ni, and Co separation from spent battery leachate was achieved by Virolainen et al. (2017) using the extractant CYANEX 272 (Bis(2,4,4-trimethylpentyl) phosphinic acid). A Li raffinate with a purity of 99.9% was produced using modified CYANEX 272 by TOA (trioctylamine, a phase modifier that helps to reduce the formation of unwanted organic phases). A 99.7% Ni aqueous solution and a 99.6% Co organic solution were recovered after additional scrubbing and stripping.

CYANEX 272 and D2HEPA (extractants) can be dissolved in kerosene (diluent) until they reach the same molar concentration of Co in the leach liquor (0.84 M). Then they can be partially saponified (65%) by adding a NaOH solution (5.0 M) while stirring. The organic phase was shaken for 5 minutes with 10 mL of purified leach liquor (50 g/L of Co, 10 g/L of Li, 5 g/L of Ni, and 1.5 g/L of Mn). All experiments used the same amount of aqueous phase but varied the amount of organic phase by 10–60 mL. Then, the ratio, O/A, between the volumes of the organic phase and aqueous phase changed from 1 to 6. Working with different O/A ratios means working under various stoichiometric conditions because the molar concentration of the extractant in the organic phase is the same as that of Co in the aqueous phase (O/A = molSLV molCo1, where molSLV are the mol of extractant in the organic phase). A separating funnel was used to separate the two phases after shaking. Leach liquor pH levels were changed by adding NaOH or H_2SO_4 solution in the range of 1–6. According to Granata et al. (2012b), all experiments were carried out in duplicate at room temperature ($25 \pm 1°C$).

The amount of extracted Co^{2+}, Li^+, Ni^{2+}, and Mn^{2+} can be determined by ICP-OES analysis of raffinates (the aqueous phase after extraction). With the same volume of the organic phase and aqueous phase (O/A = 1), stripping tests for metal recovery from the organic phase can be performed using a 4.0 M H_2SO_4 solution at $25 \pm 1°C$.

8.10.1 COBALT RECOVERY

Co can be recovered as oxide or oxalate from the Cyanex extraction's stripping solution. Co can be precipitated as hydroxide for the production of cobalt oxide, which is then produced thermally. Cobalt hydroxide exists in two polymorphic forms, – $Co(OH)_2$ and -$Co(OH)_2$, which mainly differ in the distance between hydroxide layers. Cobalt tends to form –$Co(OH)_2$, which is made up of –$Co(OH)_2$ sheets intercalated with water molecules, when precipitated directly with hydroxide. Based on earlier experimental results showing quantitative precipitation of Co at pH = 11, the pH of precipitation can be set. At 300°C, cobalt hydroxide begins to transform into cobalt oxide, but it takes until 450°C or 500°C to complete the transformation. According to Atia et al. (2019), the main components of the cobalt oxide produced at 450°C are 60 wt% Co, 1.2 wt%, 2.2 wt% Mn, 0.1 wt% Cu, and 0.5 wt% Al. Co-extracted Al, Cu, and Mn precipitated as hydroxides during Co recovery and consequently changed into oxides during final thermal treatment despite D2EHPA and CYANEX 272 treatments. As a result, the hydroxide route is not very selective because other divalent cations and anions are present. Through this process, 82%–83% pure Co_3O_4 can be produced from 85% to 86% of the Co in the LIB powder. Between 94% and 96% of the $CoC_2O_4 2H_2O$ was ultimately pure.

Both solutions from primary purification (NaOH precipitation) and secondary purification (stripping solution after SX) can be used to recover cobalt as carbonate with a 98% yield. To achieve a final pH of 9–10, purified leach liquors can be stirred while being added to a saturated Na_2CO_3 solution. Suspension can be filtered after 2 hours of stirring at room temperature for ICP-OES analysis. To remove soluble salt (such as Na_2SO_4), solid precipitates can be washed with water and then dried for 24 hours at 60°C. Cobalt carbonate samples can be dissolved in water and the resulting solutions can be examined using an ICP-OES to determine the purity of the produced products. About 47% (w/w) of Co is present in the cobalt carbonate that was recovered after secondary purification, compared to 36%–37% (w/w) in the cobalt carbonate that was recovered after primary purification. These results demonstrate that the commercial standard required for this chemical (45–47%, weight-for-weight Co content), can only be met by SX. $Co(OH)_2$ begins to convert to Co_3O_4 at a temperature of 300°C, but it takes until 450°C–500°C to complete the process.

The hydrolysis constant (pK_a value) is in the following order: $Co^{2+} > Li^+$, indicating that divalent Co is extracted preferentially over monovalent Li. As a result, extraction at $pH_{eq} = 5.0$ was optimized, allowing for the extraction of 88.2% Co with only 0.7% Li (Sattar et al., 2019).

Additionally, 0.64 M CYANEX 272 (50% saponified) in the organic phase was optimized in the range of 3:1–1:3 at $pH_{eq} = 5.0$ to achieve quantitative extraction of Co. The percentage of Co extracted increases with an increase in the organic population, reaching its maximum of more than 99.9% in a single contact when O/A is equal to 3:1. The batch counter-current study also confirmed that the experiment, which was conducted at O/A = 1:1, would require two stages for the quantitative extraction from the aqueous solution containing 9.96 g/L Co. Stripping with 2.0 M H_2SO_4 solution and contacting at O/A = 10:1 for 10 min will recover Co back into the aqueous phase from the loaded organic. Eq. 8.24 clarifies the stripping reaction with the acid solution. High-purity crystals of $CoSO_4*H_2O$ were produced by slowly evaporating a high-purity stripped solution of concentrated Co. According to the chemical analysis of the crystal, the precipitated product contained 20.54% Co (Sattar et al., 2019).

$$Co(R.RH)_{2org} + 2H_2SO_{4aq} \leftrightarrow CoSO_{4aq} + 2(RH)_{2org} \qquad (8.24)$$

8.11 LITHIUM RECOVERY

From raffinates, Li was recovered as carbonate. Raffinates were stirred while a saturated Na_2CO_3 solution was added to bring the pH up to 8. Suspensions were filtered and ICP-OES were analyzed after 2 hours of stirring. Filtration was used to remove the solid precipitates (primarily the Ni, Mn, and Cu carbonates), and the remaining solution was evaporated, leaving varying amounts of residual water (evaporation degrees tested: 70%, 80%, and 90%). Filtration can separate Li_2CO_3, which can then be dried after being washed in hot water (90°C). ICP-OES can analyze Li_2CO_3 samples that have been dissolved in water. By evaporating 80% of the water volume, Li_2CO_3 is recovered with a yield of 80% and with a purity of greater than 98%. The degree of evaporation is selected to increase Li recovery and purity. In actuality, Na_2SO_4 (formed during precipitation steps) crystallized at the same time as Li_2CO_3 did. While the sole purpose of hot washing is to remove Na_2SO_4, some small amounts of Li were also lost in the process. Over 80% was found to have little positive impact on Li recovery and a negative impact on its purity due to the continued formation of Na_2SO_4, which is difficult to remove by washing without suffering significant Li loss.

$$(2Li^+)_{aq} + (CO_3^{2-})_{aq} \rightarrow (Li_2CO_3)s\downarrow \qquad\qquad (8.25)$$

An original aspect of the hydrometallurgical route is the recovery of Na_2SO_4 before Li_2CO_3 precipitation. The attempts to recover Li_2CO_3 by carbonatation at 95°C with Li-bearing diluted solution resulted in an unsatisfactory Li recovery product that contained Na_2SO_4 and Na_2CO_3. Due to subsequent Na_2SO_4 co-precipitation upon heating and Na_2CO_3 addition, the pre-separation of Na by direct cooling to 5°C of the Li-bearing dilute solution resulted in the poor recovery of Na_2SO_4, impeding the production of pure Li_2CO_3. The process that produced the best results involved evaporating the Li-containing solution, removing the Na_2SO_4 by crystallization after cooling, and then precipitating the Li_2CO_3 by adding Na_2CO_3. It is possible to obtain a 99.6%–99.8% pure phase of zabuyelite (Li_2CO_3) (Atia et al., 2019).

8.12   NICKEL RECOVERY

After Cyanex extraction, Ni was extracted as hydroxide from the aqueous solution. To achieve a quantitative recovery of Ni and minimize Li losses (under these circumstances, LiOH begins to precipitate at a pH higher than 12), the precipitation pH was fixed at 11. Nickel hydroxide was precipitated and had its typical green hue. NiO can be produced thermally from $Ni(OH)_2$. With a recovery rate for Ni that is higher than 89%, NiO can be purified to a range of 87%–89%.

8.13   SOL-GEL METHOD

High-value elements in leachate can be separated and recovered using a variety of emerging new techniques, in addition to chemical precipitation and liquid-liquid extraction. The sol-gel method is one of them; it directly recovers cathodic materials from used-up PLSs that have been leached by batteries. The leachate initially goes through several processes before it transforms into a gel. The gel is then subjected to further thermal treatments to produce solid products. $LiCoO_2$ (Li et al., 2012a; Santana et al., 2017), $LiMn_2O_4$ (Yao et al., 2021), and NCM (Li et al., 2018; Zhang et al., 2020a) are among the recovered products that have been reported. Due to the chelating properties of organic acids, sol-gel is frequently used in conjunction with them. The environmentally friendly sol-gel method can greatly streamline the challenging separation and recovery process. However, due to the substantial consumption of organic acid and the prolonged reaction period, it is relatively expensive. To change the composition of the leachate to satisfy the requirement of three-generated cathodic substances, additional chemicals are also necessary (Dobo et al., 2023).

8.14   ELECTROCHEMICAL PROCESS

Electrochemical deposition is another technique for removing target metals from S-LIB leaching solutions and recovering them. This technique's fundamental idea is based on the metallic ion's dissimilar redox potential. The recovery of Co from used battery leachate is the main objective of studies on electrochemical deposition (Freitas and Garcia, 2007; Garcia et al., 2011; Song and Zhao, 2018). In addition, the use of electrodialysis and ion exchange has been documented (Iizuka et al., 2013; Strauss et al., 2021). Due to their complex operations and astronomical costs, the commercial application of these techniques is constrained (Dobo et al., 2023).

Shen (2002) looked into the electrowinning (EW) and H_2SO_4 acid-leaching processes for recovering Co from S-LIBs. Nearly all of the Co in the S-LIBs is dissolved under the conditions of 70°C temperature, 10.0 M H_2SO_4 acid concentration, and 1 hour retention time. By hydrolyzing deposition in the pH range of 2.0–3.0 and at 90°C, the leach liquor is purified. EW generates the cathode Co at a current density of 235 A/m². The cathode Co quality complies with GB6517–86, China's 1A# cobalt standard. Co has a net recovery rate that exceeds 93%. It was believed that this procedure could be scaled up for industrial use. The electrolysis process can produce the Co compound of very high purity from S-LIBs compared to other hydrometallurgical processes for recycling metals from S-LIBs because it does not introduce other substances and thereby avoids the introduction of impurities. The drawback of this method is that it uses excessive amounts of electricity.

8.15 SOME OF THE PREVIOUS SX STUDIES

To date, a large number of methods to recover metals from the leaching liquor of S-LIBs have been proposed by many researchers. According to the previous research studies, the main techniques have some important drawbacks of the low purity of the Co compounds. Therefore, it is urgent to exploit new recovery processes to meet the demand for metal recycling from the already complicated waste streams.

A 4.0 M solution of HCl was used in the majority of processes reported to achieve the best metal extraction from the cathode material of S-LIBs. Even in 3.0 M HCl, metal leaching was still possible in the presence of a reducing agent like H_2O_2. Co^{2+}, which was reduced from Co^{3+}, was made to dissolve more easily by H_2O_2, but Li, which was also present in the same oxide, was also made to dissolve more quickly. Metals were either selectively precipitated as $Co(OH)_2$ and Li_2CO_3 from the leach solutions or separated and recovered from the leach liquors using SX, PC-88A, CYANEX 272, etc., to produce pure metal salts. In the majority of H_2SO_4 leaching procedures, H_2O_2 was utilized as a reductant. Al could be removed in some circumstances by first leaching with alkali and then acid. Using PC-88A/P507, CYANEX 272, Acorga M5640, and precipitation processes that are very similar to those of the HCl system, metals from the leach solutions were separated and recovered by SX (Meshram et al., 2014).

After 3.0 M H_2SO_4 leach at 70°C and 200 g/L solid ratio for 240 min, Nan et al. (2005) used precipitation & SX using Acorga M5640 and CYANEX 272. 90% Co-precipitation and 97% Cu by Acorga M5640, 97% Co by CYANEX 272 were succeeded.

As part of the alkali-acid leaching procedure, the cathode was first treated with 10% (w/w) NaOH at 30°C to dissolve Al, then 97% Co and 100% Li were reductively leached with H_2SO_4 and H_2O_2 (Ferreira et al., 2009; Nan et al., 2005). 98% Cu and 97% Co, respectively, were extracted and recovered from the solutions using Acorga M5640 and CYANEX 272. With 0.5% impurities, about 90% of the Co was recovered as oxalate. Utilizing the compounds, they have covered, $LiCoO_2$ positive electrode material with good electrochemical performance was created. In their 2016 study, Wang et al. used H_2SO_4 leaching, solid-liquid separation, and SX to recover Co from S-LIBs. H_2SO_4 concentration of 3.0 M, S/L ratio of 1:7, and H_2O_2 dosage of 1.6 mL/g for 2.5 h at 70°C are the ideal leaching operation conditions. The best experimental conditions are D2EHPA and PC-88A saponification rates of 20% and 30%, respectively, sulfonated kerosene volume of 70%, oil-water (O/A) ratio of 1:1, and extraction time of 10 min. This is based on the extraction characteristics of D2EHPA and PC-88A for a specific ion in different pH value leaching solutions. Sulfonated kerosene was used as the diluent, tri-n-butylphosphate was used as the phase modifier, and NaOH solution was used to saponify the extractant PC-88A. To remove the Cu and Mn ions, two extractions are used, the first at a pH of 2.70 and the second at a pH of 2.60. Following the extraction process, PC-88A is used to further extract the leaching solution and keep the pH at 4.25, ensuring that Co and Ni ions are successfully separated. Oxalic acid is then used to separate the Co ions, yielding cobalt oxalate. Co is as pure as 99.5%.

To separate Co/Li, Dorella and Mansur (2007) used an $H_2SO_{4+}H_2O_2$ leach, NH_4OH precipitation, and liquid-liquid extraction with CYANEX 272 as the extractant agent. The primary metal species targeted in the residue were Al, Co, Pb, and Li. Al was partially separated from Co and Li during the precipitation step at pH 5, which was achieved by adding NH_4OH to the leach liquor to raise the pH. Following filtration, the aqueous solution underwent a purification step by SX with 0.72 M CYANEX 272, and approximately 85% of Co was separated. According to Dorella et al. (2016), metals were extracted using CYANEX 272 in the following order: Al^{3+} is followed by Co^{2+} and Li^+. The obtained $pH_{1/2}$ values for Al, Co, and Li are 3.0, 4.0, and 6.5, respectively. Based on the findings, extraction levels of 100% for Al, 88% for Co, and 33% for Li can be obtained if the liquor is treated at pH 5 (the pH at which the precipitation step is skipped). Although the extraction behavior for Al and Co supports earlier research, the extraction level for Li was deemed to be excessive. The isotherm of Co with CYANEX 272 with the McCabe-Thiele method also shows that a limited number of contact stages will be needed to extract Co.

In addition, 2.0 M H_2SO_4 was used to leach the waste cathodic active material produced during the production of LIBs for 30 minutes at 75°C and 100 g/L solid ratio. About 85% of the Co was recovered during the Co/Li separation in a two-step SX process with 1.5 M Cyanex 272 at O/A 1.6. At O/A 1 and pH 5.35, the remaining Co was extracted in 0.5 M CYANEX 272. $CoSO_4$ in solution was discovered to be 99.99% pure. The highest separation factor (Co/Li) of 62 was recorded by Swain et al. (2006) during extraction with saponified CYANEX 272 from a synthetic solution at pH 6.9.

Using CYANEX 272, Swain et al. (2008) produced a pure $CoSO_4$ solution (99.99%) by acid leaching using CAM and SX waste. Using the supported liquid membrane (SLM) and a mixed extractant containing CYANEX 272 and DP-8R as the mobile carrier, quantitative separation of Co^{2+}/Li_+ was reported in a different study (Swain et al., 2010). At pH:5, they were able to separate synthetic solution (10 M Co^{2+} and 20 M Li_2SO_4) at a Co/Li separation factor of 497.

Kang et al. (2010a, 2010b) investigated SX by 50% saponified 0.4 M CYANEX 272, pH: 6, O/A = 2 after precipitation at pH: 6.5 for Cu, Fe, and Al. S-LIBs were leached by 2.0 M H_2SO_4 + 6 vol.% H_2O_2. Leaching achieved more than 99% Co extraction and SX succeeded 99.9% Co in two stages. Ni extraction was about 1%. Separation factor of 750 (Co/Li) and Co/Ni at pH: 6 accomplished.

After 4.0 M H_2SO_4 + 10 vol.% H_2O_2 leaching at 85°C and 100 g/L solid ratio for 120 min, Chen et al. (2011a, 2011b) used SX by 25% P507 after precipitation by ammonium oxalate, pH:1.5. 96% Li and > 95% Co were leached out of the S-LIBs under optimal conditions, and 99% of the Fe, Mn, and Cu were precipitated from the leach liquor with only 2% loss of Co. SX removed 97% Ni and Li while achieving 98% Co. The yield and purity of the Co oxalate were both 93% in the end.

Metal separation by CYANEX 272 and the leaching of Li and Co from CAMs of used mobile phone batteries in the presence of H_2O_2 have both been reported (Jha et al., 2013a, 2013b).

Suzuki et al. (2012) created a method for recycling S-LIBs using H_2SO_4 leach and SX. Al, Co, and Li were left in the raffinate after Cu was extracted using Acorga M5640 within a pH range of 1.5–2.0. Then, Al is selectively extracted using PC-88A at a pH of 2.5–3.0. Despite Acorga M5640's higher Co selectivity, PC-88A/TOA's higher stripping efficiency (N: 98%) allows it to separate Co^{2+} and Li^+ more effectively than Acorga M5640.

Zhao et al. (2011) investigated the synergistic extraction and separation of Co^{2+}, Mn^{2+}, and Li^+ from simulated H_2SO_4 leaches of waste cathodic materials using a mixture of CYANEX 272 and PC-88A in N-heptane. The leach solutions of S-LIBs were also handled using a mixed extractant system (Pranolo et al., 2010). First, Ionquest 801 and Acorga M5640 were used to extract Fe^{3+}, Al^{3+}, and Cu^{2+}. The raffinate containing Co, Ni, and Li was then treated with 15% (v/v) CYANEX 272 to separate out Co. At pH 5.5–6.0 and A/O 1:2, Co could be separated by more than 90%. Then, Ni^{2+}/Li^+ was separated using anion-exchange resin, such as Dowex M4195. Cu underwent a significant pH isotherm shift following the addition of 2% (v/v) Acorga M5640 to 7% (v/v) Ionquest 801, resulting in a $pH_{1/2}$ of 3.45.

Chen et al. (2015b) looked into how to separate and recover metal values from mixed-type S-LIB leaching liquor. It was suggested to use a combined hydrometallurgical process. Fe, Mn, Ni, and Li were precipitated and recovered with care. Mextral 5640H was used to selectively extract and recover Cu. To recover Co, Ni-loaded Mextral 272P was used. It is possible to efficiently separate and recover all metals. Equilibrium pH of 2.06, A/O = 2/1, and 10% Mextral 5640H for a 5-minute Cu extraction; equilibrium pH of 5.50, A/O = 2/1, and 30% Mextral 272P for a 30-minute Co extraction. Finally, NaOH and Na_3PO_3 solutions were used to successively precipitate the Ni and Li ions still present in the leachate.

After filtration and drying, Ni and Li were recovered as $Ni(OH)_2$ and Li_3PO_4, respectively. Under their ideal experimental conditions, the following recovery efficiencies could be reached: 100% for Cu, 99.2% for Mn, 97.8% for Co, 99.1% for Ni, and 95.8% for Li.

40% D2EHPA was used by Yang et al. (2017) with an A:O = 1:3 and a pH of 3.5. They reported a co-extraction of 100% Mn, 99% Co, and 85% Ni into the organic phase, with approximately 70% Li remaining in the aqueous phase.

8.16 SUMMARY AND OPPORTUNITIES

For storing energy in electric vehicles, energy systems, and electronics, the LIB is a superb method. Recent decades have seen a boom in its applications, which has resulted in an increase in the number of S-LIBs. For the sake of preserving natural resources, human health, the ecosystem, and waste LIBs, a serious issue that needs to be addressed.

Aqueous solutions are used in hydrometallurgical processes to extract the desired metals from CAMs of S-LIBs. H_2SO_4/H_2O_2 has been reported as the most frequent reagent combination thus far. Numerous investigations have been conducted to identify the ideal set of circumstances for a maximum leaching rate. These factors include the amount of leaching acid present, time, solution temperature, the ratio of solids to liquids, and the addition of a reducing agent. In the majority of these studies, it was discovered that adding reductant (H_2O_2), which turns insoluble Co^{3+} materials into soluble Co^{2+}, increased leaching efficiency.

Numerous additional potential reducing agents and leaching acids have been researched. Additionally, an organic solvent may be used to perform a SX on the leached solution. Once the metals have been leached, they may be recovered using a variety of precipitation reactions that can be controlled by adjusting the pH of the solution. Li can be extracted through a precipitation reaction, forming Li_2CO_3, LiOH, or Li_3PO_4, after Co is typically extracted as the sulfate, oxalate, hydroxide, or carbonate.

Most current hydrometallurgical recycling processes aim to recover reagents used because the materials, with sufficient purity, can be re-used not just for resynthesizing the original cathode materials, but also in a range of other applications, such as the synthesis of $CoFe_2O_4$ or $MnCo_2O_4$. Following initial work focused on the leaching and remanufacturing of $LiCoO_2$, work has since moved on to strategies for new cell chemistries, which typically contain multiple transition metals. For NMC_{111} a precursor hydroxide, $Ni_xMn_yCo_z(OH)_2$ is produced.

The major problems to be addressed with all solve-metallurgical processes are the volumes of expensive solvents required, the speed of delamination, the costs of neutralization, and the likelihood of cross-contamination of materials. Although shredding is a fast and efficient method of rendering the battery materials safe, mixing the anode and cathode materials at the beginning of the recycling process complicates downstream processing. A method in which anode and cathode assemblies could be separated before mechanical or solvent-based separation would greatly improve material segregation. This is one of several key areas where designing for EoL recycling promises to have a real impact, but the historic backlog of batteries containing polyvinylidene fluoride (PVDF) as a binder will still need to be processed. It is clear that the current design of cells makes recycling extremely complex and neither hydro- nor pyrometallurgy currently provides routes that lead to pure streams of material that can easily be fed into a closed-loop system for batteries.

A number of improvements could make S-LIB recycling processes economically more efficient, such as better sorting technologies, a method for separating electrode materials, greater process flexibility, design for recycling, and greater manufacturer standardization of batteries. There is a clear opportunity for a more sophisticated approach to battery recovery through automated disassembly, smart segregation of different batteries, and the intelligent characterization, evaluation, and 'triage' of used batteries into streams for remanufacture, reuse, and recycling. The potential benefits of this are many and include reduced costs, higher value of recovered material streams, and the near elimination of the risk of harm to human workers.

The design of current battery packs is not optimized for easy disassembly. The use of adhesives, bonding methods, and fixtures do not lend themselves to easy deconstruction either by hand or machine. All reported current commercial physical cell-breaking processes employ shredding or milling with subsequent sorting of the component materials. This makes separating the components more difficult than if they were presorted and considerably reduces the economic value of waste material streams. Many of the challenges this presents to remanufacture, reuse and recycling could be addressed if considered early in the design process (Harper et al., 2019).

In this chapter, we have focused on the scientific challenges of recycling S-LIBs, but we recognize that the system performance of the S-LIB recycling industry will be strongly affected by a range of non-technical factors, such as the nature of the collection, transportation, storage, and logistics of S-LIBs at the EoL. As these vary from country to country, region to region, and company to company, it follows that different jurisdictions may arrive at different answers to the problems posed. Research is underway in the Faraday Institution ReLiB Project, UK; the ReCell Project, US; at CSIRO in Australia and at several European Union projects including ReLieVe, Lithorec, and AmplifII.

Figure 8.9 compares two possible flowsheets for LIB recycling with precipitation and precipitation & SX. Flowsheet including only precipitation is cheaper in equipment and operating costs. Flowsheet with precipitation and SX is quite expensive in solvent purchasing cost. If the yearly capacity of the plant is more than 250 t and the Li_2CO_3 price is 18 \$/kg, the payback time is only 1 year for flowsheet with only precipitation (Granata et al., 2012b).

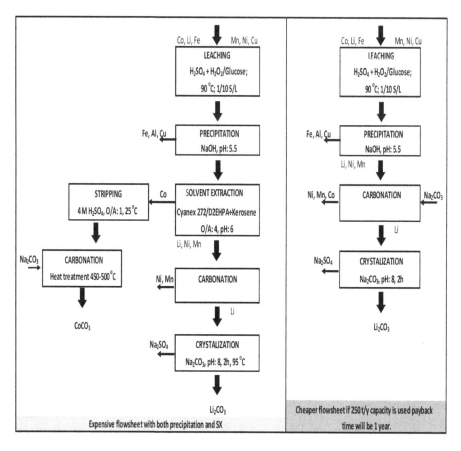

FIGURE 8.9 Comparison of two LIBs recycling flowsheets with precipitation and SX.

References

Aaltonen, M., Peng, C., Wilson, B., & Lundstrom, M., 2017. Leaching of metals from spent lithium-ion batteries, *Recycling*, 2(4), 20. https://doi.org/10.3390/recycling2040020.

Al-Shammari, H., & Farhad, S., 2021. Heavy liquids for rapid separation of cathode and anode active materials from recycled lithium-ion batteries, *Resour. Conserv. Recycl.*, 174, 105749.

Arshad, F., Li, L., Amin, K., Fan, E., Manurkar, N., Ahmad, A., Yang, J., Wu, F., & Chen, R., 2020. A comprehensive review of the advancement in recycling the anode and electrolyte from spent lithium-ion batteries, *ACS Sustain. Chem. Eng.*, 8(36), 13527–13554.

Atia, T.A., Elia, G., Hahn, R., Altimari, P., & Pagnanelli, F., 2019. Closed-loop hydrometallurgical treatment of end-of-life lithium-ion batteries: Towards zero-waste process and metal recycling in advanced batteries, *J. Energy Chem.*, 35, 220. https://doi.org/10.1016/j.jechem.2019.03.022.

Bae, H., & Kim, Y., 2021. Technologies of lithium recycling from waste lithium-ion batteries: A review. *Mater. Adv.*, 2(10), 3234–3250.

Bahaloo-Horeh, N., & Mousavi, S.M., 2017. Enhanced recovery of valuable metals from spent lithium-ion batteries through optimization of organic acids produced by Aspergillus niger, *Waste Manage.*, 60, 666–679. https://doi.org/10.1016/j.wasman.2016.10.034.

Bahgat, M., Farghaly, F., Basir, S.A., & Fouad, O., 2017. Synthesis, characterization, and magnetic properties of microcrystalline lithium cobalt ferrite from spent lithium-ion batteries, *J. Mater. Process. Technol.*, 183(1), 117–121. https://doi.org/10.1016/j.jmatprotec.2006.10.005.

Balasooriya, N.W.B., Touzain, P., & Bandaranayake, P.W.S.K., 2007. Capacity improvement of mechanically and chemically treated Sri Lanka natural graphite as an anode material in Li-ion batteries, *Ionics*, 13(5):305–309.

Bankole, O.E., Gong, C., & Lei, L., 2013. Battery recycling technologies: Recycling waste lithium-ion batteries with the impact on the environment in-view, *J. Environ. Ecol.*, 4(1), 14–28. https://doi.org/10.5296/jee.v4i1.3257.

Barik, S.P., Prabaharan, G., & Kumar, L., 2017. Leaching and separation of Co and Mn from electrode materials of spent lithium-ion batteries using hydrochloric acid: Laboratory and pilot scale study, *J. Clean. Product.*, 147, 37–43. https://doi.org/10.1016/j.jclepro.2017.01.095.

Bertuol, D.A., Toniasso, C., Jimenez, B.M., Meili, L., Dotto, G.L., Tanabe, E.H., & Aguiar, M.L., 2015. Application of spouted bed elutriation in the recycling of lithium-ion batteries, *J. Power Sources*, 275, 627–632. https://doi.org/10.1016/j.jpowsour.2014.11.036.

Bi, H., Zhu, H., Zu, L., Bai, Y., Gao, S., & Gao, Y., 2019. A new model of trajectory in eddy current separation for recovering spent lithium iron phosphate batteries, *Waste Manag.*, 100, 1–9.

Buchert, M., Jenseit, W., Merz, C., & Schüller, D., 2011. *Verbundprojekt: Entwicklung eines realisierbaren recycling-konzepts für die Hochleistungsbatterien zukünftiger Elektrofahrzeuge-LIBRI-Teilprojek: LCA der Recyclingverfahren.* Darmstadt, Germany: Öko-Institut.

Castillo, S., Ansart, F., Laberty-Robert, C., & Portal, J., 2002. Advances in the recovering of spent lithium battery compounds, *J. Power Sources*, 112, 247–254.

Chan, K. H., Anawati, J., Malik, M., & Azimi, G., 2021. Closed-loop recycling of lithium, cobalt, nickel, and manganese from waste lithium-ion batteries of electric vehicles, *ACS Sustain. Chem. Eng.*, 9(12), 4398–410.

Chehreh Chelgani, S., Rudolph, M., Kratzsch, R., Sandmann, D., & Gutzmer, J., 2015. A review of graphite beneficiation techniques, *Miner. Process. Extract. Metallur. Rev.*, 37(1), 58–68.

Chen, L., Tang, X., Zhang, Y., Li, L., Zeng, Z., & Zhang, Y., 2011a. Process for the recovery of cobalt oxalate from spent lithium-ion batteries, *Hydrometallurgy*, 108, 80–86.

Chen, Y., Tian, Q., Chen, B., Shi, X., & Liao, T., 2011b. Preparation of lithium carbonate from spodumene by a sodium carbonate autoclave process, *Hydrometallurgy*, 109, 3–46.

Chen, X., Chen, Y., Zhou, T., Liu, D., Hu, H., & Fan, S., 2015a. Hydrometallurgical recovery of metal values from sulfuric acid leaching liquor of spent lithium-ion batteries, *Waste Manag.*, 38, 349–356. https://doi.org/10.1016/j.wasman.2014.12.023.

Chen, X., Luo, C., Zhang, J., Kong, J., & Zhou, T., 2015b. Sustainable recovery of metals from spent lithium-ion batteries: A green process, *ACS Sustain. Chem. Eng.*, 3(12), 3104–3113. https://doi.org/10.1021/acssuschemeng.5b01000.

Chen, X., Xu, B., Zhou, T., Liu, D., Hu, D., & Fan, S., 2015c. Separation and recovery of metal values from leaching liquor of mixed-type of spent lithium-ion batteries, *Sep. Purif. Technol.*, 144, 197–205. https://doi.org/10.1016/j.seppur.2015.02.006.

Chen, J., Li, Q., Song, J., Song, D., Zhang, L., & Shi, X., 2016a. Environmentally friendly recycling and effective repairing of cathode powders from spent LiFePO$_4$ batteries, *Green Chem.*, 18(8), 2500–2506.

Chen, X., Fan, B., Xu, L., Zhou, T., & Kong, J., 2016b. An atom-economic process for the recovery of high value-added metals from spent lithium-ion batteries, *J. Cleaner Prod.*, 112, 3562–3570. https://doi.org/10.1016/j.jclepro.2015.10.132.

Chen, X., Ma, H., Luo, C., & Zhou, T., 2017. Recovery of valuable metals from waste cathode materials of spent lithium-ion batteries using mild phosphoric acid, *J. Hazard. Mater.*, 326, 77–86. https://doi.org/10.1016/j.jhazmat.2016.12.021.

Chen, X., Cao, L., Kang, D. Li, J., Zhou, T., & Ma, H., 2018. Recovery of valuable metals from mixed types of spent lithium-ion batteries. Part II: selective extraction of lithium, *Waste Manage.*, 80, 198–210. DOI: 10.1016/j.wasman.2018.09.013.

Chen, M., Ma, X., Chen, B., Arsenault, R., Karlson, P., & Simon, N., 2019. Recycling end-of-life electric vehicle lithium-ion batteries, *Joule*, 3, 2622–2646.

Chen, D., Rao, S., Wang, D., Cao, H., Xie, W., & Liu, Z., 2020. Synergistic leaching of valuable metals from spent Li-ion batteries using the sulfuric acid-l-ascorbic acid system, *Chem. Eng. J.*, 388, 124321. https://doi.org/10.1016/j.cej.2020.124321.

Claus, D., 2011. Materials and processing for lithium-ion batteries, *TMS*, 60(9), 43–48.

Contestabile, M., Panero, S., & Scrosati, B., 1999. A laboratory-scale lithium battery recycling process, *J Power Sourc.*, 83, 75–78.

Contestabile, M., Panero, S., & Scrosati, B., 2001. A laboratory-scale lithium-ion battery recycling process, *J. Power Sourc.*, 92, 65–69.

Crundwell, F., Moats, M., Ramachandran, V., Robinson, T., & Davenport, W.G., 2011. *Extractive Metallurgy of Nickel, Cobalt, and Platinum Group Metals*. London, UK: Elsevier.

Cuhadar, E., Mennik, F., Dinc, N.I., & Burat, F., 2023. Characterization and recycling of lithium nickel manganese cobalt oxide type spent mobile phone batteries on mineral processing technology, *J. Mater. Cycles Waste Manag.*, 25(3), 1746–1759. https://doi.org/10.1007/s10163-023-01652-5.

Cui, J., & Forssberg, E., 2003. Mechanical recycling of waste electric and electronic equipment: Are view, *J. Hazard. Mater.*, B99, 243–263.

Dai, C.Y., Wang, Z., Liu, K., Zhu, X.X., Liao, X.B., Chen, X., & Pan, Y., 2019. Effects of cycle times and C-rate on mechanical properties of copper foil and adhesive strength of electrodes in commercial LiCoO$_2$ LIBs, *Eng. Fail. Analys.*, 101, 193–205.

Danino-Perraud, R., 2020. *The Recycling of Lithium-Ion Batteries. A Strategic Pillar for the European Battery Alliance*, IFRI Center for Energy.

DeMeuse, M., 2020. Battery separators: How can the plastics industry meet the challenges? August 18, 2020. https://omnexus.specialchem.com/tech-library/article/battery-separators-how-can-the-plastics-industrymeet-current-challenges.

Diaz, F., Wang, Y., Moorthy, T., & Friedrich B., 2018. Degradation mechanism of nickel-cobalt-aluminum (NCA) cathode material from spent lithium-ion batteries in microwave-assisted pyrolysis, *Metals*, 8(8), 565.

Diekmann, J., Hanisch, C., Fröböse, L., Schälicke, G., Loellhoeffel, T., Fölster, A.-S., & Kwade, A., 2016. Ecological recycling of lithium-ion batteries from electric vehicles with a focus on mechanical processes, *J. Electrochem. Soc.*, 164(1), A6184.

Diez, E.D., Ventosa, E., Guarnieri, M., Trovò, A., Cristina Flox, C., Marcilla, R., Soavi, F., Mazur, P., Aranzabe, E., & Ferret, R., 2021. Redox flow batteries: Status and perspective towards sustainable stationary energy storage, *J. Power Sourc.*, 481, 228804. https://doi.org/10.1016/j.jpowsour.2020.228804.

Dobó, Z., Dinh, T., & Kulcsár, T., 2023. A review on recycling of spent lithium-ion batteries, *Energ. Rep.*, 9, 6362–6395. https://doi.org/10.1016/j.egyr.2023.05.264.

Dong Joon, A., & Xing Cheng, X., 2015. Enhanced rate capability of oxide coated lithium titanate within extended voltage ranges, *Front. Energ. Resour.*, 3(21), 1–9. https://doi.org/10.3389/fenrg.2015.00021.

Dorella, G., and Mansur, M.B., 2007. A study of the separation of cobalt from spent Li-ion battery residues, *J. Power Sources*, 170, 210–215. https://doi.org/10.1016/j.jpowsour.2007.04.02.

Elwert, T., Goldman, D., Römer, F., Buchert, M., Merz, C., Schueler, D., & Sutter, J., 2016. Current developments and challenges in the recycling of key components of (Hybrid) electrical vehicles, *Recycling*, 1, 25–60.

ERG Inc. and JSE Associates Reports, 2022. EPA lithium-ion battery disposal and recycling stakeholder workshop, Summary Report.

Fan, C.-L., He, H., Zhang, K.-H., & Han, S.-C., 2012. Structural developments of artificial graphite scraps in further graphitization and its relationships with discharge capacity, *Electrochim. Acta*, 75, 311–315.

Fan, B., Chen, X., Zhou, T., Zhang, J., & Xu, B., 2016. A sustainable process for the recovery of valuable metals from spent lithium-ion batteries, *Waste Manag. Res.*, 34(5), 474–481. https://doi.org/10.1177/0734242X16634454.

Feng, X., Ouyang, M., Liu, X., Lu, L., Xia, Y., & He, X., 2018. Thermal runaway mechanism of lithium-ion battery for electric vehicles: A review, *Energ. Stor. Mater.*, 10, 246–267.

Ferreira, D.A., Prados, L.M.Z., Majuste, D., & Mansur, M.B., 2009. Hydrometallurgical separation of aluminium, cobalt, copper and lithium from spent Li-ion batteries, *J. Power Sources*, 187, 238–246. https://doi.org/10.1016/j.jpowsour.2008.10.077.

Freitas, M.B.J.G., & Garcia, E.M., 2007. Electrochemical recycling of cobalt from cathodes of spent lithium-ion batteries, *J. Power Sources*, 171(2), 953–959. https://doi.org/10.1016/j.jpowsour.2007.07.002.

Frohlich, S., & Sewing, D., 1995. The BATENUS process for recycling mixed battery waste, *J. Power Sources*, 57, 27–30.

Forrest, W., Adel, G., & Yoon, R.-H., 1994. Characterizing coal flotation performance using release analysis, *Coal Prepar.*, 14(1–2), 13–27.

Fu, L.J., Liu, H., Li, C., Wu, Y.P., Rahm, E., Holze, R., & Wu, H.Q., 2006. Surface modifications of electrode materials for lithium-ion batteries, *Solid State Sci.*, 8(2), 113–128.

Gaines, L., 2018. Lithium-ion battery recycling processes: Research towards a sustainable course, *Sustain. Mater. Technol.*, 17, e00068.

Gaines, L., Dai, Q., Vaughey, J.T., & Gillard, S., 2021. Direct recycling R&D at the Recell center, *Recycling*, 6(2), 31. https://doi.org/10.3390/recycling6020031.

Ganter, M.J., Landi, B.J., Babbitt, C.W., Anctil, A., & Gaustad, G., 2014. Cathode function-alization as a lithium-ion battery recycling alternative, *J. Power Sources*, 256, 274–280. https://doi.org/10.1016/j.jpowsour.2014.01.078.

Gao, W., Zhang, X., Zheng, X., Lin, X., Cao, H., & Zhang, Y., 2017. Lithium carbonate recovery from cathode scrap of spent lithium ion battery: A closed-loop process, *Env. Sci. Technol.*, 51, 1662–1669. https://doi.org/10.1021/acs.est.6b03320.

Gao, H., Yan, Q., Xu, P., Liu, H., Li, M., Liu, P., Luo, J., & Chen, Z., 2020a. Efficient direct recycling of degraded $LiMn_2O_4$ cathodes by one-step hydrothermal relithiation, *ACS Appl. Mater. Interf.*, 12(46), 51546–51554.

Gao, Y., Li, Y., Li, J., Xie, H., & Chen, Y., 2020b. Direct recovery of $LiCoO_2$ from the recycled lithium-ion batteries via structure restoration, *J. Alloys Comp.*, 845, 156234.

Garcia, E.M., Tarôco, H.A., Matencio, T., Domingues, R.Z., dos Santos, J.A.F., & de Freitas, M.B.J.G., 2011. Electrochemical recycling of cobalt from spent cathodes of lithium-ion batteries: Its application as a coating on SOFC interconnects, *J. Appl. Electrochem.*, 41(11), 1373–1379. https://doi.org/10.1007/s10800011-0339-3.

Georgi-Maschler, T., Freiedrich, B., Weyhe, R., Heegn, H., & Rutz, M., 2012. Development of recycling process for Li-ion batteries, *J. Power Sources*, 207, 173–182.

Ghassa, S., Farzanegan, A., Gharabaghi, M., & Abdollahi, H., 2020. The reductive leaching of waste lithium-ion batteries in presence of iron ions: Process optimization and kinetics modelling, *J. of Cleaner Production*, 262, 121312, https://doi.org/10.1016/j.jclepro.2020.121312.

Ghassa, S., Farzanegan, A., Gharabaghi, M., & Abdollahi, H., 2021. Iron scrap a sustainable reducing agent for waste lithium-ion batteries leaching: An environmentlly friendly method to treating waste with waste, *Resour. Conserv. Recycl.*, 166, 105384. https://doi.org/10.1016/j.resconrec.2020.105348.

Granata, G., Moscardini, E., Pagnanelli, F., Trabucco, F., & Toro, L., 2012a. Product recovery from Li-ion battery wastes coming from an industrial pre-treatment plant: Lab scale tests and process simulations, *J. Power Sources*, 206, 393. https://doi.org/10.1016/.jpowsour.2012.01.115.

Granata, G., Pagnanelli, F., Moscardini, E., Takacova, Z., Havlik, T., & Toro, L., 2012b. Simultaneous recycling of nickel metal hydride, lithium-ion, and primary lithium batteries: Accomplishment of European Guidelines by optimizing mechanical pre-treatment and solvent extraction operations, *J. Power Sources*, 212, 205–211.

Gratz, E., Sa, Q., Apelian, D., & Wang, Y., 2014. A closed loop process for recycling spent lithium-ion batteries, *J. Power Sources*, 262, 255–262, https://doi.org/10.1016/j.jpowsour.2014.03.126.

Grützke, M., Kraft, V., Weber, W., Wendt, C., Friesen, A., Klamor, S., Winter, M., & Nowak, S., 2014. Supercritical carbon dioxide extraction of lithium-ion battery electrolytes, *J. Supercrit. Fluids*, 94, 216–222. https://doi.org/10.1016/j.supflu.2014.07.014.

Grützke, M., Mönnighoff, X., Horsthemke, F., Kraft, V., Winter, M., & Nowak, S., 2015. Extraction of lithium-ion battery electrolytes with liquid and supercritical carbon dioxide and additional solvents, *RSC Adv.*, 5, 43209–43217. https://doi.org/10.1039/C5RA04451K.

Guan, J., Li, Y.G., Guo, Y.G., Su, R.J., Gao, G.L., Song, H.X., Yuan, H., Liang, B., & Guo, Z.H., 2017. Mechanochemical process enhanced cobalt and lithium recycling from wasted lithium-ion batteries, *ACS Sustain. Chem. Eng.*, 5(1), 1026–1032.

Guo, Y., Li, F., Zhu, H., Li, G., Huang, J., & He, W., 2016. Leaching lithium from the anode electrode materials of spent lithium-ion batteries by hydrochloric acid (HCl), *Waste Manag.*, 51, 227–233. https://doi.org/10.1016/j.wasman.2015.11.036.

Guzolu, J. S., Gharabaghi, M., Mobin, M., & Alilo, H. 2017. Extraction of Li and Co from Li-ion batteries by chemical methods. *J. Inst. Eng. (India): Ser. D*, 98(1), 43–48. DOI: 10.1007/s40033-016-0114-z.

Hagelüken, C., & Grehl, M., 2012. Recycle and loop concept for a sustainable usage, In *Precious Metal Handbook*, Ed. By Sehrt, U., & Grehl, M., Germany: Hanau-Wolfgang, ISBN 978-3-8343-3259-2, pp: 35–79.

Harper, G., Sommerville, R., Kendrick, E., Driscoll, L., Slater, P., Stolkin, R., Walton, A., Christensen, P., Heidrich, O., Lambert, S., Abbott, A., Ryder, K., Graines, L., & Anderson, P., 2019. Recycling lithium-ion batteries from electric vehicles, *Nature*, 575(7781), 75–86.

He, L.P., Sun, S.Y., Song, X.F., & Yu, J.G., 2015. Recovery of cathode materials and Al from spent lithium-ion batteries by ultrasonic cleaning, *Waste Manag.*, 46, 523–528.

He, L.-P., Sun, S.-Y., Mu, Y.-Y., Song, X.-F., & Yu, J.-G., 2017a. Recovery of lithium, nickel, cobalt, and manganese from spent lithium-ion batteries using L-tartaric acid as a leachant, *ACS Sustain. Chem. Eng.*, 5(1), 714–721, https://doi.org/10.1021/acssuschemeng.6b02056.

He, L.P., Sun, S.Y., Song, X.F., & Yu, J.G., 2017b. Leaching process for recovering valuable metals from the LiNi1/3Co1/3Mn1/3O2 cathode of lithium-ion batteries, *Waste Manag.*, 64, 171–181.

He, Y., Zhang, T., Wang, F., Zhang, G., Zhang, W., & Wang, J., 2017c. Recovery of $LiCoO_2$ and graphite from spent lithium-ion batteries by Fenton reagent-assisted flotation, *J. Clean. Product.*, 143, 319–325.

He, K., Zhang, Z.Y., Alai, L.G., & Zhang, F.S., 2019. A green process for exfoliating electrode materials and simultaneously extracting electrolytes from spent lithium-ion batteries, *J. Hazard. Mater.*, 375, 43–51.

Heiskanen, S.K., Kim, J., & Lucht, B.L., 2019. Generation and evolution of the solid electrolyte interphase of lithium-ion batteries, *Joule*, 3(10), 2322–2333.

Horeh, N.B., Mousavi, S.M., & Shojaosadati, S.A., 2016. Bioleaching of valuable metals from spent lithium-ion mobile phone batteries using aspergillusniger, *J. Power Sources*, 320, 257–266.

Huang, B., Pan, Z., Su, X., & An, L., 2018. Recycling of lithium-ion batteries: Recent advances and perspectives, *J. Power Sources*, 399, 274–286.

Iizuka, A., Yamashita, Y., Nagasawa, H., Yamasaki, A., & Yanagisawa, Y., 2013. Separation of lithium and cobalt from waste lithium-ion batteries via bipolar membrane electrodialysis coupled with chelation, *Sep. Purif. Technol.*, 113, 33–41. https://doi.org/10.1016/j.seppur.2013.04.014.

Illes, I.B., & Kekesi T., 2023, Extraction of pure Co, Ni, Mn, and Fe compounds from spent Li-ion batteries by reductive leaching and combination oxidative precipitation in chloride media, *Minerals Eng.*, 201, 108169. https://doi.org/10.1016/j.mineng.2023.108169.

Jara, A.D., Betemariam, A., Woldetinsae, G., & Kim, J.Y., 2019. Purification, application and current market trend of natural graphite: A review, *Int. J. Min. Sci. Technol.*, 29(5), 671–689.

Jha, M.K., Kumari, A., Jha, A.K., Kumar, V., Hait, J., & Pandey, B.D., 2013a. Recovery of lithium and cobalt from waste lithium-ion batteries of mobile phones, *Waste Manag.*, 33(9), 1890–1897. https://doi.org/10.1016/j.wasman.2013.05.008.

Jha, A.K., Jha, M.K., Kumari, A., Sahu, S.K., Kumar, V., & Pandey, B.D., 2013b. Selective separation and recovery of cobalt from leach liquor of discarded Li-ion batteries using thiophosphinic extractant, *Sep. Purif. Technol.*, 104, 160–166.

Jo, C.H., & Myung, S.T., 2019. Efficient recycling of valuable resources from discarded lithium-ion batteries, *J. Power Sources*, 426, 259. https://doi.org/10.1016/j.jpowsour.2019.04.048.

Joulie, M., Loucournet, K., & Billy, E., 2014. Hydrometallurgical process for the recovery of high value metals from spent lithium nickel cobalt aluminum oxide based lithium-ion batteries. *J. Power Sources*, 247, 551–555. https://doi.org/10.1016/j. jpowsour.2013.08.128.

Kang, J., Senanayake, G., Sohn, J., & Shin, S.M., 2010a. Recovery of cobalt sulfate from spent lithium-ion batteries by reductive leaching and solvent extraction with Cyanex 272, *Hydrometallurgy*, 100(3–4), 168–171. https://doi.org/10.1016/j.hydromet.2009.10.010.

Kang, J., Sohn, J., Chang, H., Senanayake, G., & Shin, S.M., 2010b. Preparation of cobalt oxide from concentrated cathode material of spent lithium-ion batteries by hydrometallurgical method, *Adv. Powder Technol.*, 21(2), 175–179. https://doi.org/10.1016/j.apt.2009.10.015.

Kar, U., 2023. Recycling of lithium-ion Batteries-Aspect of Mineral Engineering, M.Sc. Thesis, Eskisehir Osmangazi University, 109 p.

Kar, U., Sponik, T., & Kaya, M., 2023. Comparison of different pre-treatment methods for lithium-ion battery recycling, VI Mineral Engineering Conference (MEC-2023), Wista-Poland (https://www.researchgate.net/publication/372021862).

Kaya, M., 2022. State-of-the-art lithium-ion battery recycling technologies, *Circ. Econ.*, 1(2), 100015. https://doi.org/10.1016/j.cec.2022.100015.

Kepler, K.D., Tsang, F., Vermeulen, R., & Hailey, P., 2016. Process for recycling electrode materials from lithium-ion batteries, Google Patents. US 2016/0072162 A1.

Kim, Y., Matsuda, M., Shibayama, A., and Fujita, T., 2004. Recovery of LiCoO$_2$ from wasted lithium-ion batteries by using mineral processing technology, *Resour. Process.*, 51(1), 37.

Kim, S., Bang, J., Yoo, J., Shin, Y., Bae, J., Jeong, J., Kim, K., Dong, P., & Kwon, K., 2021. A comprehensive review on the pretreatment process in lithium-ion battery recycling, *J. Clean. Product.*, 294, 126329.

Krüger, S., Hanisch, C., Kwade, A., Winter, M., & Nowak, S., 2014. Effect of impurities caused by a recycling process on the electrochemical performance of Li[Ni$_{0.33}$Co$_{0.33}$Mn$_{0.33}$]O$_2$, *J. Electroanal. Chem.*, 726, 91–96. https://doi.org/10.1016/j.jelechem.2014.05.017.

Ku, H., Jung, Y., Jo, M., Park, S., Kim, S., Yang, D., Rhee, K., An, E.-M., Sohn, J., & Kwon, K., 2016. Recycling of spent lithium-ion battery cathode materials by ammoniacal leaching, *J. Hazard. Mater.*, 313, 138–146. https://doi.org/10.1016/j.jhazmat.2016.03.062.

Lee, C.K., & Rhee, K.I., 2002. Preparation of LiCoO$_2$ from spent lithium-ion batteries, *J. Power Sources*, 109(1), 17–21.

Lee, C.K., & Rhee, K.-I., 2003. Reductive leaching of cathodic active materials from lithium-ion battery wastes, *Hydrometallurgy*, 68(1–3), 5–10. https://doi.org/10.1016/S0304-386X(02)00167-6.

Li, J., Shi, P., Wang, Z., Chen, Y., & Chang, C.-C., 2009. A combined recovery process of metals in spent lithium-ion batteries, *Chemosphere*, 77(8), 1132–1136. https://doi.org/10.1016/j.chemoshere.2009.08.040.

Li, L., Ge, J., Chen, R., Chen, S., & Wu, B., 2010a. Recovery of cobalt and lithium from spent lithium ion batteries using organic citric acid as leachant, *J. Hazard. Mater.*, 176(1–3), 288–293.

Li, L., Ge, J., Chen, R., Wu, F., Chen, S., & Zhang, X., 2010b. Environmental friendly leaching reagent for cobalt and lithium recovery from spent lithium-ion batteries, *Waste Manage.*, 30(12), 2615–2621. https://doi.org/10.1016/j.wasman.2010.08.008.

Li, L., Chen, R.J., Sun, F., Wu, F., Liu, J.R., 2011. Preparation of LiCoO$_2$ films from spent lithium-ion batteries by a combined recycling process, *Hydrometallurgy*, 108, 220–225.

Li, L., Chen, R., Zhang, X., Wu, F., Ge, J., & Xie, M., 2012a. Preparation and electrochemical properties of re-synthesized LiCoO$_2$ from spent lithium-ion batteries, *Chinese Sci. Bull.*, 57(32), 4188–4194. https://doi.org/10.1007/s11434-012-5200-5.

Li, L., Lu, J., Ren, Y., Zhang, X.X., Chen, R.J., & Wu, F., 2012b. Ascorbic-acid-assisted recovery of cobalt and lithium from spent Li-ion batteries, *J. Power Sci.*, 218, 21–27. https://doi.org/10.1016/j.jpowsour.2012.06.068.

Li, L., Dunn, J.B., Zhang, X.X., Graines, L., Chen, R.J., & Wu, F., 2013. Recovery of metals from spent lithium-ion batteries with organic acids as leaching reagents and environmental assessment, *J. Power Sci.*, 233, 180–189. https://doi.org/10.1016/j.jpowsour.2012.12.089.

Li, L., Zhai, L., Zhang, X., Lu, J., Chen, R., Wu, F., & Amine, K., 2014. Recovery of valuable metals from spent lithium-ion batteries by ultrasonic-assisted leaching process, *J. Power Sources*, 262, 380–385. https://doi.org/10.1016/j.jpowsour.2014.04.013.

Li, L., Qu, W.J., Zhang, X.X., Lu, J., Chen, R.J., Wu, F., & Amine, K., 2015. Succinic acid-based leaching system: A sustainable process for recovery of valuable metals from spent Li-ion batteries, *J. Power Sources*, 282, 544–551.

Li, J., Wang, G.X., & Xu, Z.M., 2016a. Environmentally friendly oxygen-free roasting/wet magnetic separation technology for in situ recycling cobalt, lithium carbonate, and graphite from spent $LiCoO_2$/graphite lithium batteries, *J. Hazard. Mater.*, 302, 97–104.

Li, J., Wang, G., & Xu, Z., 2016b. Generation and detection of metal ions and volatile organic compounds (VOCs) emissions from the pretreatment processes for recycling spent lithium-ion batteries, *Waste Manag.*, 52, 221–227.

Li, H., Xing, S., Liu, Y., Li, F., Guo, F., & Kuang, G., 2017a. Recovery of lithium, iron, and phosphorus from spent $LiFePO_4$ batteries using stoichiometric sulfuric acid leaching system, *ACS Sustain. Chem. Eng.*, 5(9), 8017–8024. https://doi.org/10.1021/acssuschemeng.7b01594.

Li, L., Fan, E., Guan, Y., Zhang, X., Xue, Q., Wei, L., Wu, F., & Chen, R., 2017b. Sustainable recovery of cathode materials from spent lithium-ion batteries using lactic acid leaching system, *ACS Sustain. Chem. Eng.*, 5(6), 5224–5333. https://doi.org/10.1021/acssuschemeng.7b00571.

Li, X.L., Zhang, J., Song, D.W., Song, J.S., & Zhang, L.Q., 2017c. Direct regeneration of recycled cathode material mixture from scrapped $LiFePO_4$ batteries, *J. Power Sources*, 345, 78–84.

Li, L., Bian, Y., Zhang, X., Guan, Y., Fan, E., Wu, F., & Chen, R., 2018a. Process for recycling mixed-cathode materials from spent lithium-ion batteries and kinetics of leaching, *Waste Manag.*, 71, 362–371. https://doi.org/10.1016/j.wasman.2017.10.028.

Li, L., Lu, J., Zhai, L., Zhang, X., Curtiss, L., Jin, Y., Wu, F., Chen, R., & Amine, K., 2018b. A facile recovery process for cathodes from spent lithium iron phosphate batteries by using oxalic acid, *CSEE J. Power. Energy Syst.*, 4(2), 219–225.

Li, L.R., Zheng, P.N., Yang, T.R., Sturges, R., Ellis, M.W., & Li, Z., 2019. Disassembly automation for recycling end-of-life lithium-ion pouch cells, *J. Occup. Med.*, 71(12), 4457–4464.

Li, J., He, Y., Fu, Y., Xie, W., Feng, Y., & Alejandro, K., 2021. Hydrometallurgical enhanced liberation and recovery of anode material from spent lithium-ion batteries, *Waste Manag.*, 126, 517–526.

Liu, Y.J., Hu, Q.Y., Li, X.H., Wang, Z.X., & Guo, H.J., 2006. Recycle and synthesize $LiCoO_2$ from incisors bound of Li-ion batteries, *Trans. Nonferrous Met. Soc. China*, 16, 956–959.

Liu, J., Wang, H., Hu, T., Bai, X., Wang, S., Xie, W., Hao, J., & He, Y., 2020. Recovery of $LiCoO_2$ and graphite from spent lithium-ion batteries by cryogenic grinding and froth flotation, *Miner. Eng.*, 148, 106223.

Liu, P., Zhang, Y., Dong, P., Zhang, Y., Meng, Q., Zhou, S., Yang, X., Zhang, M., & Yang, X., 2021. Direct regeneration of spent $LiFePO_4$ cathode materials with pre-oxidation and V-doping, *J. Alloys Compounds*, 860, 157909.

Lu, M., Zhang, H., Wang, B., Zheng, X., & Dai, C., 2013. The re-synthesis of $LiCoO_2$ from spent lithium-ion batteries separated by the vacuum-assisted heat-treating method, *Int. J. Electrochem. Sci.*, 8(6), 8201–8209.

Lupi, C., Pasquali, M., & Dell'Era, A., 2005. Nickel and cobalt recycling from lithium-ion batteries by electrochemical process, *Waste Manag.*, 25(2), 215–220.

Mackenzie, A., 2020. Australian battery recycling challenge and opportunities, 2020 Lithium & Battery Technology Proceedings, Australia, 147 154.

Madrigal-Arias, J.E., Argumedo, D.R., Alarcon, A, Mendoza, L.M.R., Berrades, D.C., Cruz, S.J.S., Ferrara, C.R., Fernandez, M., 2015. Bioleaching of gold, copper, and nickel from waste cellular phone PCBs and computer gold finger motherboards by two Aspergillusnigerstrains, *Brazilian J. Microbiol.*, 46(3), 707–713. https://doi.org/10.1590/S1517-838246320140256.

Meng, Q., Zhang, Y.J., & Dong, P., 2017. Use of glucose as a reductant to recover Co from spent lithium-ion batteries, *Waste Manag.*, 64, 214–218.

Meshram, P., Pandey, B.D., & Mankhand, T.R., 2014. Extraction of lithium from primary and secondary sources by pre-treatment, leaching, and separation: A comprehensive review, *Hydrometallurgy*, 150, 192–208, https://doi.org/10.1016/j.hydromet.2014.10.012.

Meshram, P., Pandey, B.D., & Mankhand, T.R., 2015a. Hydrometallurgical processing of spent lithium-ion batteries (LIBs) in the presence of a reducing agent with emphasis on the kinetics of leaching, *Chem. Eng. J.*, 281, 418–427. http://doi.org/10.1016/j.cej.2015.06.071.

Meshram, P., Pandey, B., & Mankhand, T., 2015b. Recovery of valuable metals from cathodic active material of spent lithium-ion batteries: Leaching and kinetic aspects, *Waste Manag.*, 45, 306–313. https://doi.org/10.1016/j.wasman.2015.05.027.

Mishra, D., Kim, D.-J., Ralph, D.E., Ahn, J.G., & Rhee, Y.H., 2008. Bioleaching of metals from spent lithium ion secondary batteries using *Acidithiobacillus ferrooxidans*, *Waste Manag.*, 28, 333–338.

Nan, J., Han, D., & Zuo, X., 2005. Recovery of metal values from spent lithium-ion batteries with deposition and solvent extraction, *J. Power Sources*, 152, 278–284.

Nan, J.M., Han, D.M., Yang, M.J., Cui, M., & Hou, X.L., 2006. Recovery of metal values from a mixture of spent lithium-ion batteries and nickel-metal hydride batteries, *Hydrometallurgy*, 84, 75–80.

Natarajan, S., Boricha, A.B., & Bajaj, H.C., 2018. Recovery of value-added products from cathode and anode material of spent lithium-ion batteries, *Waste Manag.*, 77, 455–465.

Nayaka, G.P., Manjanna, J., Pai, K.V., Vadavi, R., Keny, S.J., & Tripathi, V.S., 2015. Recovery of valuable metal ions from the spent lithium-ion battery using an aqueous mixture of mild organic acids as an alternative to mineral acids, *Hydrometallurgy*, 151, 73–77. https://doi.org/10.1016/j.hydromet.2014.11.006.

Nayaka, G.P., Pai, K.V., Manjanna, J., & Keny, S.J., 2016. Use of mild organic acid reagents to recover the Co and Li from spent Li-ion batteries, *Waste Manag.*, 51, 234–238. https://doi.org/10.1016/j.wasman.2015.12.008.

Nayl, A.A., Hamed, M.M., & Rizk, S.E., 2015. Selective extraction and separation of metal values from leach liquor of mixed spent Li-ion batteries, *J. Taiwan Inst. Chem. Eng.*, 55, 119–125. https://doi.org/10.1016/j.jtice.2015.04.006.

Nayl, A.A., Elkhashab, R.A., Badawy, S.M., & El-Khateeb, M.A., 2017. Acid leaching of mixed spent Li-ion batteries, *Arab. J. Chem.*, 10, S3632–S3639. http://doi.org/10.1016/j.arabjc.2014.04.001.

Nestoridi, M., & Barde, H., 2017. Beyond lithium-ion: Lithium-sulfur battery for space, ESPC-2016, *E3S Web Conf.*, 16, 08005. https://doi.org/10.1051/e3sconf/20171608005.

Nitta, N., Wu, F., Lee, J.T., & Yushin, G., 2015. Li-ion battery materials: present and future, *Mater. Today*, 18(5), 252–264. https://doi.org/10.1016/j.mattod.2014.10.040.

Niu, Z., Zou, Y., Xin, B., Chen, S., Liu, C., & Li, Y., 2014. Process controls for improving bioleaching performance of both Li and Co from spent lithium-ion batteries at high pulp density and its thermodynamics and kinetics exploration, *Chemosphere*, 109, 92–98. https://doi.org/10.1016/j.chemosphere.2014.02.059.

Ojanen, S., Lundström, M., Santasalo-Aarnio, A., & Serna-Guerrero, R., 2018. Challenging the concept of electrochemical discharge using salt solutions for lithium-ion battery recycling, *Waste Manag.*, 76, 242–249.

Olushola, S.A., Folahan, A.A., Alafara, A.B., Bhekumusa, J.X., & Olalekan, S.F., 2013. Application of Cyanexextractant in Cobalt/Nickel separation process by solvent extraction, *Int. J. Phys. Sci.*, 8(3), 89–97. https://doi.org/10.5897/IJPS12.135.

Ordonez, J., Gago, E.J., & Girard, A., 2016. Processes and technologies for the recycling and recovery of spent lithium-ion batteries, *Renew. Sustain. Energ. Rev.*, 60, 195–205. https://doi.org/10.1016/j.rser.2015.12.363.

Pagnanelli, F., Moscardini, E., Granata, G., Cerbelli, S., Agosta, L., Fieramosca, A., & Toro, L., 2014. Acid reducing leaching of cathodic powder from spent lithium-ion batteries: glucose oxidative pathways and particle area evolution, *J. Ind. Eng. Chem.*, 20(5), 3201–3207. https://doi.org/10.1016/j.jiec.2013.11.066.

Pagnanelli, F., Moscardini, E., Altimari, P., Abo Atia, T., & Toro, L., 2016. Cobalt products from real waste fractions of end-of-life lithium-ion batteries, *Waste Manag.*, 51, 214–221. https://doi.org/10.1016/j.wasman.2015.11.003.

Pan, S., 2020. Natural graphite battles for market share in battery anodes, *Ind. Miner.*, 617, 3.

Parker, J.F., Chervin, C.N., Pala, I.R., Machler, M., Burz, M.F., Long, J.W., & Rolison, D.R., 2017. Rechargeable nickel-3D zinc batteries: An energy-dense, safer alternative to lithium-ion, *Science*, 356(6336), 415–418. https://doi.org/10.1126/science.aak9991.

Partinen, J., Halli, P., Wilson, B.P., & Lundstrom, M., 2023. The impact of chlorides on NMC leaching in hydrometallurgical battery recycling, *Minerals Eng.*, 202, 108244. https://doi.org/10.1016/j.mineng.2023.108244.

Parvali, A., Aaltonen, M., Velazquez-Martinez, S., Eronen, O., Liu, E., Wilson, F., Serna, B.P., & Lundstrom, M., 2019. Mechanical and hydrometallurgical processes in HCl media for the recycling of valuable metals from Li-ion battery, Resour. Conserv. Recycl., 142, 257–266.

Peng, W., Wang, C., Hu, Y., & Song, S., 2016. Effect of droplet size of the emulsified kerosene on the floatation of amorphous graphite, *J. Dispersion Sci. Tech.*, 38(6), 889–894.

Peng, C., Liu, F., Wang, Z., Wilson, B.P., & Lundström, M., 2019. Selective extraction of lithium (Li) and preparation of battery-grade lithium carbonate (Li_2CO_3) from spent Li-ion batteries in nitrate system, *J. Power Sources*, 415, 179–188. https://doi.org/10.1016/j.jpowsour.2019.01.072.

Pinegar, H., & Smith, Y.R., 2019. Recycling of end-of-life lithium-ion batteries, Part I: Commercial processes, *J. Sustain. Met.*, 5(3), 402–416.

Pinna, E.G., Ruiz, M.C., Ojeda, M.W., & Rodriguez, M.H., 2017. Cathodes of spent lithium-ion batteries: Dissolution with phosphoric acid and recovery of lithium and cobalt from leach iquors, *Hydrometallurgy*, 167, 66–71. https://doi.org/10.1016/j.hydromet.2016.10.024.

Pranolo, Y., Zhang, W., & Cheng, C.Y., 2010. Recovery of metals from spent lithium-ion battery leach solutions with a mixed solvent extractant system, *Hydrometallurgy*, 102, 37–42.

Prazanova, A., Knap, V., & Stroe, D.I., 2022. Literature review, recycling of lithium-ion batteries from electric vehicles, part I: Recycling technology, *Energies*, 15(3), 1086.

Pudas, J., Erkkila, A., & Viljamaa, J., 2015. Battery recycling method, Google Patents.

Qadir, R., Gulshan, F., 2018. Reclamation of lithium cobalt oxide from waste lithium ion batteries to be used as recycled active cathode materials, *Mater. Sci. Appl.*, 9(1).

Rothermel, S., Evertz, M., Kasnatscheew, J., Qi, X., Grützke, M., Winter, M., & Nowak, S., 2016. Graphite recycling from spent lithium-ion batteries, *Chem. Sustain. Chem.*, 9, 3473–3484. https://doi.org/10.1002/cssc.201601062.

Ruismäki, R., Rinne, T., Dańczak, A., Taskinen, P., Serna-Guerrero, R., & Jokilaakso, A., 2020. Integrating flotation and pyrometallurgy for recovering graphite and valuable metals from battery scrap, *Metals*, 10(5), 680.

Sa, Q., Gratz, E., He, M.N., Lu, W.Q., Apelian, D., & Wang, Y., 2015. Synthesis of high-performance LiNi$_{1/3}$Mn$_{1/3}$Co$_{1/3}$O$_2$ from lithium-ion battery recovery stream, *J. Power Sources*, 282, 140–145. https://doi.org/10.1016/j.jpowsour.2015.02.046.

Saeki, S., Lee, J., Zhang, Q.W., & Saito, F., 2004. Co-grinding LiCoO$_2$ with PVC and water leaching of metal chlorides formed in ground product, *Int. J. Miner. Process.*, 74S, S373–S378. https://doi.org/10.1016/j.minpro.2004.08.002.

Santana, I.L., Moreira, T.F.M., Lelis, M.F.F., & Freitas, M.B.J.G., 2017. Photocatalytic properties of Co$_3$O$_4$/LiCoO$_2$ recycled from spent lithium-ion batteries using citric acid as a leaching agent, *Mater. Chem. Phys.*, 190, 38–44. https://doi.org/10.1016/j.matchemphys.2017.01.003.

Sattar, R., Ilyas, S., Bhatti, H.N., & Ghaffar, A., 2019. Resource recovery of critically rare metals by hydrometallurgical recycling of spent lithium-ion batteries, *Sep. Purif. Technol.*, 209, 725. https://doi.org/10.1016/j.seppur.2018.09.019.

Schulz, K.J., DeYoung, J.H., Seal, R.R., & Bradley, D.C., 2018. Critical mineral resources of the United States: Economic and environmental geology and prospects for future supply, Geological Survey.

Shaibani, M., Eshraghi, N., & Majumder, M., 2020, Lithium Battery Ande R&D Trends, Alta 2020 Lithium & Battery Technology Proceedings, 127–137.

Shaw, S., 2013. Graphite demand growth: The future of lithium-ion batteries in EVs and HEVs. Proceedings of 37th ECGA General Assembly, Roskill Report, 30: 2015.

Shaw-Stewart, J.; Alvarez-Reguera, A.; Greszta, A.; Marco, J.; Masood, M.; Sommerville, R.; Kendrick, E., 2019. Aqueous solution discharge of cylindrical lithium-ion cells, *Sustain. Mater. Technol.*, 22, e00110. https://doi.org/10.1016/j.susmat.2019.e00110.

Shen, Y., 2002. Chinese nonferro, *Metals*, 54, 69–71 (in Chinese).

Shi, Y., Chen, G., & Chen, Z., 2018. Effective regeneration of LiCoO$_2$ from spent lithium-ion batteries: A direct approach towards high-performance active particles, Green *Chem.*, 20(4), 851–862.

Shi, Q., Liang, X., Feng, Q., Chen, Y., & Wu, B., 2015. The relationship between the stability of emulsified diesel and flotation of graphite, *Miner. Eng.*, 78, 89–92.

Shin, S.M., Kim, N.H., Sohn, J.S., Yang, D.H., & Kim, Y.H., 2005. Development of a metal recovery process from Li-ion battery wastes, *Hydrometallurgy*, 79(3–4), 172–181.

Shin, H., Zhan, R., Dhindsa, K.S., Pan, L., & Han, T., 2020. Electrochemical performance of recycled cathode active materials using froth flotation-based separation process, *J. Electrochem. Soc.*, 167(2), 020504.

Shuva, M. A. H., & Kurny, A. S.W., 2013a. Hydrometallurgical recovery of value metals from spent lithium ion batteries. *Am. J. Mater. Eng. Technol.*, 1(1), 8–12. doi: 10.12691/materials-1-1-2.

Shuva, M. A. H., & Kurny, A. S. W., 2013b. Dissolution kinetics of cathode of spent lithium ion battery in hydrochloric acid solutions. *J. Inst. Eng. (India): Ser. D*, 94(1), 13–16. doi: 10.1007/s40033-013-0018-0.

Sloop, S., Crandon, L., Allen, M., Koetje, K., Reed, L., & Gaines, L., 2020. A direct recycling case study from a lithium-ion battery recall, *Sustain. Mater. Technol.*, 25, e00152. https://doi.org/10.1016/j.susmat.2020.e00152.

Sohn, J.-S., Shin, S.-M., Yang, D.-H., Kim, S.-K., & Lee, C.-K., 2006a. Comparison of two acidic leaching processes for selecting the effective recycle process of spent lithium-ion battery, *Geosyst. Eng.*, 9(1), 1–6.

Sohn, J.-S., Yang, D.-H., Shin, S.-M., & Kang, J.-G., 2006b. Recovery of cobalt in sulfuric acid leaching solution using oxalic acid, *Geosyst. Eng.*, 9(3), 81–86. https://doi.org/10.1080/12269328.2006.10541259.

Sojka, R., Pan, Q., & Billmann, L., 2020, Comparative study on Li-ion battery recycling processes, Accurec Recycling GmnH Report, 53 p.

Sommerville, R., Shaw-Stewart, J., Goodship, V., Rowson, N., & Kendrick, E., 2020. A review of physical processes used in the safe recycling of lithium ion batteries, *Sustain. Mater. Technol.*, 25, e00197.

Sommerville, R., Zhu, P., Rajaeifar, M.A., Heidrich, O., Goodship, V., & Kendrick, E., 2021. A qualitative assessment of lithium-ion battery recycling processes, *Resourc. Conserv. Recycl.*, 165, 105219.

Song, D.W., Wang, X.Q., Zhou, E.L., Hou, P.Y., Guo, F.X., & Zhang, L.Q., 2013. Recovery and heat treatment of the $Li(Ni_{1/3}Co_{1/3}Mn_{1/3})O_2$ cathode scrap material for lithium-ion battery, *J. Power Sources*, 232, 348–352.

Song, D.W., Wang, X.Q., Nie, H.H., Shi, H., Wang, D.G., Guo, F.X., Shi, X.X., & Zhang, L.Q., 2014. Heat treatment of $LiCoO_2$ recovered from cathode scraps with the solvent method, *J. Power Sources*, 249, 137–141.

Song, Y., & Zhao, Z., 2018. Recovery of lithium from spent lithium-ion batteries using precipitation and electrodialysis techniques, *Sep. Purif. Technol.*, 206, 335–342. https://doi.org/10.1016/j.seppur.2018.06.022.

Spangerberger, J., 2018. *Introduction to Lithium-Ion Batteries*. Argonne National Laboratory.

Strauss, M.L., Diaz, L.A., McNally, J., Klaehn, J., & Lister, T.E., 2021. Separation of cobalt, nickel, and manganese in leach solutions of waste lithium-ion batteries using dowex M4195 ion exchange resin, *Hydrometallurgy*, 206, 105757. https://doi.org/10.1016/j.hydromet.2021.105757.

Suarez, D. S., Pinna, E. G., Rosales, G. D., & Rodriguez, M. H., 2017. Synthesis of lithium fluoride from spent lithium ion batteries. Minerals, 7(5), 81. Doi: 10.3390/min7050081.

Sun, L., & Qiu, K., 2011. Vacuum pyrolysis and hydrometallurgical process for the recovery of valuable metals from spent lithium-ion batteries, *J. Hazard. Mater.*, 194, 378–384. https://doi.org/10.1016/j.jhazmat.2011.07.114.

Sun, L., & Qiu, K., 2012. Organic oxalate as leachant and precipitant for the recovery of valuable metals from spent lithium-ion batteries, *Waste Manag.*, 32(8), 1575–1582. https://doi.org/10.1016/j.wasman.2012.03.027.

Sun, Z., Cao, H., Xiao, Y., Sietsma, J., Jin, W., Agterhuis, H., & Yang, Y., 2017. Toward sustainability for recovery of critical metals from electronic waste: The hydrochemistry processes. *ACS Sustain. Chem. Eng.*, 5, 21. https://doi.org/10.1021/acssuschemeng.6b00841.

Suzuki, K., Hamada, T., & Sugiura, T., 1999. Effect of graphite surface structure on initial irreversible reaction in graphite anodes, *J. Electrochem. Soc.*, 146(3), 890.

Suzuki, T., Nakamura, T., Inoue, Y., Niinae, M., & Shibata, J., 2012. A hydrometallurgical process for the separation of aluminum, cobalt, copper, and lithium in acidic sulfate media, *Sep. Purif. Technol.*, 98, 396–401.

Swain, B., Jeong, J., Lee, J.C., & Lee, G.H., 2006. Separation of cobalt and lithium from mixed sulphate solution using Na-Cyanex 272, *Hydrometallurgy*, 84, 130–138.

Swain, B., Jeong, J., Lee, J.-C.; Lee, G.-H., & Sohn, J.-S., 2007. Hydrometallurgical process for recovery of cobalt from waste cathodic active material generated during manufacturing of lithium-ion batteries, *J. Power Sources*, 167(2), 536–544. https://doi.org/10.1016/j.jpowsour.2007.02.046.

Swain, B., Jeong, J., Lee, J.C., & Lee, G.H., 2008. Development of process flow sheet for recovery of high pure cobalt from sulfate leach liquor of LIB industry waste: a mathematical model correlation to predict optimum operational conditions, *Sep. Purif. Technol.*, 63, 360–369.

Swain, B., Jeong, J., Yoo, K., & Lee, J.C., 2010. Synergistic separation of Co(II)/Li(I) for the recycling of LIB industry wastes by supported liquid membrane using Cyanex 272 and DR-8R, *Hydrometallurgy*, 101, 20–27.

Swoffer, W.N.S.A.S., 2013. Recovery of Lithium Ion Batteries, I. Toxco, Anaheim, CA (US), US 8,616,475 B1.

Takacova, Z., Havlik, T., Kukurugya, F., & Orac, D., 2016. Cobalt and lithium recovery from active mass of spent Li-ion batteries: Theoretical and experimental approach. *Hydrometallurgy*, 163, 9–17. https://doi.org/10.1016/j.hydromet.2016.03.007.

Talbot, P., 2020. Establishing Australia's battery industry supply chain, Alta-2020, Lithium & Battery Technology Proceedings.

Tanii, T., Tsuzuki, S., Honmura, S., Kamimura, T., Sasaki, K., Yabuki, M., & Nishida, K., 2003. US Patent n6,524,737.

Tedjar, F., & Foudraz, J.-C., 2010. Method for the mixed recycling of lithium-based anode batteries and cells. Google Patents.

Tiilhonen, M., Haavanlammi, L., Kinnunen, S., & Kolehmainen, E., 2020. Outotec lithium hydroxide process—A progress update. ALTA Lithium & battery Technology Proceedings, 94–102.

Toshiba, 2017. https://www.global.toshiba/ww/news/corporate/2017/10/pr0301.html.

Traore, N., & Kelebek, S., 2023. Characteristics of spent lithium-ion batteries and their recycling potential using flotation separation: A review, *Miner. Process. Extract. Metal. Rev.*, 44(3), 231–259.

Tytgat, J., 2015. Li-ion and NiMH battery recycling at Umicore: strategic choices. ESA Report. Göteborg, Sweden: Chalmers University.

Ubbelohde, A.R., & Lewis, F.A., 1960. *Graphite and Its Crystal Compounds*. London, UK: Clarendon Press.

Umicore. 2019. Our recycling process, an internationally recognized. Avaliable at https://csm. umicore.com

Wang, F., Sun, R., Xu, J., Chen, Z., & Kang, M., 2016a. Recovery of cobalt from spent lithium-ion batteries using sulphuric acid leaching followed by solid-liquid separation and solvent extraction, *RSC Adv.*, 88(6), 85303–85311. https://doi.org/10.1039/C6RA16801A.

Wang, M.-M., Zhang, C.-C., & Zhang, F.-S., 2016b. An environmentally benign process for cobalt and lithium recovery from spent lithium-ion batteries by mechanochemical approach, *Waste Manag.*, 51, 239–244.

Wang, H., & Whitacre, J.F., 2018. Direct recycling of aged $LiMn_2O_4$ cathode materials used in Aqueous Lithium-ion Batteries: Processes and sensitivities, *Energy Technol.*, 6(12), 2429–2437.

Wang, F., Zhang, T., He, Y., Zhao, Y., Wang, S., Zhang, G., Zhang, Y., & Feng, Y., 2018. Recovery of valuable materials from spent lithium-ion batteries by mechanical separation and thermal treatment, *J. Clean. Product.*, 185, 646–652.

Wang, H.F., Liu, J.S., Bai, X.J., Wang, S., Yang, D., Fu, Y.P., & He, Y.Q., 2019a. Separation of the cathode materials from the Al foil in spent lithium-ion batteries by cryogenic grinding, *Waste Manag.*, 91, 89–98.

Wang, M.M., Tan, Q.Y., Liu, L.L., & Li, J.H., 2019b. Efficient separation of aluminum foil and cathode materials from spent lithium-ion batteries using a low-temperature molten salt, *ACS Sustain. Chem. Eng.*, 7(9), 8287–8294.

Wicser, M., 2023. The future direction of lithium-ion chemistry, Energy next-2023. www. anteotech.com/energy.

Widijatmoko, S.D., Fu, G., Wang, Z., & Hall, P., 2020a. Recovering lithium cobalt oxide, aluminum, and copper from spent lithium-ion batteries via attrition scrubbing, *J. Clean. Product.*, 260, 120869. https://doi.org/10.1016/j.jclepro.2020.120869.

Widijatmoko, S.D., Gu, F., Wang, Z., & Hall, P., 2020b. Selective liberation in dry-milled spent lthium-ion batteries, *Sustain. Mater. Technol.*, 23, e00134.

Wills, B.A., & Finch, J., 2015. *Wills' Mineral Processing Technology: An Introduction to the Practical Aspects of Ore Treatment and Mineral Recovery*. Oxford: Butterworth Heinemann. https://doi.org/10.1016/B978-008097053-0.00001-7.

Wu, Z., Zhu, H., Bi, H., He, P., & Gao, S., 2021. Recycling of electrode materials from spent lithium-ion power batteries via thermal and mechanical treatments, *Waste Manag. Researc.*, 39(4), 607–619.

Vanderbruggen, A., Salces, A., Fereira, A., Rudolph, M., & Serna-Guerrero, R., 2022. Improving separation efficiency in end-of-life lithium ion batteries flotation using attrition pre-treatment, *Minerals*, 12, 72. https://doi.org/10.3390/min12010072.

Venkatraman, S., Choi, J., & Manthiram, A., 2004. Factors influencing the chemical lithium extraction rate from layered $LiNi_{1-y-z}Co_yMn_zO_2$ cathodes, *Electrochem. Commun.*, 6(8), 832–837. https://doi.org/10.1016/j.elecom.2004.06.004.

Verma, A., Kore, R., Corebin, D.R., & Shiflett, M.B., 2019. Metal recovery using oxalate chemistry, a technical review, *Ind. Eng. Chem. Resour.*, 58(34), 15381–15393. https://doi.org/10.1021/acs.iecr.9b02598.

Virolainen, S., Fallah Fini, M., Laitinen, A., & Sainio, T., 2017. Solvent extraction fractionation of Li-ion battery leachate containing Li, Ni, and Co, *Sep. Purif. Technol.*, 179, 274–282. https://doi.org/10.1016/j.seppur.2017.02.010.

Xiao, J., Guo, J., Zhan, L., & Xu, Z., 2020. A cleaner approach to the discharge process of spent lithium-ion batteries in different solutions, *J. Clean. Product.*, 255, 120064.

Xu, J., Thomas, H.R., Francis, R.W., Lum, K.R., Wang, J., & Liang, B., 2008. A review of processes and technologies for the recycling of lithium-ion secondary batteries, *J. Power Sources*, 177(2), 512–527, https://doi.org/10.1016/j.jpowsour.2007.11.074.

Xu, Y.N., Song, D.W., Li, L., An, C.H., Wang, Y.J., Jiao, L.F., & Yuan, H.T., 2014. A simple solvent method for the recovery of $LixCoO_2$ and its applications in alkaline rechargeable batteries, *J. Power Sources*, 252, 286–291. https://doi.org/10.1016/j.jpowsour.2013.11.052.

Xuan, W., Otsuki, A., & Chagnes, A., 2019. Investigation of the leaching mechanism of NMC 811 ($LiNi_{0.8}Mn_{0.1}Co_{0.1}O_2$) by hydrochloric acid for recycling lithium ion battery cathodes. *RSC Adv.*, 9, 38612. https://doi.org/10.1039/c9ra06686a.

Xuan, W., de Souza Braga, A., Korbel, C., & Chagnes, A., 2021. New insights in the leaching kinetics of cathodic materials in acidic chloride media for lithium-ion battery recycling. *Hydrometallurgy*, 204. https://doi.org/10.1016/j.hydromet.2021.105705.

Yang, L., Xi, G., & Xi, Y., 2015. Recovery of Co, Mn, Ni, and Li from spent lithium-ion batteries for the preparation of $LiNixCoyMnzO_2$ cathode materials, *Ceram. Int.*, 41(9), 11498–11503. https://doi.org/10.1016/j.ceramint.2015.05.115.

Yang, Y., Huang, G.Y., Xu, S.M., He, Y.H., & Liu, X., 2016. Thermal treatment process for the recovery of valuable metals from spent lithium-ion batteries, *Hydrometallurgy*, 165, 390–396.

Yang, Y., Xu, S., & He, Y., 2017. Lithium recycling and cathode material regeneration from acid leach liquor of spent lithium-ion battery via facile co-extraction and co-precipitation processes, *Waste Manag.*, 64, 219–227. https://doi.org/10.1016/j.wasman.2017.03.018.

Yang, Y., Song, S., Lei, S., Sun, W., Hou, H., Jiang, F., Ji, X., Zhao, W., & Hu, Y., 2019. A process for combination of recycling lithium and regenerating graphite from spent lithium-ion battery, *Waste Manage.*, 85, 529. https://doi.org/10.1016/j.wasman.2019.01.008.

Yang, T., Lu, Y., Li, L., Ge, D., Yang, H., Leng, W., Zhou, H., Han, X., Schmidt, N., & Ellis, M., 2020. An effective relithiation process for recycling lithium-ion battery cathode materials, *Adv. Sustain. Syst.*, 4(1), 1900088.

Yang, H., Deng, B., Jing, X., Li, W., & Wang, D., 2021. Direct recovery of degraded $LiCoO_2$ cathode material from spent lithium-ion batteries: Efficient impurity removal toward practical applications, *Waste Manag.*, 129, 85–94.

Yao, L., Feng, Y., & Xi, G., 2015. A new method for the synthesis of $LiNi_{1/3}Co_{1/3}Mn_{1/3}O_2$ from waste lithium-ion batteries, *RSC Adv.*, 5(55), 44107–44114. https://doi.org/10.1039/C4RA16390G.

Yao, L., Yao, H., Xi, G., & Feng, Y., 2016. Recycling and synthesis of $LiNi_{1/3}Co_{1/3}Mn_{1/3}O_2$ from waste lithium-ion batteries using D, L-malic acid, *RSC Adv.*, 6(22), 17947–17954. https://doi.org/10.1039/C5RA25079J.

Yao, Y., Zhu, M., Zhao, Z., Tong, B., Fan, Y., & Hua, Z., 2018. Hydrometallurgical processes for recycling spent lithium-ion batteries: A critical review, *ACS Sustain. Chem. Eng.*, 6(11), 13611–13627. https://doi.org/10.1021/acssuschemeng.8b03545.

Yao, L., Xi, Y., Han, H., Li, W., Wang, C., & Feng, Y., 2021. $LiMn_2O_4$ prepared from waste lithium-ion batteries through sol-gel process, *J. Alloys Compd.*, 868, 159222. https://doi.org/10.1016/j.jallcom.2021.159222.

Yu, J.D., He, Y.Q., Li, H., Xie, W.N., & Zhang, T., 2017. Effect of the secondary product of semi-solid phase Fenton on the flotability of electrode material from spent lithium-ion battery, *Powder Technol.*, 315, 139–146.

Yu, J.D., He, Y.Q., Ge, Z.Z., Li, H., Xie, W.N., & Wang, S., 2018. A promising physical method for recovery of $LiCoO_2$ and graphite from spent lithium-ion batteries: Grinding flotation, *Sep. Purif. Technol.*, 190, 45–52.

Yu, H., Dai, H., Zhu, Y., Hu, H., Zhao, R., Wu, B., & Chen, D., 2021. Mechanistic insights into the lattice reconfiguration of the anode graphite recycled from spent high-power lithium-ion batteries, *J. Power Sources*, 481, 229159.

Zeng, G., Deng, X., Luo, S., Luo, X., & Zou, J., 2012. A copper-catalyzed bioleaching process for enhancement of cobalt dissolution from spent lithium-ion batteries, *J. Hazard. Mater.*, 199–200. https://doi.org/10.1016/j.jhazmat.2011.10.063.

Zeng, G., Luo, S., Deng, X., Li, L., & Au, C., 2013. Influence of silver ions on bioleaching of cobalt from spent lithium batteries, *Miner. Eng.*, 49, 40–44. https://doi.org/10.1016/j.mineng.2013.04.021.

Zeng, X., Li, J., & Singh, N., 2014. Recycling of spent lithium-ion battery: A critical review, *Crit. Rev. Environ. Sci. Technol.*, 44(10), 1129–1165.

Zeng, X., & Li, J., 2014. Innovative application of ionic liquid to separate Al and cathode materials from spent high-power lithium-ion batteries, *J. Hazard. Mater.*, 271, 50–56. https://doi.org/10.1016/j.jhazmat.2014.02.001.

Zeng, X., Li, J., & Shen, B., 2015a. Noval approach to recover cobalt and lithium from spent lithium-ion battery using oxalic acid, *J. Hazard. Mater.*, 295, 112–118. https://doi.org/10.1016/j.jhazmat.2015.02.064.

Zeng, X., Wei, S., Sun, L., Jacques, D.A., Tang, J., Lian, M., Ji, Z., Wang, J., Zhu, J., & Xu, Z., 2015b. Bioleaching of heavy metals from contaminated sediments by the Aspergillusnigerstrain SY1, *J. Soils Sediments*, 15(4), 1029–1038. https://doi.org/10.1007/s11368-015-1076-8.

Zhan, R.T., Oldenburg, Z., & Pan, L., 2018. Recovery of active cathode materials from lithium-ion batteries using froth flotation, *Sustain. Mater. Technol.*, 17, e00062.

Zhan, R., Payne, T., Leftwich, T., Perrine, K., & Pan, L., 2020a. Deagglomeration of cathode composites for direct recycling of Li-ion batteries, *Waste Manag.*, 105, 39–48.

Zhan, R., Yang, Z., Bloom, I., & Pan, L., 2020b. Significance of a solid electrolyte interphase on separation of anode and cathode materials from spent Li-ion batteries by froth flotation, *ACS Sustain. Chem. Eng.*, 9(1), 531–540. https://doi.org/10.1021/acssuschemeng.0c07965.

Zhang, P., Yokoyama, T., Habashi, O., Suzuki, T.M., & Inoue, K., 1998. Hydrometallurgical process for recovery of metal values from spent lithium-ion secondary batteries, *Hydrometallurgy*, 47, 259–271.

Zhang, S.S., 2006. A review on electrolyte additives for lithium-ion batteries, *J. Power Sources*, 162(2), 1379–1394.

Zhang, T., He, Y., Ge, L., Fu, R., Zhang, X., & Huang, Y., 2013. Characteristics of wet and dry crushing methods in the recycling process of spent lithium-ion batteries, *J. Power Sources*, 240, 766–771. https://doi.org/10.1016/j.jpowsour.2013.05.009.

Zhang, T., He, Y., Wang, F., Li, H., Duan, C., & Wu, C., 2014a, Surface analysis of cobalt-enriched crushed products of spent lithium-ion batteries by X-ray photoelectron spectroscopy, *Sep. Purif. Tech.*, 138(10), 21–28. https://doi.org/10.1016/j.seppur.2014.09.033.

Zhang, T., He, Y., Wang, F., Ge, L., Zhu, X., & Li, H., 2014b. Chemical and process mineralogical characterizations of spent lithium-ion batteries: An approach by multi-analytical techniques, *Waste Manag.*, 34(6), 1051–1058.

Zhang, X., Xie, Y., Cao, H., Nawaz, F., & Zhang, Y., 2014c. A novel process for recycling and resynthesizing $LiNi_{1/3}Co_{1/3}Mn_{1/3}O_2$ from the cathode scraps intended for lithium-ion batteries, *Waste Manag.*, 34(9), 1715–1724. https://doi.org/10.1016/j.wasman.2014.05.023.

Zhang, X., Hongbib, C., Pengge, N., Haixia, Y., & Faheem, N., 2015. A closed-loop process for recycling $LiNi_{1/3}Co_{1/3}Mn_{1/3}O_2$ from the cathode scraps of lithium-ion batteries: Process optimization and kinetics analysis, *Sep. Purif. Technol.*, 150, 186–195. https://doi.org/10.1016/j.seppur.2015.07.003.

Zhang, G.W., He, Y.Q., Feng, Y.W., Wang, H.F., & Zhu, X.N., 2018a. Pyrolysis-ultrasonic-assisted flotation technology for recovering graphite and $LiCoO_2$ from spent lithium-ion batteries, *ACS Sustain. Chem. Eng.*, 6(8), 10896–10904.

Zhang, G., He, Y., Feng, Y., Wang, H., Zhang, T., Xie, W., & Zhu, X., 2018b. Enhancement in the liberation of electrode materials derived from spent lithium-ion battery by pyrolysis, *J. Clean. Product.*, 199, 62–68.

Zhang, Y., He, Y., Zhang, T., Zhu, X., Feng, Y., Zhang, G., & Bai, X., 2018c. Application of Falcon centrifuge in the recycling of electrode materials from spent lithium-ion batteries, *J. Clean. Product.*, 202, 736–747.

Zhang, X., Bian, Y., Xu, S., Fan, E., Xue, Q., Guan, Y., Wu, F., Li, L., & Chen, R., 2018d. Innovative application of acid leaching to regenerate $Li(Ni_{1/3}Co_{1/3}Mn_{1/3})O_2$ cathodes from spent lithium ion batteries, *ACS Sustain. Chem. Eng.*, 6(5), 5959–5968.

Zhang, G., Du, Z., He, Y., Wang, H., Xie, W., & Zhang, T., 2019. A sustainable process for the recovery of anode and cathode materials derived from spent lithium-ion batteries, *Sustainability*, 11, 2363. https://doi.org/10.3390/su11082363.

Zhang, G.W., He, Y.Q., Wang, H.F., Feng, Y.W., Xie, W., & Zhu, X.N., 2020a. Removal of organics by pyrolysis for enhancing liberation and flotation behavior of electrode materials derived from spent lithium-ion batteries, *ACS Sustain. Chem. Eng.*, 8, 2205–2214.

Zhang, G., Yuan, X., He, Y., Wang, H., Xie, W., & Zhang, T., 2020b. Organics removal combined with in situ thermal reduction for enhancing the liberation and metallurgy efficiency of $LiCoO_2$ derived from spent lithium-ion batteries, *Waste Manag.*, 115, 113–120.

Zhang, Z., Qiu, J., Yu, M., Jin, C., Yang, B., & Guo, G., 2020c. Performance of Al-doped $LiNi_{1/3}Co_{1/3}Mn_{1/3}O_2$ synthesized from spent lithium-ion batteries by sol-gel method, *Vacuum*, 172, 109105. https://doi.org/10.1016/j.vacuum.2019.109105.

Zhao, H., Ren, J., He, X., Li, J., Jiang, C., & Wan, C., 2008. Modification of natural graphite for lithium-ion batteries, *Solid State Sci.*, 10(5), 612–617.

Zhao, J.M., Shen, X.Y., Deng, F.L., Wang, F.C., Wu, Y., & Liu, H.Z., 2011. Synergistic extraction and separation of valuable metals from waste cathodic material of lithium-ion batteries using Cyanex 272 and PC-88A, *Sep. Purif.Technol.*, 78(3), 345–351.

Zhao, J., Qu, X., Qu, J., Zhang, B., Ning, Z., Xie, H., Zhou, X., Song, Q., Xing, P., Yin, H., 2019. Extraction of Co and Li_2CO_3 from cathode materials of spent lithium-ion batteries through a combined acid-leaching and electrode oxidation approach, *J. Hazard. Mater.*, 379, 120817. https://doi.org/10.1016/j.jhazmat.2019.120817.

Zhao, C., & Zhang, X., 2020. Retracted: Reverse flotation process for the recovery of pyrolytic $LiFePO_4$, *Colloids Surf. A Physicochem. Eng. Asp.*, 596, 124741.

Zheng, Y., Long, H., Zhou, L., Wu, Z., Zhou, X., You, L., Yang, Y., & Liu, J., 2016. Leaching procedure and kinetic studies of cobalt in cathode materials from spent lithium-ion batteries using organic citric acid as leachant, *Int. J. Env. Resour.*, 10(1), 159–168. https://doi.org/10.22059/ijer.2016.56898.

Zhong, X., Liu, W., Han, J., Jiao, F., Qin, W., Liu, T., & Zhao, C., 2019. Pyrolysis and physical separation for the recovery of spent LiFePO$_4$ batteries, *Waste Manag.*, 89, 83–93. https://doi.org/10.1016/j.wasman.2019.03.068.

Zhou, X., He, W.Z., Li, G.M., Zhang, X.J., Huang, J.W., & Zhu, S.G., 2010. Recycling of electrode materials from spent lithium-ion batteries. Proceedings of the 4th international conference on Bioinformatics and biomedical engineering. Chengdu, China.

Zhu, S.-G., He, W.-Z., Li, G.-M., Zhou, X., Zhang, X.-J., & Huang, J.-W., 2012. Recovery of Co and Li from spent lithium-ion batteries by combination method of acid leaching and chemical precipitation, *Trans. Nonferrous Met. Soc. China*, 22(9), 2274–2281. https://doi.org/10.1016/S1003-6326(11)61460-X.

Zou, H., Gratz, E., Apelian, D., & Wang, Y., 2013. A novel method to recycle mixed cathode materials for lithium-ion batteries, *Green Chem.*, 15, 1183.

WEB SITES

https://www.alibaba.com/product-detail/Environmental-big-lithium-battery-recycling machine_1600525922332.html?spm=a2700.wholesale.0.0.5e0b 46e2V40a43

https://americanbatterytechnology.com.

https://americanbatterytechnology.com/solutions/lithium-ion-battery-recycling

https://americanbatterytechnology.com/projects/usabc-project/

https://americanmanganeseinc.com/recyclico-battery-materials-advances-to-lithium-recovery-stage-of-demonstration-plant-project/ü

https://analyticalscience. wiley.com/do/10.1002/gitlab.15680/full/

https://analyticalscience.wiley.com/do/10.1002/gitlab.10220

https://www.argusmedia.com/en//news/

https://www.argusmedia.com/en//news/2473890-chinas-battery-firms-accelerate-overseas-expansions?

https://www.argusmedia.com/en//news/2475368-chinas-yahua-extends-li-hydroxide-supplies-to-tesla

https://www.argusmedia.com/en/news/2458422-chinas-gem-expands-battery-recycling-production

https://ascendelements.com/

https://ascendelements.com/innovation/

www.ateotech.com/energy

https://arstechnica.com/science/2022/04/lithium-costs-a-lot-of-money-so-why-arent-we-recycling-lithium-batteries/

https://batrec.ch/en/

https://www.bestmag.co.uk/aqua-metals-in-tie-up-with-south-korean-storage-solution-and-battery-materials-company-yulho/

https://www.chemistryworld.com/features/the-drive-to-recycle-lithium-ionbatteries/4012222.article.

https://www.chargecccv.com/technology/innovation

https://www.chargecccv.com/technology/portfolio

https://www.duesenfeld.com/recycling_en.html

https://www.electrive.com/2021/05/21/oxis-energy-is-facing-bankruptcy/

https://en.brunp.com.cn/intro/14.html

https://envirostream.com.au/

https://www.epa.gov/recycle/used-lithium-ion-batteries#single-use.

https://www.highpowertech.com/li-ion-li-polymer-rechargeable-batteries

https://www.isc.fraunhofer.de/en/press-and-media/press-releases/efficient-recycling-of-lithium-ionbatteries-launch-of-research-project-new-bat.html

https://www.iwks.fraunhofer.de/en/competencies/Separation-and-SortingTechnologies/
Physical-Separation-and-Sorting-Technologies/Electro-hydraulic-Fragmentation-EHF.html
https://www.kemetco.com/battery-research.html
https://www.kobar.co.kr/
https://nickelhuette-aue.de/en/about
https://www.neometals.com.au/
https://www.nobelprize.org/prizes/chemistry/2019/press-release/
https://onto-technology.com/).
https://polyplus.com/glass-protected-lithium-battery/#future
https://pubs.acs.org/doi/10.1021/acs.energyfuels.1c02489
https://polyplus.com/product-pipeline/
https://recyclico.com/
https://www.sungeelht.com/en
https://www.tes-amm.com/it-services/commercial-battery-recycling#!/
https://www.r4-innovation.de/de/new-bat.html
https://www.r4-innovation.de/de/new-bat.htmlhttps:/www.r4-innovation.de/de/n
https://www.simslifecycle.com/ resources/tip-sheet-recycling/
https://www.simslifecycle.com/blog/2019/%20guide-how-to-responsibly-dispose-of-lithium-
    ion-batteries/
https://sionpower.com/
https://sionpower.com/products/
https://steatite-batteries.co.uk/partners/oxis-energy-batteries/
https://westwaterresources.net/minerals-portfolio/graphite-market/
https://www.wikipedia.org/

Index

Printed in the United States
by Baker & Taylor Publisher Services